Extracellular ATP and Adenosine as Regulators of Endothelial Cell Function

Evgenia Gerasimovskaya · Elzbieta Kaczmarek
Editors

Extracellular ATP and Adenosine as Regulators of Endothelial Cell Function

Implications for Health and Disease

Springer

Editors
Dr. Evgenia Gerasimovskaya
University of Colorado Denver
Department of Pediatrics
Critical Care Medicine
Pediatric Clinical Services
12700 East 19th Avenue
Aurora CO 80045
USA

Dr. Elzbieta Kaczmarek
Harvard Medical School
Beth Israel Deaconess
Medical Center
99 Brookline Ave.
Boston MA 02215
USA

ISBN 978-90-481-3434-2 e-ISBN 978-90-481-3435-9
DOI 10.1007/978-90-481-3435-9
Springer Dordrecht Heidelberg London New York

Library of Congress Control Number: 2010922800

Printed on acid-free paper

Springer is part of Springer Science+Business Media (www.springer.com)

Preface

Purine and pyrimidine nucleotides and nucleosides function as components of DNA and RNA, are central in cellular energy metabolism, directly interact with mutliple proteins and participate in over 500 enzymatic reactions including phosphorylations, glycosylations, and phospholipid biosynthesis. Perhaps because of these well studied functions of intracellular purine and pyrimidine compounds, the view of extracellular nucleotides as important signaling molecules that are secreted from both resting and activated cells and exert various autocrine/paracrine effects, was not accepted readily.

The breakthrough in the purinergic signaling field was achieved by identification and cloning of the nucleotide targets, P2 purinergic receptors, followed by extensive characterization of two groups of these receptors, P2Y (metabotropic) and P2X (ionotropic). These advances led to numerous studies that elucidated extracellular nucleotide-initiated signaling pathways and their outcomes in different cell types and tissues. Moreover, it is well accepted now that purinergic signaling is regulated by the levels/availability of their specific ligands, which, in turn, depend on the net balance between nucleotide release and their metabolism by nucleotide- and nucleoside- converting ecto-enzymes.

One of the cell types frequently exposed to extracellular nucleotides is the endothelium. In addition, endothelial cells themselves release nucleotides, therefore they represent both the source and the target for adenine nucleotides. It is well established that the endothelium plays a central role in regulation of vascular tone and permeability, blood clotting, inflammation, interaction with blood cells and angiogenesis. However, despite the progress in our understanding of purinergic regulation of vascular function, the role of extracellular purines and pyrimidines in regulation of endothelial cell function remains incomplete.

The editors recognize that covering all available data on purinergic regulation of endothelial function under various physiological and pathological conditions in one book would be a considerbale challenge. However, facing the lack of any compendium on this subject, we have chosen to compile topics that represent recent advances on the effects of extracellular nucleotides and adenosine in endothelial cells. These topics, presented in chapters written by experts in their respective fields, address roles of purinergic signaling in regulation of vascular tone, tissue perfusion and cell barrier, as well as in angiogenesis, hypoxia, wound healing, inflammation, coagulation, diabetes, and pregnancy. Moreover, interactions of endothelial cells originating from different vascular beds, such as lung, skin, cornea, pancreas and uterus with host tissues are discussed.

We believe that this book should be of broad interest to basic science researchers, clinicians and students interested in biology of the endothelium and thus curious about the mechanisms of purinergic signaling in endothelial cells. Finally, we would like to thank all contributors for their effort to launch this up to date book.

Aurora, CO, USA Evgenia Gerasimovskaya
Boston, MA, USA Elzbieta Kaczmarek

Contents

Contributors

Aftab Ahmad Department of Pediatrics, National Jewish Health, 1400 Jackson Street, Denver, CO 80206, USA, ahmada@njhealth.org

Shama Ahmad Department of Pediatrics, National Jewish Health, 1400 Jackson Street, Denver, CO 80206, USA, ahmads@njhealth.org

Ian M. Bird Perinatal Research Laboratories, Department OBGyn, University Wisconsin School Medicine and Public Health, 7E Meriter Hospital, 202 South Park Street, Madison, WI 53715, USA, imbird@wisc.edu

Derek S. Boeldt Perinatal Research Laboratories, Department OBGyn, University Wisconsin School Medicine and Public Health, 7E Meriter Hospital, 202 South Park Street, Madison, WI 53715, USA, dsboeldt@wisc.edu

Geert Bultynck Laboratory of Molecular and Cellular Signaling, Department of Molecular Cell Biology, KULeuven, Campus Gasthuisberg O&N1, B-3000, Leuven, Belgium, Geert.Bultynck@med.kuleuven.be

John D. Catravas Vascular Biology Center, Medical College of Georgia, Augusta, GA 30912, USA, jcatrava@mcg.edu

Sean P. Colgan Departments of Medicine and Anesthesiology and the Mucosal Inflammation Program, University of Colorado Denver, Aurora, CO, USA, sean.colgan@ucdenver.edu

Catheleyne D'hondt Laboratory of Molecular and Cellular Signaling, Department Molecular Cell Biology, KULeuven, Campus Gasthuisberg O&N1, B-3000 Leuven, Belgium, Catheleyne.Dhondt@med.kuleuven.be

Holger K. Eltzschig Departments of Medicine and Anesthesiology and the Mucosal Inflammation Program, University of Colorado Denver, Aurora, CO, USA, holger.eltzschig@ucdenver.edu

David Erlinge Department of Cardiology, Lund University Hospital, SE-22185 Lund, Sweden, david.erlinge@med.lu.se

Evgenia V. Gerasimovskaya Department of Pediatrics, Critical Care Medicine, University of Colorado Denver, Aurora, CO 80045, USA, evgenia.gerasimovskaya@ucdenver.edu

Priya Gomes Physiology lab, KULeuven, subfaculty Kortrijk, E. Sabbelaan 53, B-8500, Kortrijk, Belgium, Priya.Gomes@med.kuleuven.be

Eva Grapengiesser Department of Medical Cell Biology, Uppsala University, Uppsala, Sweden, eva.grapengiesser@mcb.uu.se

Bo Hellman Department of Medical Cell Biology, Uppsala University, Uppsala, Sweden, bo.hellman@mcb.uu.se

Bernard Himpens Laboratory of Molecular and Cellular Signaling, Department Molecular Cell Biology, KULeuven, Campus Gasthuisberg O&N1, B-3000 Leuven, Belgium, Bernard.Himpens@med.kuleuven.be

Leif Jansson Department of Medical Cell Biology, Biomedical Centre, Uppsala University, Husargatan 3, Box 571, SE-751 23 Uppsala, Sweden, leif.jansson@mcb.uu.se

Agnieszka Jezierska Vascular Biology Center, Medical College of Georgia, Augusta, GA 30912, USA, ajezierska@mcg.edu

Elzbieta Kaczmarek Center for Vascular Biology Research, Surgery Department, Microvascular Surgery Division, Beth Israel Deaconess Medical Center, Harvard Medical School, 99 Brookline Avenue, Boston, MA, USA, ekaczmar@bidmc.harvard.edu

Irina A. Kolosova Department of Anesthesiology and Critical Care Medicine, Johns Hopkins University School of Medicine, Baltimore, MD 21205, USA, ikoloso1@jhmi.edu

Rudolf Lucas Vascular Biology Center, Medical College of Georgia, Augusta, GA 30912, USA, rlucas@mcg.edu

Christopher F. MacManus Mucosal Inflammation Program, 12700 E. 19th Ave, Aurora, CO 80045, USA, christopher.macmanus@ucdenver.edu

M. Carmen Montesinos Department de Farmacologia, Universitat de València, Ave. Vicent Andrès Estellès s/n, 46100 Burjassot, Valencia, Spain; Instituto de Reconocimiento Molecular y Desarrollo Tecnológico, Centro Mixto Universidad Politécnica de Valencia, Universidad de Valencia, Spain; Department of Medicine, New York University School of Medicine, 550 First Ave, New York, NY 10016, USA, m.carmen.montesinos@uv.es

Yvonne M. Mowery Department of Pathology, Duke University, DUMC Box 3712, Durham, NC 27710, USA, ymm3@duke.edu

Salvatore V. Pizzo Department of Pathology, Duke University, DUMC Box 3712, Durham, NC 27710, USA, pizzo001@mc.duke.edu

Raf Ponsaerts Laboratory of Molecular and Cellular Signaling, Department Molecular Cell Biology, KULeuven, Campus Gasthuisberg O&N1, B-3000 Leuven, Belgium, raf.ponsaerts@med.kuleuvan.be

Cheikh I. Seye Department of Cellular and Integrative Physiology, Indiana University School of Medicine, 635 Barnhill Drive, MS 332, Indianapolis, Indiana 46202, cseye@iupui.edu

Sangly P. Srinivas Indiana University, School of Optometry, Bloomington, IN 47405, USA, srinivas@indiana.edu

Kurt R. Stenmark Department of Pediatrics, Critical Care Medicine, University of Colorado Denver, Aurora, CO 80045, USA, kurt.stenmark@ucdenver.edu

Nagavedi S. Umapathy Vascular Biology Center, Medical College of Georgia, Augusta, GA 30912, USA, usiddaramappa@mcg.edu

María D. Valls Department de Farmacologia, Universitat de València, Ave. Vicent Andrès Estellès s/n, 46100 Burjassot, Valencia, Spain, mvn204@yahoo.com

Johan Vereecke Laboratory of Molecular and Cellular Signaling, Department Molecular Cell Biology, KULeuven, Campus Gasthuisberg O&N1, B-3000 Leuven, Belgium, Johan.Vereecke@med.kuleuven.be

Alexander D. Verin Vascular Biology Center, Medical College of Georgia, Augusta, GA 30912, USA, averin@mcg.edu

Carl W. White Department of Pediatrics, National Jewish Health, 1400 Jackson Street, Denver, CO 80206, USA, whitec@njhealth.org

Gary A. Weisman Department of Biochemistry, 540E Life Sciences Center, University of Missouri, 1201 Rollins Road, Columbia, MO 65211–7310, USA, weismang@missouri.edu

Gennady G. Yegutkin MediCity Research Laboratory, University of Turku, Turku, 20520, Finland, genyeg@utu.fi

FuXian Yi Perinatal Research Laboratories, Department OBGyn, University Wisconsin School Medicine and Public Health, 7E Meriter Hospital, 202 South Park Street, Madison, WI 53715, USA, fuxianyi@wisc.edu

Evgeny A. Zemskov Vascular Biology Center, Medical College of Georgia, Augusta, GA 30912, USA, ezemskov@mcg.edu

Chapter 1
Purinergic and Pyriminergic Activation of the Endothelium in Regulation of Tissue Perfusion

David Erlinge

Abstract The nucleotides ATP, ADP, UTP and UDP acting on P2 receptors stimulate endothelium dependent dilatation mediated via nitric oxide, endothelium derived hyperpolarising factors, and prostacyclins (to a lesser extent). They are crucial for shear stress-mediated increases in blood flow and for the increase of tissue perfusion during exercise and hypoxia. The endothelium can also release potent contractile tetraphosphate nucleotides (AP_4A and UP_4A). The purinergic signaling has been implicated in blood pressure regulation, development of myocardial infarction, heart failure, pulmonary hypertension, circulatory shock and regulation of blood flow in most organs of the body. Purinergic and pyriminergic endothelium-mediated effects are important for physiological and patophysiological regulation of tissue perfusion.

Keywords Endothelium · ATP · UTP · ADP · UDP · Nitric oxide · EDHF · AP4A · UP4A · Red blood cell · Spreading dilatation · Vasodilatation · Vasoconstriction · Ectonucleotidase · Hypertension · Congestive heart failure · Reactive hyperemia · Pulmonary hypertension · Hypoxia

1.1 Introduction

Effects purines on tissue perfusion were recognized already in 1929 by Drury and Szent-Györgyi [25]. Since then, it has become increasingly clear that, in addition to functioning as an intracellular energy source, as DNA or RNA, the purines and pyrimidines ATP, ADP, UTP and UDP can serve as important extracellular signaling molecules acting on fifteen P2 receptor subtypes [9, 11, 16, 19, 52]. To terminate signaling, ectonucleotidases present in the circulation and on cell surfaces, rapidly

D. Erlinge (✉)
Department of Cardiology, Lund University Hospital, SE-22185 Lund, Sweden
e-mail: david.erlinge@med.lu.se

E. Gerasimovskaya, E. Kaczmarek (eds.), *Extracellular ATP and Adenosine as Regulators of Endothelial Cell Function*, DOI 10.1007/978-90-481-3435-9_1,

degrade extracellular ATP into ADP, AMP and adenosine [30,83]. Experimental studies on the vascular effects of purines have been of crucial importance for revealing the complexities of this system and for the classification of the receptors [18, 20, 27, 49, 51]. Purines and pyrimidines stimulate vasoconstriction and vasodilatation, are involved in blood pressure regulation, development of myocardial infarction, heart failure, blood flow regulation during exercise and hypoxia, pulmonary hypertension and regulation of blood flow in most organs of the body. This chapter is an attempt to summarize the endothelium-mediated effects of purinergic and pyriminergic regulation on tissue perfusion.

1.2 Regulation of Vascular Tone: Balance Between Contractile and Dilatory Effects

The net effect on tissue perfusion is a balance between contractile and dilatory responses. Vasoconstriction produced by ATP released as a cotransmitter with noradrenaline from perivascular sympathetic nerves was recognised early [10, 14]. However, following the seminal discovery of endothelium dependent vasodilatation by Furchgott in the early 1980s, it was shown that ATP acted on endothelial cells (EC) to cause endothelial dependent vasodilatation [12] and dual purinergic neural and endothelial control of vascular tone was established [13, 15]. Extracellular nucleotides (ATP, ADP, UTP and UDP) cause endothelial-dependent relaxation, while adenosine dilates blood vessels by direct action on vascular smooth muscle cells (VSMC), mainly via A_{2a} receptors [51] (Fig. 1.1).

In some situations, adenosine does not solely dilate via direct action on the VSMC, but may in part activate the endothelium and release nitric oxide (NO).

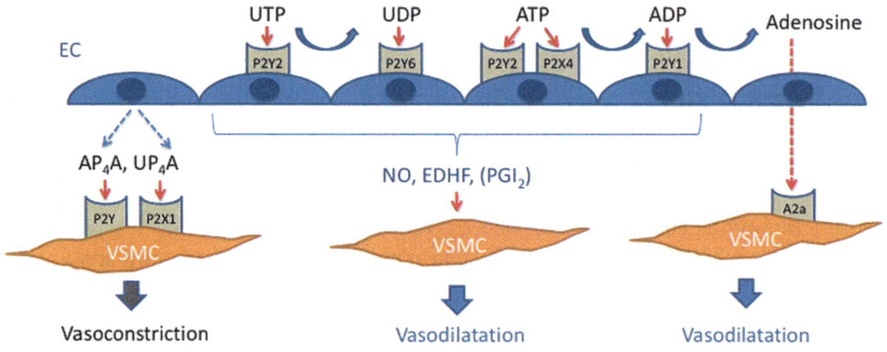

Fig. 1.1 Purines and pyrimidines regulate vascular tone via endothelial cells. ATP, ADP, UTP and UDP dilate via release of NO and endothelium derived hyperpolarising factor (EDHF). Adenosine acts mainly via direct effect on the vascular smooth muscle cells (VSMC), but may also in some situations cause dilatation via NO-release. The endothelium can release the contractile nucleotides UP_4A and AP_4A, which cause contraction of the VSMC

This is exemplified by the human forearm circulation where adenosine have been shown also to cause NO dependent increases in blood flow [61].

If the endothelium is damaged or if extracellular nucleotides are administered on the adventitial side, the vessel constricts as a result of activation of P2Y and P2X receptors on the VSMC.

1.3 Nucleotide Release in the Vasculature

1.3.1 Endothelial Cells

Shear stress and hypoxia are an important stimuli of both ATP and UTP release from endothelial cells [17]. The nucleotides can be released from intracellular sources, but there is also evidence for cell surface ATP synthase involved in shear stress-induced ATP release [78]. Immunofluorescence staining of human pulmonary arterial ECs showed that cell surface ATP synthase is distributed in lipid rafts/caveole and critical for shear stress-induced ATP release. Interestingly, Planck et al., applied a mathematical model demonstrating that the area directly after a stenotic lesion with reduced shear stress will have lower levels of extracellular ATP and thereby reduced endothelium stimulation [50], indicating that lack of ATP mediated endothelial stimulation might contribute to atherosclerosis development.

1.3.2 Red Blood Cells

The matching of oxygen supply with demand requires a mechanism that increases blood flow in response to decreased tissue oxygen levels. There is a growing literature suggesting that the red blood cell (RBC) acts as a sensor for hypoxia and different mechanisms have been suggested by which the deoxygenated RBC stimulates vasodilatation [22, 26, 46, 82]. The intracellular levels of ATP in RBCs are at millimolar levels and on the inside of the membrane are abundant glycolytic enzymes generating new ATP from adenosine [2, 62, 64]. ATP is released in response to reductions in oxygen tension and pH [26, 82]. It has been shown in vitro that vessels dilate in response to low O_2 levels only when blood vessels are perfused with RBCs [68]. ATP is released in working human skeletal muscle circulation depending on the number of unoccupied hemoglobin O_2 binding sites [26, 82]. The released ATP then binds to P2Y receptors on the endothelium and stimulates vasodilatation. Thus, the RBC may function as an O_2 sensor, contributing to the regulation of blood flow and O_2 delivery, by releasing ATP depending on the oxygenation state of hemoglobin (Fig. 1.2).

When ATP is degraded to ADP it can activate $P2Y_{13}$ receptors as a negative feedback pathway for ATP release from human RBCs [73]. Blood consists of approximately 40% RBCs, containing a 1000-fold higher ATP concentration than plasma (mmol/L vs. μmol/L). With this gradient, even a minor release of ATP from the high intracellular concentrations raises local extracellular ATP levels

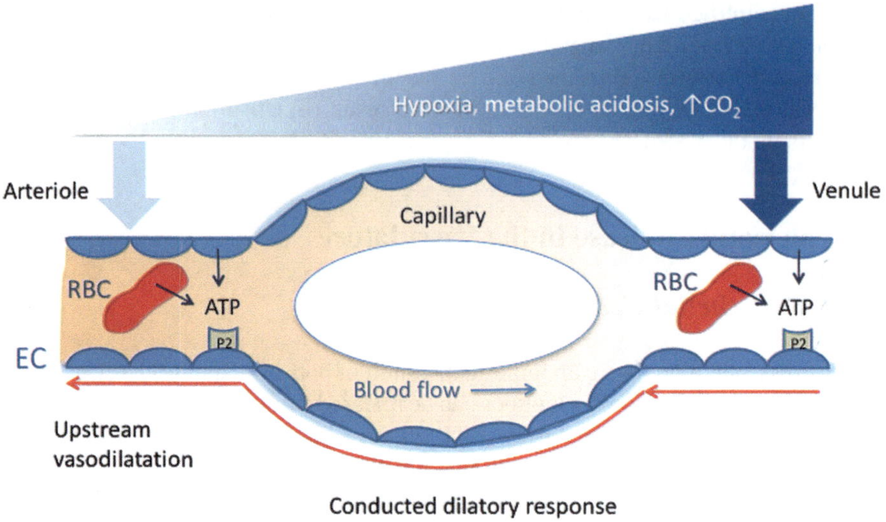

Fig. 1.2 Regulation of tissue perfusion. Hypoxia, metabolic acidosis and increased CO_2 cause release of intraluminal ATP from red blood cells (RBC) and endothelial cells (EC). ATP activates P2-receptors on endothelial cells and induces dilatation via NO and EDHF. This result in both a local dilatation and a retrograde spreading dilatation giving a conducted response resulting in increased blood flow to the tissue. The retrograde spreading dilation is important since the most pronounced hypoxia, metabolic acidosis and increase in CO_2 is found on the venular side

substantially and could have major circulatory effects. A negative feedback system may therefore be of great physiological importance to mitigate ATP release.

1.4 Role of Ectonucleotidases in Regulation of Tissue Perfusion

Extracellular ATP in the circulation is rapidly degraded into ADP, AMP and adenosine by ectonucleotidases. Vascular NTPDase1 (CD39) is an endothelial cell membrane protein with both ecto-ATPase and ecto-ADPase activities [38, 83]. Ectonucleotidases are released by shear stress from endothelial cells [81], and from sympathetic nerves together with its substrate ATP, as a termination mechanism for the signalling [66]. In exercising humans when skeletal muscle blood flow increases, ectonucleotidases are released and it may be speculated that this could contribute to the benefits of physical activity [82].

The importance of ectonucleotidases for tissue perfusion has been established in several other pathophysiological situations. Endothelial ADPase activity is lost following ischemia-reperfusion injury, xenograft rejection and inflammation resulting in increased levels of ATP and ADP [38, 54]. This can cause platelet aggregation and microvascular obstruction. Infusion of systemic apyrase inhibits platelet aggregation and prolongs xenograft survival [40]. This has also been shown with adenoviral transfer of NTPDase1 [36]. NTPDase1 is also lost in vascular cardiac grafts subjected to oxidant stress.

1.5 Receptor Subtypes Involved in Endothelial Relaxation

The prototypical endothelial receptor, since the definition of the P2Y and P2X receptor subclasses, is the $P2Y_1$ receptor, and still, even after cloning of eight different P2Y receptors, it seems to be of main importance in most vascular beds [51]. Pharmacology of vasodilatation and mRNA quantification in man indicates that $P2Y_1$, $P2Y_2$ and $P2Y_6$ are the most important endothelial P2Y-receptors [72, 74]. Knockout mice experiments recently confirmed this picture and demonstrated that the $P2Y_4$ receptor does not mediate dilatation [33]. Knockout of the $P2Y_2$ and $P2Y_6$ receptor reveals that a major part of the UTP mediated dilatation is in fact mediated by the UDP receptor $P2Y_6$ [1, 33], after degradation of UTP to UDP by ectonucleotidases.

The ATP receptor $P2Y_{11}$ is one of the transcriptionally highest expressed P2Y receptors in the endothelium, but it does not seem to be involved in dilatory mechanisms [72]. Instead, it could stimulate proinflammatory responses.

The $P2X_1$ receptor is not expressed on the human endothelium and the evidence for other P2X receptors have been scarce, except for the $P2X_4$ receptor which is the highest expressed P2 receptor in endothelium [72, 77]. Using antisense oligonucleotides the $P2X_4$ receptor was shown to be important for shear stress dependent Ca^{2+}-influx via an ATP dependent mechanism [76]. This indicates that ATP and P2 receptors may be of importance for shear stress mediated effects which is in agreement with the well established release of ATP from endothelial cells during shear stress [3]. Vessel dilation induced by acute increases in blood flow is markedly suppressed in $P2X_4$-/-mice. Thus, endothelial $P2X_4$ channels are crucial to flow-sensitive mechanisms that regulate blood pressure and vascular remodeling [79].

The importance of the presence of ATP for flow-induced vasodilation has been confirmed rat mesenteric arteries [41]. If ATP was not present in the lumen, the flow-induced vasodilation was reduced by half. Interestingly, the effect of ATP could be mimicked by UTP, indicating that P2Y receptors could be of similar importance as the $P2X_4$ receptor for shear induced vasodilatation.

In summary, endothelium dependent dilatation is stimulated directly by ATP, ADP, UTP and UDP by activation of $P2Y_1$, $P2Y_2$ and $P2Y_6$ receptors and shear stress mediated vasodilatation is dependent on ATP activation of $P2X_4$ receptors, but also by UTP.

1.6 Mediators of Endothelial Dependent Vasodilation: NO, EDHF and PGI_2

ATP and ADP release prostacyclin (PGI_2) from endothelial cells [6]. PGI_2 is a dilator in some blood vessel but usually its contribution to endothelial dependent dilatation is small or absent. It may have a more important role as a potent platelet inhibitor, preventing thrombus formation in the microcirculation.

Vasodilatation and decreased blood pressure by release of NO has been demonstrated in several studies [24, 52]. NO is especially important in the regulation of medium to large arteries. In smaller arteries and vascular beds the dilatory response has not been possible to block with NO-inhibitors. This has led to the definition of a third endothelial dilatory factory called endothelium derived hyperpolarizing factor (EDHF). The exact chemical entity is not fully understood but EDHF is defined as resistant to NO and prostaglandin inhibitors, it hyperpolarises VSMC and it is blocked by potassium channel blockers.

Extracellular nucleotides acting on P2Y receptors have been shown to mediate a major part of the dilatation via endothelium derived hyperpolarizing factor (EDHF) [44, 45, 47]. This has been confirmed in human blood vessels examined in vitro [74]. The importance of EDHF becomes even more pronounced in vivo, e.g. in the mesenteric vascular bed were NO played a minor role and EDHF was the dominating factor mediating ATP, ADP and UTP induced dilatations [43]. This was even more pronounced in man where both UTP and ATP reduced forearm vascular resistance in a prostaglandin and NO independent way [35], indicating an important role for EDHF in P2 receptor mediated vasodilatation [35].

1.7 Mediators of Endothelial Dependent Vasoconstriction: AP_4A, UP_4A

The endothelial can also release substances that stimulate vasoconstriction, as exemplified by the peptide endothelin which play an important role in pulmonary hypertension. Recently, the two tetraphosphate nucleotides AP_4A, UP_4A have been shown to be released from the endothelium and to mediate potent contractile effects by activation of $P2X_1$ and P2Y receptors [37, 67]. They seem to be especially important in the renal vascular regulation and infusion of AP_4A results in nanomolar concentration results in mean arterial pressure.

1.8 Extracellular Nucleotides in Physiological Vascular Regulation

1.8.1 Retrograde Spreading Dilatation

Blood flow to a tissue is a complex process designed to provide adequate amounts of oxygen to meet the changing metabolic needs of the tissue. Hypoxia and acidosis that causes release of dilatory ATP, occurs on the venular side of the capillary bed, but to increase blood flow to the tissue the arterioles upstream need to be dilated. Ellsworth have demonstrated that addition of ATP to the venular side binds to purinergic receptors located on the vascular endothelium and induces a vasodilation that is conducted upstream increasing oxygen supply to the region of tissue supplied by the vessel [21].

In an elegant experiment, Winter and Dora, used triple cannulation of isolated arteries to enable focal application of purine and pyrimidine nucleotides to the endothelium, avoiding potential complicating actions of these agents on the smooth muscle [75]. Nucleotides were locally infused through one branch of a bifurcation, causing near maximal local dilatation attributable to EDHF. Dilatation then spread rapidly backwards to the adjacent feed artery and upstream against the direction of luminal flow, and increases flow into the feed artery. The data demonstrate that direct luminal stimulation of P2Y receptor on the endothelium of rat mesenteric arteries leads to marked spreading retrograde dilatation and thus suggests that circulating purines and pyrimidines may act as important regulators of blood flow [75]. In summary, ATP and UTP are released by endothelial and red blood cells in tissues in need of increased blood flow and stimulate an EDHF-dependent dilatation that spreads in a retrograde direction to the supplying artery to increase blood flow (Fig. 1.2).

1.8.2 Sympathicolytic Effects Mediated Via Endothelial Activation

Sympathetic nerves express inhibitory P2Y and P1 (A1) receptors indicating a P2-receptor mediated negative feedback loop both directly by ATP and its degradation product adenosine [5, 23]. An important balancing mechanism is the endothelium dependent sympathicolytic effect seen in exercising skeletal muscle. This mechanism is mimicked by injection in the arterial lumen of ATP and UTP, inducing a vasodilatation that overrides sympathetic vasoconstrictor activity in human skeletal muscle, an effect not obtained by injection of adenosine, indicating involvement of $P2Y_2$ receptors [55, 56]. Thus, endothelial P2 receptors mediate increases in skeletal blood flow during exercise both via direct dilatory effects, but also via a sympathicolytic effect.

1.8.3 Reactive Hyperemia

Reactive hyperemia is the massive increase in blood flow that starts when a blood vessel is opened after a period of ischemia. It is well known that ATP is released during ischemia [30, 70] and ADP has been shown to mediate the midportion and peak of hyperemia via endothelial $P2Y_1$ receptors [48]. ADP is then degraded to adenosine mediating the late phase hyperemia via A_{2A} receptors on SMC [57]. It is possible that ATP mediates dilatation during the first phase.

1.8.4 The Role of P2 Receptors in Circulatory Shock and Sepsis

It is possible that in extreme conditions such as circulatory or septic shock with acidosis and hypoxia, high ATP levels could be deleterious, leading to a drop in blood pressure [69]. At levels above 100 μM, ATP concentrations may exceed the catalytic capacity of ectonucleotidases and, could in fact, stimulate ATP release by increasing permeability of the RBC [69], probably via $P2X_7$ receptors [60]. ATP can also release ATP from endothelial cells [4]. At high concentrations of ATP, a self sustaining process may thus be instigated which may contribute to the irreversible stage of circulatory shock that can develop rapidly in severely ill patients. Similar mechanisms may be of importance in malaria because induction of the osmolyte permeability in Plasmodium-infected erythrocytes involves purinoceptor signalling [65].

Recently, a role for $P2X_7$ receptors in alpha-hemolysin induced hemolysis was suggested. The pore formation triggers purinergic receptor activation to mediate the full hemolytic action. These findings potentially have clinical perspectives as P2 antagonists may ameliorate symptoms during sepsis with hemolytic bacteria [59].

1.8.5 Hypertension

Purinergic blood pressure regulation is the net result of the balancing contractile and dilatory effects described above. ATP and UTP released on the luminal side from endothelial cells and erythrocytes stimulates vasodilatation in contrast to release from nerves on the adventitial side which results in vasoconstriction.

In spontaneously hypertensive rats (SHR), the importance of ATP as a sympathetic cotransmitter is increased [7, 71]. The renal vasculature shows an enhanced responsiveness to ab-meATP in perfused rat kidneys from SHR [28], while mesenteric vascular contractile reactivity to ATP via $P2X_1$ and $P2Y_2$ receptors is not altered in DOCA-salt hypertension [29], and in the aorta of SHR, the endothelium-dependent relaxation to ATP is impaired because of the simultaneous generation of an endothelium-derived contracting factor (EDCF). Vasoconstriction to ATP is potentiated in SHR aorta [80].

Diadenosine polyphosphates such as AP_4A, AP_5A and AP_6A are combinations of two adenosine molecules connected with 4–6 phosphate groups. They have been identified as vasocontractile agents [58], via actions on $P2X_1$ and $P2Y_2$ receptors. AP_5A and AP_6A are stored at higher levels in platelets from patients with hypertension and may contribute to their increased peripheral vascular resistance [34]. UP_4A is a novel endothelium derived vasoconstrictive factor more potent than endothelin in renal vasoconstriction [37]. It is released upon stimulation of the endothelium by acetylcholine, thrombin and mechanical stress and can be cleaved into either ATP or UTP to stimulate both $P2X_1$ and $P2Y_2$ receptors on VSMC resulting in increased blood pressure [37]. $P2X_4$-/- mice demonstrated a higher blood pressure and excrete smaller amounts of NO products in their urine than do wild-type mice [79]. $P2X_4$ receptor protein is up-regulated in the placenta in preeclampsia [53]. Interestingly, some of the most commonly used medicines to lower blood pressure, beta-blockers, may in part mediate their dilatory effects via release of ATP from endothelial cells [39].

1.8.6 Pulmonary Hypertension

Erythrocyte release of ATP regulates pulmonary resistance [63], and patients with pulmonary hypertension has impaired release of ATP from red blood cells [64]. Endothelium-dependent relaxation to ATP has been demonstrated in human pulmonary arteries [31]. Infusions of ATP-$MgCl_2$ have been claimed to be clinically useful in the treatment of children with pulmonary hypertension [8]. The endothelium derived vasoconstrictive factor UP_4A is a potent vasoconstrictor in pulmonary arteries and may have a similar effect as endothelin in pulmonary hypertension [32].

1.8.7 Congestive Heart Failure

Congestive heart failure is accompanied by impaired peripheral blood flow and endothelial dysfunction with decreased release of NO. In a model of non-atherosclerotic congestive heart failure there was a minor decrease in P2Y-mediated total dilatation and a marked down-regulation of the NO-mediated dilatation, while the EDHF-dilatation was up-regulated. Increased EDHF-activity in CHF may represent a compensatory response to decreased NO-activity to preserve endothelial function and tissue perfusion [42].

1.9 Conclusion

Purines and pyrimidines are released into the vascular lumen from red blood cells and endothelial cells in situations when there is a need for increased tissue perfusion. They activate several P2-receptors on endothelial cells stimulating vasodilatation through release of NO and EDHF. A mechanism has been demonstrated of retrograde spreading of the dilatation to the supplying artery, thereby enabling improved tissue perfusion. This system is ubiquitous and could be important for normal physiology and cardiovascular disease.

References

1. Bar I, Guns PJ, Metallo J, Cammarata D, Wilkin F, Boeynams JM, Bult H, Robaye B. (2008) Knockout mice reveal a role for P2Y6 receptor in macrophages, endothelial cells, and vascular smooth muscle cells. Mol Pharmacol 74:777–84.
2. Bergfeld GR, Forrester T. (1992) Release of ATP from human erythrocytes in response to a brief period of hypoxia and hypercapnia. Cardiovasc Res 26:40–7.
3. Bodin P, Bailey D, Burnstock G. (1991) Increased flow-induced ATP release from isolated vascular endothelial cells but not smooth muscle cells. Br J Pharmacol 103:1203–5.
4. Bodin P, Burnstock G. (1996) ATP-stimulated release of ATP by human endothelial cells. J Cardiovasc Pharmacol 27:872–5.
5. Boehm S, Huck S. (1997) Receptors controlling transmitter release from sympathetic neurons in vitro. Progress in Neurobiology 51:225–42.

6. Boeynaems JM, Galand N. (1983) Stimulation of vascular prostacyclin synthesis by extracellular ADP and ATP. Biochem Biophys Res Commun 112:290–6.
7. Brock JA, Van Helden DF. (1995) Enhanced excitatory junction potentials in mesenteric arteries from spontaneously hypertensive rats. Pflugers Arch 430:901–8.
8. Brook MM, Fineman JR, Bolinger AM, Wong AF, Heymann MA, Soifer SJ. (1994) Use of ATP-MgCl2 in the evaluation and treatment of children with pulmonary hypertension secondary to congenital heart defects. Circulation 90:1287–93.
9. Burnstock G. (1972) Purinergic nerves. Pharmacological Reviews 24:509–81.
10. Burnstock G. (1976) Do some nerve cells release more than one transmitter? Neuroscience 1:239–48.
11. Burnstock G. (1978) A basis for distinguishing two types of purinergic receptor. In: Cell membrane receptors for drugs and hormones: A multidiciplinary approach. New York, Raven press.
12. Burnstock G. (1987) Local control of blood pressure by purines. Blood Vessels 24:156–60.
13. Burnstock G. (1988) Local purinergic regulation of blood pressure (The First John T. Shepherd Lecture). In: Vasodilatation: Vascular Smooth Muscle, Peptides, Autonomic Nerves, and Endothelium. New York, Raven Press.
14. Burnstock G. (1988) Sympathetic purinergic transmission in small blood vessels. Trends Pharmacol Sci 9:116–17.
15. Burnstock G. (1990) Local mechanisms of blood flow control by perivascular nerves and endothelium. J Hypertens Suppl 8:S95–S106.
16. Burnstock G. (1997) The past, present and future of purine nucleotides as signalling molecules. Neuropharmacology 36:1127–39.
17. Burnstock G. (1999) Release of vasoactive substances from endothelial cells by shear stress and purinergic mechanosensory transduction. J Anat 194 (Pt 3):335–42.
18. Burnstock G. (2002) Purinergic signalling and vascular cell proliferation and death. Arteriosclerosis, Thrombosis Vasc Biol 22:364–73.
19. Burnstock G. (2004) Introduction: P2 receptors. Curr Top Med Chem 4:793–803.
20. Burnstock G, Kennedy C. (1986) A dual function for adenosine 5′-triphosphate in the regulation of vascular tone. Excitatory cotransmitter with noradrenaline from perivascular nerves and locally released inhibitory intravascular agent. Circul Res 58:319–30.
21. Collins DM, McCullough WT, Ellsworth ML. (1998) Conducted vascular responses: communication across the capillary bed. Microvasc Res 56:43–53.
22. Cosby K, Partovi KS, Crawford JH, Patel RP, Reiter CD, Martyr S, Yang BK, Waclawiw MA, Zalos G, Xu X, Huang KT, Shields H, Kim-Shapiro DB, Schechter AN, Cannon RO, 3rd, Gladwin MT. (2003) Nitrite reduction to nitric oxide by deoxyhemoglobin vasodilates the human circulation. Nat Med 9:1498–505.
23. Cunha R. (2001) Adenosine as a neuromodulator and as a homeostatic regulator in the nervous system: different roles, different sources and different receptors. Neurochem Int 38:107–25.
24. De Mey JG, Vanhoutte PM. (1981) Role of the intima in cholinergic and purinergic relaxation of isolated canine femoral arteries. J Physiol 316:347–55.
25. Drury AN, Szent-Gyorgyi A. (1929) The physiological activity of adenine compounds with especial reference to their action upon the mammalian heart. J Physiol 68:213–37.
26. Ellsworth ML, Forrester T, Ellis CG, Dietrich HH. (1995) The erythrocyte as a regulator of vascular tone. Am J Physiol 269:H2155–61.
27. Erlinge D. (1998) Extracellular ATP: a growth factor for vascular smooth muscle cells. Gen Pharmacol 31:1–8.
28. Fernandez O, Wangensteen R, Osuna A, Vargas F. (2000) Renal vascular reactivity to P(2)-purinoceptor activation in spontaneously hypertensive rats. Pharmacology 60:47–50.
29. Galligan JJ, Hess MC, Miller SB, Fink GD. (2001) Differential localization of P2 receptor subtypes in mesenteric arteries and veins of normotensive and hypertensive rats. J Pharmacol Exp Ther 296:478–85.
30. Gordon JL. (1986) Extracellular ATP: effects, sources and fate. Biochem J 233:309–19.

31. Greenberg B, Rhoden K, Barnes PJ. (1987) Endothelium-dependent relaxation of human pulmonary arteries. Am J Physiol 252:H434–8.
32. Gui Y, Walsh MP, Jankowski V, Jankowski J, Zheng XL. (2008) Up4A stimulates endothelium-independent contraction of isolated rat pulmonary artery. Am J Physiol Lung Cell Mol Physiol 294:L733–8.
33. Guns PJ, Korda A, Crauwels HM, Van Assche T, Robaye B, Boeynaems JM, Bult H. (2005) Pharmacological characterization of nucleotide P2Y receptors on endothelial cells of the mouse aorta. Br J Pharmacol 146:288–95.
34. Hollah P, Hausberg M, Kosch M, Barenbrock M, Letzel M, Schlatter E, Rahn KH. (2001) A novel assay for determination of diadenosine polyphosphates in human platelets: studies in normotensive subjects and in patients with essential hypertension. J Hypertens 19:237–45.
35. Hrafnkelsdottir T, Erlinge D, Jern S. (2001) Extracellular nucleotides ATP and UTP induce a marked acute release of tissue-type plasminogen activator in vivo in man. Thromb Haemost 85:875–81.
36. Imai M, Takigami K, Guckelberger O, Kaczmarek E, Csizmadia E, Bach FH, Robson SC. (2000) Recombinant adenoviral mediated CD39 gene transfer prolongs cardiac xenograft survival. Transplantation 70:864–70.
37. Jankowski V, Tolle M, Vanholder R, Schonfelder G, van der Giet M, Henning L, Schluter H, Paul M, Zidek W, Jankowski J. (2005) Uridine adenosine tetraphosphate: a novel endothelium- derived vasoconstrictive factor. Nat Med 11:223–7.
38. Kaczmarek E, Koziak K, Sevigny J, Siegel JB, Anrather J, Beaudoin AR, Bach FH, Robson SC. (1996) Identification and characterization of CD39/vascular ATP diphosphohydrolase. J Biol Chem 271:33116–22.
39. Kalinowski L, Dobrucki LW, Szczepanska-Konkel M, Jankowski M, Martyniec L, Angielski S, Malinski T. (2003) Third-generation beta-blockers stimulate nitric oxide release from endothelial cells through ATP efflux: a novel mechanism for antihypertensive action. Circulation 107:2747–52.
40. Koyamada N, Miyatake T, Candinas D, Hechenleitner P, Siegel J, Hancock WW, Bach FH, Robson SC. (1996) Apyrase administration prolongs discordant xenograft survival. Transplantation 62:1739–43.
41. Liu C, Mather S, Huang Y, Garland CJ, Yao X. (2004) Extracellular ATP facilitates flow-induced vasodilatation in rat small mesenteric arteries. Am J Physiol Heart Circ Physiol 286:H1688–95.
42. Malmsjo M, Bergdahl A, Zhao XH, Sun XY, Hedner T, Edvinsson L, Erlinge D. (1999) Enhanced acetylcholine and P2Y-receptor stimulated vascular EDHF-dilatation in congestive heart failure. Cardiovasc Res 43:200–9.
43. Malmsjo M, Chu ZM, Croft K, Erlinge D, Edvinsson L, Beilin LJ. (2002) P2Y receptor-induced EDHF vasodilatation is of primary importance for the regulation of perfusion pressure in the peripheral circulation of the rat. Acta Physiol Scand 174:301–9.
44. Malmsjo M, Edvinsson L, Erlinge D. (1998) P2U-receptor mediated endothelium-dependent but nitric oxide-independent vascular relaxation. Br J Pharmacol 123:719–29.
45. Malmsjo M, Erlinge D, Hogestatt ED, Zygmunt PM. (1999) Endothelial P2Y receptors induce hyperpolarisation of vascular smooth muscle by release of endothelium-derived hyperpolarising factor. Eur J Pharmacol 364:169–73.
46. McMahon TJ, Moon RE, Luschinger BP, Carraway MS, Stone AE, Stolp BW, Gow AJ, Pawloski JR, Watke P, Singel DJ, Piantadosi CA, Stamler JS. (2002) Nitric oxide in the human respiratory cycle. Nat Med 8:711–7.
47. Mistry H, Gitlin JM, Mitchell JA, Hiley CR. (2003) Endothelium-dependent relaxation and endothelial hyperpolarization by P2Y receptor agonists in rat-isolated mesenteric artery. Br J Pharmacol 139:661–71.
48. Olivecrona GK, Gotberg, M., Harnek, J., Wang, L., Jacobson, K. A. & Erlinge, D. (2006) Coronary artery reperfusion: The ADP receptor P2Y1 mediates early reactive hyperemia in vivo in pigs. Purinergic Signalling 1:59–64.

49. Olsson RA, Pearson JD. (1990) Cardiovascular purinoceptors. Physiol Rev 70:761–845.
50. Plank M, Wall D, David T. (2006) Atherosclerosis and calcium signalling in endothelial cells. Prog Biophys Mol Biol 91:287–313.
51. Ralevic V, Burnstock G. (1991) Roles of P2-purinoceptors in the cardiovascular system. Circulation 84:1–14.
52. Ralevic V, Burnstock G. (1998) Receptors for purines and pyrimidines. Pharmacol Rev 50:413–92.
53. Roberts VH, Webster RP, Brockman DE, Pitzer BA, Myatt L. (2006) Post-Translational Modifications of the P2X(4) Purinergic Receptor Subtype in the Human Placenta are Altered in Preeclampsia. Placenta.
54. Robson SC, Kaczmarek E, Siegel JB, Candinas D, Koziak K, Millan M, Hancock WW, Bach FH. (1997) Loss of ATP diphosphohydrolase activity with endothelial cell activation. J Exp Med 185:153–63.
55. Rosenmeier JB, Hansen J, Gonzalez-Alonso J. (2004) Circulating ATP-induced vasodilatation overrides sympathetic vasoconstrictor activity in human skeletal muscle. J Physiol 558: 351–65.
56. Rosenmeier JB, Yegutkin GG, Gonzalez-Alonso J. (2008) Activation of ATP/UTP-selective receptors increases blood flow and blunts sympathetic vasoconstriction in human skeletal muscle. J Physiol 586:4993–5002.
57. Saito D, Steinhart CR, Nixon DG, Olsson RA. (1981) Intracoronary adenosine deaminase reduces canine myocardial reactive hyperemia. Circ Res 49:1262–7.
58. Schluter H, Offers E, Bruggemann G, van der Giet M, Tepel M, Nordhoff E, Karas M, Spieker C, Witzel H, Zidek W. (1994) Diadenosine phosphates and the physiological control of blood pressure. Nature 367:186–8.
59. Skals M, Jorgensen NR, Leipziger J, Praetorius HA. (2009) Alpha-hemolysin from Escherichia coli uses endogenous amplification through P2X receptor activation to induce hemolysis. Proc Natl Acad Sci USA 106:4030–5.
60. Sluyter R, Shemon AN, Barden JA, Wiley JS. (2004) Extracellular ATP increases cation fluxes in human erythrocytes by activation of the P2X7 receptor. J Biol Chem 279: 44749–55.
61. Smits P, Williams SB, Lipson DE, Banitt P, Rongen GA, Creager MA. (1995) Endothelial release of nitric oxide contributes to the vasodilator effect of adenosine in humans. Circulation 92:2135–41.
62. Sprague RS, Ellsworth ML, Stephenson AH, Lonigro AJ. (2001) Participation of cAMP in a signal-transduction pathway relating erythrocyte deformation to ATP release. Am J Physiol Cell Physiol 281:C1158–64.
63. Sprague RS, Olearczyk JJ, Spence DM, Stephenson AH, Sprung RW, Lonigro AJ. (2003) Extracellular ATP signaling in the rabbit lung: erythrocytes as determinants of vascular resistance. Am J Physiol Heart Circ Physiol 285:H693–700.
64. Sprague RS, Stephenson AH, Ellsworth ML, Keller C, Lonigro AJ. (2001) Impaired release of ATP from red blood cells of humans with primary pulmonary hypertension. Exp Biol Med (Maywood) 226:434–9.
65. Tanneur V, Duranton C, Brand VB, Sandu CD, Akkaya C, Kasinathan RS, Gachet C, Sluyter R, Barden JA, Wiley JS, Lang F, Huber SM. (2006) Purinoceptors are involved in the induction of an osmolyte permeability in malaria-infected and oxidized human erythrocytes. Faseb J 20:133–5.
66. Todorov LD, Mihaylova-Todorova S, Westfall TD, Sneddon P, Kennedy C, Bjur RA, Westfall DP. (1997) Neuronal release of soluble nucleotidases and their role in neurotransmitter inactivation. Nature 387:76–9.
67. Tolle M, Jankowski V, Schuchardt M, Wiedon A, Huang T, Hub F, Kowalska J, Jemielity J, Guranowski A, Loddenkemper C, Zidek W, Jankowski J, van der Giet M. (2008) Adenosine 5′-tetraphosphate is a highly potent purinergic endothelium-derived vasoconstrictor. Circ Res 103:1100–8.

68. Touchman JW, Anikster Y, Dietrich NL, Maduro VV, McDowell G, Shotelersuk V, Bouffard GG, Beckstrom-Sternberg SM, Gahl WA, Green ED. (2000) The genomic region encompassing the nephropathic cystinosis gene (CTNS): complete sequencing of a 200-kb segment and discovery of a novel gene within the common cystinosis-causing deletion. Genome Res 10:165–73.
69. Trams EJ KH, Burnstock GG. (1980) A proposal for the role of ecto-enzymes and adenylates in traumatic shock. J Theor Biol 87:609–21.
70. Vassort G. (2001) Adenosine 5′-triphosphate: a P2-purinergic agonist in the myocardium. Physiol Rev 81:767–806.
71. Vidal M, Hicks PE, Langer SZ. (1986) Differential effects of alpha-beta-methylene ATP on responses to nerve stimulation in SHR and WKY tail arteries. Naunyn Schmiedebergs Arch Pharmacol 332:384–90.
72. Wang L, Karlsson L, Moses S, Hultgardh-Nilsson A, Andersson M, Borna C, Gudbjartsson T, Jern S, Erlinge D. (2002) P2 receptor expression profiles in human vascular smooth muscle and endothelial cells. J Cardiovasc Pharmacol 40:841–53.
73. Wang L, Olivecrona G, Gotberg M, Olsson ML, Winzell MS, Erlinge D. (2005) ADP acting on P2Y13 receptors is a negative feedback pathway for ATP release from human red blood cells. Circ Res 96:189–96.
74. Wihlborg AK, Malmsjo M, Eyjolfsson A, Gustafsson R, Jacobson K, Erlinge D. (2003) Extracellular nucleotides induce vasodilatation in human arteries via prostaglandins, nitric oxide and endothelium-derived hyperpolarising factor. Br J Pharmacol 138:1451–8.
75. Winter P, Dora KA. (2007) Spreading dilatation to luminal perfusion of ATP and UTP in rat isolated small mesenteric arteries. J Physiol 582:335–47.
76. Yamamoto K, Korenaga R, Kamiya A, Ando J. (2000) Fluid shear stress activates Ca(2+) influx into human endothelial cells via P2X4 purinoceptors. Circ Res 87:385–91.
77. Yamamoto K, Korenaga R, Kamiya A, Qi Z, Sokabe M, Ando J. (2000) P2X(4) receptors mediate ATP-induced calcium influx in human vascular endothelial cells. Am J Physiol Heart Circ Physiol 279:H285–92.
78. Yamamoto K, Shimizu N, Obi S, Kumagaya S, Taketani Y, Kamiya A, Ando J. (2007) Involvement of cell surface ATP synthase in flow-induced ATP release by vascular endothelial cells. Am J Physiol Heart Circ Physiol 293:H1646–53.
79. Yamamoto K, Sokabe T, Matsumoto T, Yoshimura K, Shibata M, Ohura N, Fukuda T, Sato T, Sekine K, Kato S, Isshiki M, Fujita T, Kobayashi M, Kawamura K, Masuda H, Kamiya A, Ando J. (2006) Impaired flow-dependent control of vascular tone and remodeling in P2X4-deficient mice. Nat Med 12:133–7.
80. Yang D, Gluais P, Zhang JN, Vanhoutte PM, Feletou M. (2004) Endothelium-dependent contractions to acetylcholine, ATP and the calcium ionophore A 23187 in aortas from spontaneously hypertensive and normotensive rats. Fundam Clin Pharmacol 18:321–6.
81. Yegutkin G, Bodin P, Burnstock G. (2000) Effect of shear stress on the release of soluble ecto-enzymes ATPase and 5′-nucleotidase along with endogenous ATP from vascular endothelial cells. Br J Pharmacol 129:921–6.
82. Yegutkin GG, Samburski SS, Mortensen SP, Jalkanen S, Gonzalez-Alonso J. (2007) Intravascular ADP and soluble nucleotidases contribute to acute prothrombotic state during vigorous exercise in humans. J Physiol 579:553–64.
83. Zimmermann H. (2006) Ectonucleotidases in the nervous system. Novartis Found Symp 276:113–28; discussion 28–30, 233–7, 75–81.

Chapter 2
Nucleotides and Novel Signaling Pathways in Endothelial Cells: Possible Roles in Angiogenesis, Endothelial Dysfunction and Diabetes Mellitus

Elzbieta Kaczmarek

Abstract In this chapter, we will focus on the signal transduction pathways in endothelial cells (ECs) studied by us and other groups. We will present data showing nucleotide-mediated activation of focal adhesion kinase (FAK), AMP-activated protein kinase (AMPK) and endothelial nitric oxide synthase (eNOS), and document the role of Ca^{2+} in these pathways. Activation by extracellular nucleotides of mitogen-activated protein kinases (MAPK), such as extracellular signal-regulated kinase (ERK), p38, and JUN NH_2-terminal kinase (JNK), phosphatidylinositol-3 kinase (PI3K), as well as the mammalian target of rapamycin (mTOR) pathway will be also discussed. Our data indicate that extracellular nucleotides activate FAK and eNOS, and modulate α_v integrin expression, EC cytoskeletal rearrangements and migration, functions associated with angiogenesis. Activation of eNOS and AMPK suggests that nucleotides acting through P2 receptors may exert anti-inflammatory, anti-oxidative, pro-proliferative, and anti-apoptotic effects in the endothelium. Sustaining of eNOS activation in ECs exposed to high glucose concentrations supports our hypothesis that extracellular nucleotides play a protective role against endothelial dysfunction observed in diabetes. We have also evidence that purine nucleotides increase intracellular energy levels, possibly protecting ECs from stress-related loss of intracellular ATP. Numerous nucleotide-mediated effects on the endothelium remain to be elucidated.

Keywords Endothelial cells · P2Y receptors · P2X receptors · ATP · UTP · ADP · Adenosine · Calcium flux · Focal adhesion kinase (FAK) · α_v integrin · AMP-activated protein kinase (AMPK) · Endothelial nitric oxide synthase (eNOS) · Nitric oxide (NO) · Protein kinase C delta (PKCδ) · Calcium/calmodulin-activated kinase II · LKB1 · Extracellular signal-regulated kinase (ERK) · Mammalian target of rapamycin (mTOR) · Nuclear factor kappa B (NF-κB) · Diabetes

---

E. Kaczmarek (✉)
Center for Vascular Biology Research, Surgery Department, Microvascular Surgery Division, Beth Israel Deaconess Medical Center, Harvard Medical School, 99 Brookline Avenue, Boston, MA, USA
e-mail: ekaczmar@bidmc.harvard.edu

E. Gerasimovskaya, E. Kaczmarek (eds.), *Extracellular ATP and Adenosine as Regulators of Endothelial Cell Function*, DOI 10.1007/978-90-481-3435-9_2,

2.1 Introduction

For decades, the endothelium was recognized only as a barrier between blood and tissues located beneath it. Data collected during the last several years provided evidence that ECs actively participate in many physiological and pathological events. The endothelium comprises a continuous, stable and quiescent, single cell layer that lines the vasculature. A growing body of experimental data indicates a central role for the endothelium in the regulation of vascular homeostasis, local immune or inflammatory reactions associated with platelet deposition and responses to vascular injury [27].

Effects of extracellular nucleotides in the vasculature, e.g., modulation of blood vessel relaxation, have been recognized for many years. However, the molecular mechanisms underlying the endothelial responses have not been fully elucidated. Recent advances in the purinergic field were enabled by cloning of P2 receptors and their detailed pharmacological characterization, followed by identification of signal transduction pathways initiated by extracellular nucleotides in various tissues/cell types. The role of particular P2 receptors was further elucidated in studies with P2 receptor-knockout mice. Published results indicate that extracellular nucleotides activate several signaling pathways in ECs with outcomes similar to those caused by other regulatory molecules, such as VEGF or thrombin [16, 47, 79].

There are several mechanisms proposed whereby nucleotides are released from tissues into extracellular fluids in response to cell activation by various stimuli, including hypoxia and stress. These mechanisms include exocytosis from nucleotide-containing granules [11], efflux through a membrane transport systems (such as ATP-binding cassette transporters) [9, 83], release by connexin or pannexin hemichannels [48, 89, 93, 101, 103] (also discussed in Chapter 10) and by plasmalemmal voltage-dependent channels [73, 96]. In addition, nucleotides can be released as a consequence of cell death or tissue damage. Extracellular nucleotide levels are also regulated by ecto-enzymes (addressed in Chapter 5), such as ATP-synthase, nucleoside diphosphate kinase and adenylate kinase [80, 126, 127] and ecto-NTPDases, which, in concert with CD73 (5′-nucleotidase), metabolize extracellular nucleotides to adenosine [50, 63, 66, 129].

Extracellular nucleotides act as paracrine or autocrine mediators via activation of purinergic P2 receptors [15, 16, 39, 90]. Pharmacological, functional, and molecular cloning data have facilitated the classification of P2 receptors into two main groups: P2Y, G protein-coupled receptors and P2X, ligand-gated ion channels [17, 18]. Eight P2Y receptor subtypes (P2Y1, P2Y2, P2Y4, P2Y6, P2Y11-P2Y14) and seven P2X receptors subtypes (P2X1-P2X7) have been described to date. While activation of P2Y receptors results in Ca^{2+} release from intracellular stores, P2X receptors facilitate the entry of extracellular Ca^{2+} [13]. P2X receptors are activated by ATP, whereas P2Y receptors respond to both purine (ATP, ADP) and pyrimidine (UTP, UDP, UDP-glucose) nucleotides. Specifically, in human tissues ATP is an agonist for P2Y2 and P2Y11 receptors, UTP for P2Y2 and P2Y4 receptors, ADP for P2Y1, P2Y12 and P2Y13 receptors, UDP for the P2Y6 receptor, and UDP-glucose for the P2Y14 receptor [17]. Activation of P2Y1, P2Y2, P2Y4, P2Y6, and P2Y11 receptors (coupled directly to Gq protein) stimulates phospholipase C, followed by an increase in inositol 1,4,5 trisphosphate and release of Ca^{2+} from intracellular stores [120]. P2Y11 receptors additionally mediate a stimulation of adenylate cyclase to generate cAMP. In contrast, activation of the P2Y12, P2Y13, and P2Y14 receptors (coupled to Gi protein) causes an inhibition of adenylate cyclase activity and a decrease in intracellular cAMP levels [12].

It is clear now, that extracellular nucleotides, acting via P2 receptors and adenosine acting via P1/adenosine receptors (A1, A2a, A2b, A3), activate multiple signaling pathways in various cells, including ECs.

2.2 P2 Receptors in ECs and Calcium Responses to Extracellular Nucleotides

Human umbilical vein endothelial cells (HUVECs) express multiple P2X and P2Y receptor subtypes, including P2X1, P2X2, P2X4, P2X6, P2X7, P2Y1, P2Y2, P2Y4, P2Y6, and P2Y11 receptors, as determined by PCR and/or quantitative PCR and partially confirmed by Western blot analysis when reliable antibodies were available ([55, 94, 121] and our unpublished data). P2Y1, P2Y2 and P2X4 receptors are highly expressed in ECs, compared to other P2 receptor subtypes.

Early cellular responses to extracellular nucleotides include elevation of $[Ca^{2+}]_i$ due to calcium influx from the extracellular space or release from intracellular stores [24]. Calcium ions function as second messengers, regulating numerous cellular processes. Brief spikes and waves in the intracellular free calcium ion concentration ($[Ca^{2+}]_i$) regulate the expression of selected genes [37, 71]. We analyzed calcium responses by confocal microscopy in HUVECs loaded with fluo-3, a fluorescent Ca^{2+} indicator. ATP, ADP, UTP and their slowly hydrolysable analogs (ATPγS and ADPβS) mediated significant increases in $[Ca^{2+}]_i$. However, UDP (P2Y6 receptor ligand), BzATP (P2X7 and P2Y11 receptor ligand) and α-β-meATP (a ligand for P2X receptors) in the concentration range studied (1–200 μM) did not induce increases in $[Ca^{2+}]_i$, suggesting that only P2Y1, P2Y2 and possibly P2Y4 receptors are functionally active in HUVECs [62]. Moreover, the nucleotide-induced increase in $[Ca^{2+}]_i$ was inhibited by BAPTA (an intracellular Ca^{2+} chelator) and thapsigargin (which depletes Ca^{2+} from intracellular stores by inhibiting endoplasmic reticulum Ca^{2+} ATPase) but not by EGTA (a chelator of extracellular Ca^{2+}) [62]. This indicates that ATP, ADP and UTP-induced changes in $[Ca^{2+}]_i$ depend on Ca^{2+} release from intracellular stores and therefore P2Y rather than P2X receptors are responsible for nucleotide-mediated increases in $[Ca^{2+}]_i$, and most likely for Ca^{2+}-related signaling events in HUVEC. This is confirmed by other studies suggesting that P2Y, and not P2X, receptors are functionally active in the endothelium [115].

In contrast to these results, the roles of some P2X receptor subtypes have been reported in ECs. This includes P2X4 receptors, whose involvement in the EC response to shear stress is recognized. In human pulmonary artery ECs, ATP released in response to shear stress mediates Ca^{2+} influx, which was inhibited by calcium chelator or apyrase, an enzyme that hydrolyzes ATP and ADP to AMP [125]. Studies with P2X1- and P2X4-knockout mice, also suggest that these receptors are active in the endothelium [56, 57, 124].

There are also reports that P2X7 receptors are expressed in ECs [22, 91]. We confirmed P2X7 receptor expression in HUVECs at the mRNA [119] and protein levels (data not shown). Moreover, we localized these receptors not only to the plasma membrane but also to the nuclear envelope, as analyzed by immunocytochemistry and Western blot analysis using various cell fractions (data not shown). P2X7 receptor expression in the nuclear membrane was earlier shown in neurons, smooth muscle cells and epithelial cells of visceral organs of the guinea pig [6, 77]. The function of P2X7 receptors in the nuclear membrane could be to open ion channels that increase the nuclear Ca^{2+} concentration, thereby affecting intra-nuclear processes, such as the regulation of gene expression. Therefore, the discrepancy between detection of P2X7 receptors in ECs and the lack of a calcium response to BzATP (an agonist of this receptor) needs to be resolved.

Multiple functions for P2Y1, P2Y2 and possibly P2Y4 receptors in the vasculature are well documented. These receptors have been implicated in the regulation of angiogenesis [99], vascular tone [53], cell migration [8, 62], and mitogenic responses [46, 117].

Although we were not able to demonstrate functional expression of P2Y6 receptors in HUVECs (*i.e.*, no changes in $[Ca^{2+}]_i$ were seen in response to UDP), there are data showing that this receptor is functional in ECs originating from other vascular beds. The endothelium-dependent relaxation of the aorta mediated by UDP was not apparent in P2Y6-knockout mice [10]. Accordingly, the contractile effect of UDP on the aorta, observed when eNOS was blocked, was also abolished in P2Y6-null mice.

Our unpublished data demonstrate that populations of HUVECs (passage 1–4) stimulated with extracellular nucleotides respond with increases in $[Ca^{2+}]_i$. Remarkably, nucleotide-induced changes in $[Ca^{2+}]_i$ at the level of the single cell were heterogeneous. Single cells stimulated sequentially with ADP, UTP and ATP respond either to all three nucleotides, or only to one or two of these ligands. The intensity of these responses, as well as the number and size of the calcium spikes were also different between individual cells. Interestingly, with increased passage number (from 1 to 4), HUVEC responses to ADP diminish, suggesting that either the expression of the P2Y1 receptor, its functionality or both are decreased with time in culture. This observation could be explained by the in vivo data showing a decrease in P2Y1 expression in aging (19-month) rat aortic and carotid arteries, as compared to younger (2-month) animals [78]. In contrast to HUVECs, single cells of the EC line, EAhy926, respond to selected extracellular nucleotides (ATP, UTP and ADP) in exactly the same way as the whole cell population. Immunocytochemical staining of HUVECs with available antibodies to P2 receptors indicate that different cells express varying levels of the same type of P2 receptor. We hypothesize that ECs may exhibit heterogeneity in the context of P2 receptor subtype expression and/or functionality. The physiological importance of this heterogeneity remains to be elucidated. Our hypothesis regarding P2 receptor heterogeneity in ECs is supported by

Buvinic *et al.* who demonstrated that functional expression levels of P2Y receptors along the cord, superficial chorionic vessels and cotyledons of the human placenta differ [20]. P2Y1 and P2Y2 receptors were not only unevenly distributed along the human placental vascular tree but also were coupled to different signaling pathways in the cord/chorionic vessels versus the cotyledon leading to opposing vasomotor responses.

2.3 Extracellular Nucleotide-Initiated Signaling Pathways

2.3.1 Extracellular Nucleotides Activate FAK and Induce Cytoskeletal Changes and EC Migration

We have demonstrated that stimulation of HUVECs with ATP or UTP induces phosphorylation of FAK, paxillin and $p130^{cas}$, molecules that were reported to regulate cell adhesion, spreading and migration [95, 97]. Indeed, we have further shown that exposure of HUVECs to extracellular UTP or ATP induces actin filament formation and cytoskeletal rearrangements resulting in the spreading of HUVECs, and enhances cell migration [62]. Since chelation of intracellular calcium with BAPTA inhibited UTP-induced FAK phosphorylation, as well as EC migration, these responses are likely linked to Ca^{2+}originating from intracellular stores, and therefore to activation of P2Y2 and possibly P2Y4 receptors. Migration of HUVECs in response to ADP, a ligand for P2Y1 receptors, was much less significant than that observed with ATP or UTP. However, in another study, involvement of P2Y1 receptors (via MAPK activation) in EC migration has been documented [109]. Moreover, we showed that the mechanism of nucleotide-induced migration of HUVECs depends upon PI3K activation, as inhibitors of PI3K, wortmannin and LY294002, attenuate migration of HUVEC in response to ATP or UTP [62].

It has been reported that $\alpha_v\beta_3$ integrin promotes cell adhesion to the extracellular matrix, cell migration and angiogenesis [67]. Integrins are also involved in activation of the FAK signaling pathway [108], and an increase in $[Ca^{2+}]_i$ is necessary to induce the interaction of integrins with the cytoskeleton [84]. Moreover, the integrin-binding motif RGD has been identified in P2Y2 receptors and shown to promote co-localization of this receptor with $\alpha_v\beta_3$ integrin, enabling nucleotides to activate integrin signaling pathways, cytoskeletal rearrangements and cell migration [8, 43, 72]. All these data prompted us to investigate effects of extracellular nucleotides on integrin expression. Indeed, we have observed an increase in the expression of α_v integrin in HUVECs stimulated with UTP or ATP. The mechanism of this up-regulation of α_v integrin expression by extracellular nucleotides still needs to be elucidated. A scheme of extracellular nucleotide effects on FAK, PI3K and integrin activation is presented in Fig. 2.1.

Phosphorylation of FAK and paxillin has been observed previously in ECs treated with VEGF [1, 7], and P2Y2 receptors have been recently reported to transactivate

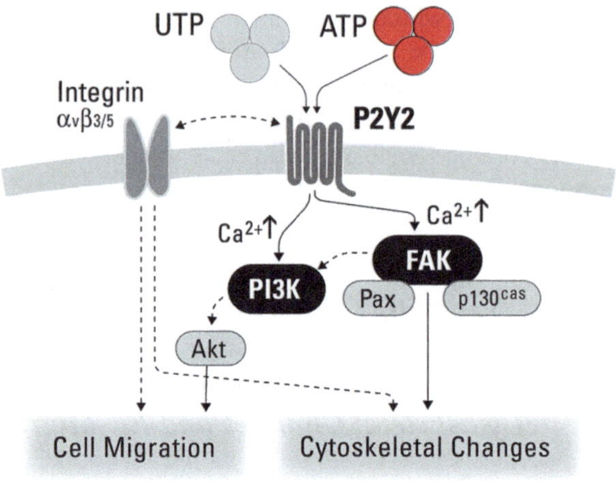

Fig. 2.1 A scheme of the P2 receptor-initiated effects (activation of FAK and PI3K, and interaction with integrin) leading to EC migration and cytoskeletal changes

VEGF receptor-2 (VEGFR-2) [99, 100, 107] (discussed in Chapter 4 of this book). Whether described here nucleotide-mediated effects observed in HUVECs are VEGFR-2-dependent or not, remain to be determined.

2.3.2 Extracellular Nucleotides and Adenosine Activate AMPK in HUVECs

AMPK is a heterotrimeric Ser/Thr kinase consisting of a catalytic α subunit and regulatory β and γ subunits. AMPK is activated in response to various stimuli that induce an increase in the intracellular AMP:ATP ratio caused by intracellular ATP depletion. This decrease in intracellular ATP levels can originate from pathological cellular stresses, such as heat shock, hypoxia or ischemia, and from physiological exercise-induced skeletal muscle contraction [54]. AMPK can also be phosphorylated and activated by the mechanism independent of changes in the intracellular AMP:ATP ratio [32]. AMPK is activated up to 5-fold by allosteric modification by AMP, and 50–100-fold by phosphorylation of Thr-172 on the α subunit by upstream kinases, including the tumor suppressor kinase, LKB1 and calcium/calmodulin-dependent protein kinase kinase (CaMKK) [58, 113]. AMPK has been localized to skeletal muscle, brain, liver, pancreas, as well as ECs [30]. However, mechanisms of AMPK activation, and the function of this kinase in the endothelium still are not fully characterized. We aimed to determine whether extracellular nucleotides are involved in the activation of AMPK.

Studies from our laboratory indicate that the extracellular nucleotides, ATP, UTP and ADP, as well as adenosine, induce phosphorylation of AMPK and its downstream target, acetyl-CoA carboxylase (ACC), in time- and nucleotide concentration-dependent manners. Further, by pharmacological analysis, we showed that P2Y1, P2Y2 and possibly P2Y4 receptors are involved in the activation of AMPK, while we excluded participation of P2Y6, P2Y11, P2X7, and other P2X receptors [28].

Additionally, we confirmed that Ca^{2+} released from the endoplasmic reticulum plays an important role in ATP/UTP-induced phosphorylation of AMPK, as a chelator of intracellular free calcium ions, BAPTA, almost completely attenuated AMPK phosphorylation. Moreover, we identified CaMKK as the kinase responsible for nucleotide-induced AMPK activation in ECs, whereas we excluded contributions from LKB1, PI3K, protein kinase C (PKC) and CaMK II [28].

As mentioned earlier in this chapter, ECs express CD39 (NTPDase I), an NTPDase that hydrolyzes nucleoside tri- and diphosphates to mono-phosphates, as well as 5′-nucleotidase (CD73), which can further hydrolyze AMP to adenosine [63, 129]. Therefore, we investigated whether phosphorylation of AMPK induced by nucleotides was related to generation of extracellular adenosine. Inhibition of 5′-nucleotidase with α,β-methylene ADP (AOPCP) had no effect on ATP/UTP/ADP-induced AMPK phosphorylation, indicating that adenosine generation is not involved in AMPK activation induced by extracellular nucleotides.

Extracellular adenosine can exert cellular responses by two mechanisms, either through activation of cell surface P1 receptors or via uptake by nucleoside transporters whereupon intracellular conversion to adenine nucleotides can affect cellular functions. We demonstrated that adenosine but not its metabolites, inosine or hypoxanthine, can activate AMPK. Furthermore, we established that the effect of adenosine was not mediated by P1 receptors, but by adenosine uptake facilitated by nucleoside transporters. Finally, we clarified that the intracellular conversion of adenosine to AMP is important for AMPK activation. We hypothesize that intracellular AMP generated from adenosine by adenosine kinase, allosterically activates AMPK so that it is more sensitive to further activation by upstream kinases. While searching for the upstream kinases, we excluded CaMKK, as well as PI3K, and showed that adenosine-induced phosphorylation of AMPK was dependent on activation of LKB1. Therefore, we conclude that the pathway of adenosine-induced activation of AMPK differs from the one identified for extracellular nucleotides [28].

We expected that AMPK activation induced by adenosine, which is converted to AMP within the cell, would be sensitive to changes in the AMP:ATP ratio. We observed an increase in intracellular AMP levels in response to extracellular adenosine, however, this increase was accompanied by a corresponding increase in ATP levels. Therefore, the ratio of AMP to ATP did not change significantly in response to extracellular adenosine, as compared to other studies showing 6 to 25-fold increases in the AMP:ATP ratio in adenosine treated cells [26, Hardie, 1998 #734]. We assume that the actual level of AMP is more important for activation of AMPK than the AMP:ATP ratio [122]. Therefore, we conclude that adenosine-induced activation of AMPK is linked to LKB1 activation and is enhanced by increases in the intracellular AMP level.

In summary, our findings demonstrate two distinct, but converging pathways for AMPK activation in HUVECs. One pathway induced by extracellular nucleotides is linked to activation of P2Y1, P2Y2 and possibly P2Y4 receptors, and is dependent on $[Ca^{2+}]_i$ and CaMKK. The second pathway induced by extracellular adenosine is

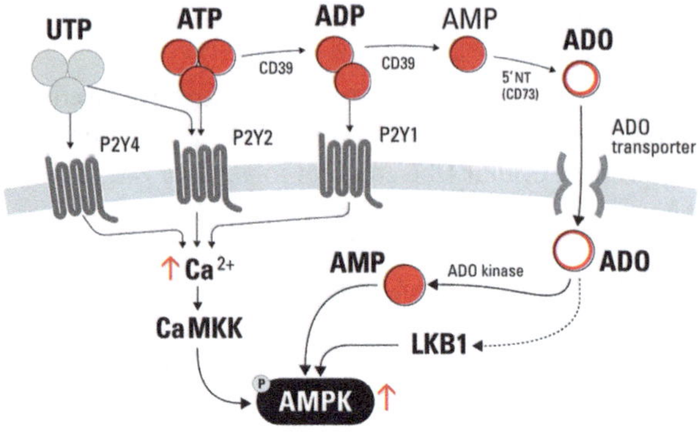

Fig. 2.2 A scheme of two signaling pathways initiated by extracellular nucleotides and adenosine, leading to AMPK activation in EC

dependent upon adenosine uptake followed by the generation of intracellular AMP that promotes activation of AMPK via LKB1. A scheme of these two pathways is presented in Fig. 2.2. Extracellular nucleotides and adenosine whose local concentrations under pathophysiological conditions can be significantly increased, could play an important role in maintaining cellular energy homeostasis by AMPK activation that modulates glucose uptake and fatty acid metabolism. We propose that P2Y receptors and nucleoside transporters could be novel targets for a pharmacological intervention to regulate AMPK and hence EC metabolism under pathological conditions.

2.3.3 eNOS Activation by Extracellular Nucleotides

NO is an important signaling molecule regulating inflammation, angiogenesis and thrombosis [74]. NO generated in the endothelium promotes vasorelaxation, protects ECs from apoptosis, inhibits vascular smooth muscle cell proliferation, blocks platelet activation and enhances vascular growth. As such, NO maintains vascular homeostasis. A decrease in NO production is implicated in endothelial dysfunction. In ECs, NO synthesis requires the activation of eNOS. The expression and function of eNOS is regulated at multiple levels: transcriptional, post-transcriptional (mRNA stability) and post-translational (O-glycosylation, myristoylation, palmitoylation and phosphorylation). Additional regulation occurs through eNOS dimerization and interaction with regulatory proteins, such as calmodulin (CaM), caveolin-1 and heat-shock protein 90 (HSP90), as well as eNOS subcellular targeting to caveolae, other cellular membranes or the cytoplasm [40, 45].

Basal production of NO is maintained through allosteric activation of eNOS by Ca^{2+}/CaM, secondary to activation of phospholipase C gamma (PLCγ), an increase in $[Ca^{2+}]_i$ and activation of CaM. However, to reach high activity, eNOS has to be phosphorylated at Ser-1177 [36]. Several kinases, including CaMKII, PI3K/Akt, AMPK, ERK, p38 MAPK, PKCα, and protein kinase A (PKA), have been reported to phosphorylate Ser-1177 and thus activate eNOS [5, 14, 36, 81, 85, 102, 104, 116, 128].

Regulation of vessel contraction/relaxation by extracellular nucleotides has been well known for many years and is associated with NO generation [19] in response to a conformational activation of eNOS by Ca^{2+}/CaM [64, 75]. However, until recently the impact of extracellular nucleotides on eNOS phosphorylation at Ser-1177 had not been fully elucidated. We showed that ADP, ATP, and UTP induce Ser-1177-eNOS phosphorylation, indicating that P2Y1, P2Y2 and possibly P2Y4 receptors are the main receptors involved in purinergic activation of eNOS in ECs [29]. We excluded P2Y6 and P2Y11 receptors, as their respective ligands, UDP and BzATP, did not induce eNOS phosphorylation.

Our results also demonstrate that nucleotide-induced eNOS phosphorylation is Ca^{2+}-dependent, as chelation of intracellular Ca^{2+} with BAPTA completely inhibited Ser-1177 phosphorylation. However, activation of Ca^{2+}-dependent kinases upstream of eNOS activation, including CaMKII and CaMKK [102] is not involved in nucleotide-induced eNOS phosphorylation on Ser-1177, even though extracellular nucleotides can activate CaMKK in ECs [28].

We also excluded AMPK, a reported upstream activator of eNOS [116] that is activated by extracellular nucleotides in ECs [28]. Inhibition of AMPK with a pharmacological inhibitor, Compound C, or by expression of dominant-negative AMPKα2 did not decrease eNOS phosphorylation induced by extracellular nucleotides [29].

Likewise, pharmacological inhibitors of PI3K/Akt, ERK, p38, PKA, PKG, and PKCα did not affect nucleotide-induced eNOS phosphorylation. In contrast, pharmacological inhibitors of PKCδ or suppression of PKCδ expression with siRNA decreased nucleotide-induced eNOS activation in ECs. Additionally, we confirmed that nucleotide-induced eNOS phosphorylation correlates with increases in cGMP levels, likely caused by NO generation. Accordingly, we propose that extracellular nucleotides induce eNOS activation, through Ca^{2+}-dependent conformational changes and PKCδ–dependent phosphorylation of eNOS (Fig. 2.3).

In contrast to our results, Brown et al. showed that eNOS is activated by P2Y receptors via the PKCα isoform [14]. Results from Ian Bird's group suggest roles for Ca^{2+} and ERK1/2 in pregnancy-related eNOS activation in uterine arteries [35]. In other studies, P2X1 [56] and P2X4 [124] receptor activation has been implicated in NO-dependent vasodilation.

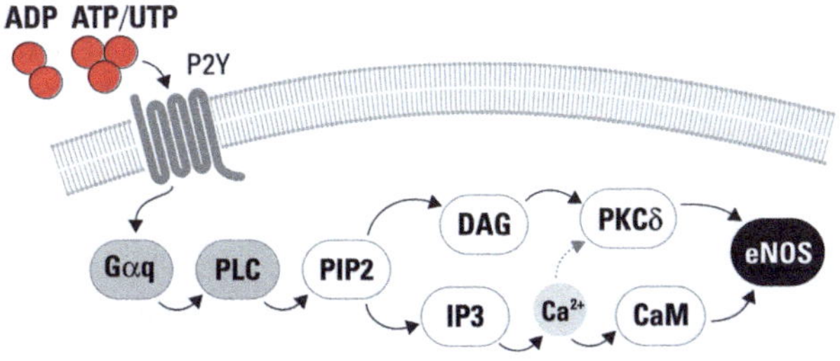

Fig. 2.3 A scheme of the proposed signaling pathway of the extracellular nucleotide-induced phosphorylation of eNOS in EC

2.3.4 Effects of Extracellular Nucleotides on ECs Exposed to High Glucose

Maintenance of eNOS activity and NO bioavailability in the endothelium has been proved to be of great importance. In certain diseases, including atherosclerosis and diabetes mellitus, NO production is impaired as a result of attenuation of eNOS expression and activity [105]. Reduced NO production, resulting from high glucose-induced inactivation of eNOS, is one of the key pathophysiological causes of EC dysfunction and vascular complications in diabetic patients.

It is documented that NO exerts anti-inflammatory effects in ECs through inhibition of nuclear factor kappa B (NF-κB) activation by various mechanisms, including stabilization of inhibitor κBα (IκBα), nitrosylation of p50, and quenching of reactive oxygen species (ROS) [25]. NO also protects ECs from apoptosis by reducing oxidative damage [60], promotes angiogenesis by increasing VEGF production [41] and inhibits platelet aggregation leading to homeostasis of thrombotic events [92].

Relevant to decreased NO bioavailability in diabetes, high glucose/ROS decrease eNOS activity by activation of PKCβII (which negatively regulates eNOS activity through its phosphorylation at Thr-495 [23, 76]), post-translational O-glycosylation of the Ser-1177 residue (which decreases phosphorylation of this residue that is required for eNOS activity [38, 44]), and inhibition of the PI3K/Akt pathway upstream of eNOS [42].

We hypothesized that the Akt-independent activation of eNOS by PKCδ in ECs treated with extracellular nucleotides [29] may be resistant to the damaging effects of high glucose. Indeed, we verified this hypothesis by showing that, in contrast to PI3K/Akt-dependent eNOS activation by VEGF that is inhibited by high glucose, 20–30 mM D-glucose did not affect nucleotide-induced phosphorylation and activation of eNOS (data not shown).

2.3.5 Effects of ATP in Diabetic ApoE-Knockout Mice

To evaluate the in vivo effects of extracellular nucleotides, we carried out pharmacokinetic studies in C57BL/6 mice. First, we verified the bioavailability of ATP in mouse blood after a single intraperitoneal (i.p.) injection of this nucleotide (1 mmol/kg body weight). ATP levels in the serum showed a 7-fold increase (from 0.7 ± 0.13 to 5.3 ± 1.1 μM) 4 hours after the ATP injection, in agreement with previously published data [69].

Then, we investigated the effects of ATP administration in atherosclerosis-prone apolipoprotein E (ApoE)-deficient mice (ApoE-KO) rendered diabetic with streptozotocin. ApoE-KO mice, control and diabetic, were injected daily with saline or ATP (1 mmol/kg) i.p. from 12 to 20 weeks of age. Twenty-week old mice were sacrificed and their aortic arches and hearts were stained with hematoxilin/eosin (H/E) or analyzed by immunohistochemistry (IHC). Our preliminary data for a small animal group ($n = 3$) show greater atherosclerotic lesion areas in aortic arches of saline-treated diabetic ApoE-KO mice than ATP-treated animals. Atherosclerotic lesions in ATP-treated diabetic ApoE-KO mice exhibited a decrease in immunostaining for the pro-inflammatory proteins, p65/NF-κB, VCAM-1, and P-selectin, as compared to saline-treated ApoE-KO mice. Administration of ATP also attenuated expression of the NADPH oxidase subunit $p22^{phox}$, which should result in decreased NADPH oxidase activity and reduced ROS formation (data not shown). These data need further confirmation in a larger animal group of experimental animals, however they are very promising, and suggest that administration of P2 receptor agonists could represent a novel therapeutic approach to protect vascular endothelium from accelerated atherosclerosis in diabetes.

In a similar, but not diabetic in vivo model, atherosclerotic lesions were analyzed in ApoE-KO mice and mice deficient in both ApoE and the P2Y1 receptor [59]. Atherosclerotic lesions in the aortic sinus and entire aorta were smaller in ApoE/P2Y1-KO mice, compared with ApoE-KO mice. Moreover, in ApoE/P2Y1-KO mouse aorta, pro-inflammatory VCAM-1 expression and macrophage infiltration were reduced, compared with ApoE-KO mice. However, detailed experiments with bone marrow transplantation into ApoE-KO and ApoE/P2Y1-KO mice indicated that non-hematopoietic-derived P2Y1 receptors, most likely originating from the endothelial or smooth muscle cells, contribute to development of atherosclerotic lesions. These data are contradictory to our results, showing rather anti-atherosclerotic and anti-inflammatory effects of administered ATP in diabetic ApoE-KO mice. However, as these two animal systems differ, direct comparisons of the data are difficult.

2.3.6 Other Signaling Pathways Activated by Extracellular Nucleotides in ECs

We have characterized in detail signaling pathways activated by extracellular nucleotides, leading to activation of FAK, AMPK and eNOS. However, published and unpublished results indicate that extracellular nucleotides also activate other signaling pathways in ECs. These include ERK1/2 [4, 86, 109], p38 MAPK [68, 107, 109] and JNK [109], a family of Ser/Thr protein kinases that transduces signals from the cell membrane to the nucleus in response to various stimuli and are implicated in the expression of genes that regulate EC proliferation and apoptosis. In one study, P2Y receptors were shown to activate ERK in HUVECs, independent of calcium, through a pathway involving PI3K, 3-phosphoinositide-dependent protein kinase-1 (PDK1), and PKCζ [79]. In another study, PKCα was shown to be responsible for ERK activation in response to ATP and UTP in the EAhy926 endothelial cell line [51].

It was also reported that ATP released by oxidative stress induces P2Y receptor-dependent activation of cell survival proteins, PI3K and ERK1/2 in human lung microvascular endothelial cells (HLMVECs), a prime target of hyperoxic injury. Moreover, under hypoxia, ATP activates mTOR-dependent pathways and increases mTOR-mediated uptake of glucose, in keeping with the importance of glucose metabolism in protection from hyperoxia [2, 3].

Our preliminary data indicate that adenine nucleotides and adenosine induce a significant increase (~1.8-fold) in the levels of intracellular ATP (ATPi; manuscript under preparation). This increase in ATPi levels occurred independently of AMPK activation. Overexpression in ECs of a dominant-negative form of AMPK α2 using a recombinant adenovirus or incubation of ECs with Compound C, a pharmacological inhibitor of AMPK, did not inhibit the rise in ATPi levels following treatment with adenosine. We assume that a significant increase in ATPi could be highly relevant for the protection of the endothelium from stress signals. This includes oxidative stress observed when ECs are exposed to high glucose levels, which has a strong negative impact on ATPi levels [61, 118]. This "surplus" of ATPi induced by extracellular adenine nucleotides or adenosine could be useful in cell protection from apoptosis or could serve to regulate cellular functions, such as cell growth or migration. Further studies should elucidate the importance of this observation. Presently, we can only speculate about possible effects of increased ATPi levels on the regulation of intracellular proteins that utilize ATPi. These include heat shock proteins (HSP; including HSP70 and HSP90), mTOR, as well as P2X7 receptors expressed in the nuclear membrane, to name a few.

One of the signaling pathways of potential importance for cell growth and proliferation, and therefore angiogenesis, is the mTOR pathway. It was reported that in ECs isolated from the pulmonary artery vasa vasorum (VVECs) ATP, ADP and MeSADP, acting via P2Y receptors, induced activation of several kinases, including ERK1/2, PI3K/Akt, mTOR and its downstream effector p70S6K. P2Y receptor activation was associated with increased DNA synthesis, cell migration and tube formation in Matrigel. Further analysis showed that the PI3K/Akt and mTOR/p70S6K pathways were not involved in extracellular ATP-induced tube formation. Interestingly, these effects were cell-specific, as they were observed only in pulmonary artery vasa vasorum EC (VVECs), but not in pulmonary artery ECs (MPAECs) or aortic ECs (AOECs) obtained from the same animals [47]. Moreover, ATP potentiated the effect of the angiogenic factors, VEGF and bFGF, on the stimulation of DNA synthesis in VVECs, but not in MPAECs and AOECs. These data suggest a mechanistic explanation for the role of extracellular ATP in the expansion of the vasa vasorum that can be observed in hypoxic conditions. At the same time, this study demonstrates cell specificity in the response to extracellular nucleotides, as well as interplay between P2 receptor signaling and other pathways.

2.3.7 Cross-talk Between Purinergic and Other Signaling Pathways

Several recent papers have reported cross-talk between purinergic and other signaling pathways. Some of these pathways seem to have great importance for EC functions.

In the EAhy926 EC line, UTP, ATP, but not UDP, inhibited TNFα-induced stimulation of the stress-activated protein kinases, JNK, p38, and its downstream target, MAPKAP kinase-2, which was mediated by Ca^{2+}-independent isoforms of PKC. These results reveal a novel cross-talk between P2Y receptors and the TNFα-stimulated p38 pathway in ECs [87].

Interaction between P2Y receptors and VEGF signaling pathways has been also documented [107]. Activation of the P2Y2 receptor with its ligand, UTP, up-regulated the expression of pro-inflammatory vascular cell adhesion molecule-1 (VCAM-1) in human coronary artery ECs (HCAECs). This effect of UTP was mediated via P2Y2 receptor-dependent phosphorylation of VEGF receptor-2 (VEGFR-2) and activation of its signaling pathway with involvement of Src homology-3-binding sites in the C-terminus of the P2Y2 receptor. The small GTPases Rho (RhoA) were implicated in this pathway, as a dominant negative form of RhoA inhibited P2Y2 receptor-mediated VCAM-1 expression. Moreover, it was established that VEGFR-2 and the P2Y2 receptor co-localize upon UTP stimulation.

In another study, Rumjahn et al. demonstrated that activation of P2Y receptors promotes EC angiogenesis. Activated P2Y1 receptors transactivate VEGFR-2 in the absence of VEGF, which stimulates EC tubulogenesis. This P2Y1 receptor-VEGFR-2 interaction and resulting signal transduction may affect vascular homoeostasis and tumor-mediated angiogenesis [100].

Interplay between P2Y receptor-induced signaling and growth factor receptor has been reported to modulate activity of MAPK. Specifically, it was documented that UTP transiently activated ERK in PC12 cells via P2Y2 receptor-mediated transactivation of the epidermal growth factor receptor, downstream at the level of the related adhesion focal tyrosine kinase (RAFTK) [111]. It remains to be determined if similar cross-talk between tyrosine kinases and purinergic receptors signaling exists in ECs.

There is also evidence of cross-talk between P2Y2 receptors and integrins [8, 43, 62]. This interaction affects cell chemotaxis and migration and therefore might be associated with regulation of angiogenesis by nucleotide-induced signaling.

On the basis of established data, we hypothesize that purinergic signaling may influence other signaling pathways. One of the novel roles of AMPK is to prevent the activation of NF-κB in ECs exposed to the fatty acid palmitate or the cytokine TNFα. The increases in NFκB activation and VCAM-1 expression induced by palmitate were markedly inhibited by AICAR, the AMPK activator, and by expression of a constitutively active AMPK [21]. The mechanism responsible for this observation remains to be elucidated, however it is possible that interplay between extracellular nucleotide-induced activation of AMPK and the TNF/NF-κB signaling pathway could play a role in the anti-inflammatory effects of purines in the endothelium.

Likewise, NO exerts anti-inflammatory effects in cytokine-stimulated human saphenous vein endothelial cells, by inhibiting NFκB-mediated up-regulation of VCAM, E-selectin and to some extent, ICAM-1 [34, 88]. Therefore, we predict that extracellular nucleotides by activation of eNOS and induction of NO generation may negatively affect the NFκB signaling pathway.

2.4 Conclusions, Hypotheses, and Controversies

We have described three extracellular nucleotide-initiated signaling pathways, leading to activation of FAK, AMPK and eNOS. Moreover, we showed that purine nucleotides and adenosine increase levels of ATPi in EC. Therefore we hypothesize that extracellular nucleotides may regulate various EC functions, including cell migration, proliferation and apoptosis (Fig. 2.4).

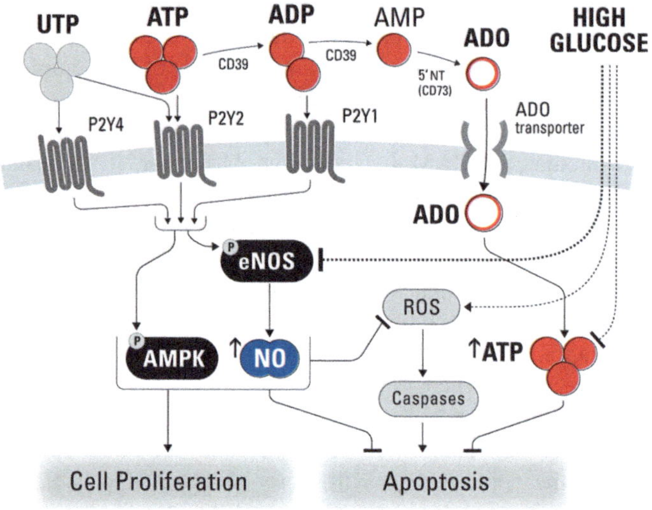

Fig. 2.4 A scheme of the extracellular nucleotide/adenosine-induced signaling pathways leading to activation of AMPK and eNOS, and an increase in the intracellular ATP (ATPi) concentration, which may regulate EC proliferation and apoptosis

These three signaling pathways are linked by their dependence on $[Ca^{2+}]_i$. Indeed, chelation of intracellular Ca^{2+} inhibits nucleotide-induced activation of FAK, AMPK, and eNOS. However, these signaling pathways diverge downstream, possibly due to compartmentalization of purinergic signaling. Otherwise, AMPK, which has been shown by other studies to be an upstream kinase that activates eNOS, would also phosphorylate eNOS on Ser-1177 in our experimental system.

Our results obtained with pharmacological ligands and antagonists indicate that P2Y1 and P2Y2 receptors are the main functional P2 receptors in HUVECs, however, these data need to be validated by additional experiments with genetic effectors, *i.e.*, specific siRNA to suppress expression of individual P2 receptor subtypes. Our studies have been restricted to HUVECs and HCAECs, however, as suggested by other reports, ECs originating from other vascular beds may express different repertoires of P2 receptor subtypes and thus respond to extracellular nucleotides differently. This intriguing aspect of purinergic signaling awaits further investigations as well.

Currently available data on the signaling effects of extracellular nucleotides were obtained from studies of ECs from different species, such as human, bovine, pig, rat, mouse, rabbit and sheep, originating from different vascular beds, i.e., umbilical vein [28], coronary artery [106], aorta [117], pulmonary artery [112], pulmonary artery VVEC [47], corneal vessels [49], uterine artery [123], brain capillary [110], renal artery [65], retinal microvasculature [114], and saphenous vein [34]. However, in these ECs, different purinergic effects and signaling pathways were studied, making comparisons of the results difficult. Therefore, effects of extracellular nucleotides in these different ECs should be systematically evaluated.

The outcomes of purinergic signaling in the endothelium also need further evaluation. We demonstrated that nucleotide-induced activation of FAK/Pax/p130 and up-regulation and transactivation of integrin α_V were followed by cytoskeletal changes and PI3K-dependent EC migration, indicating that P2 receptors may be involved in the regulation of angiogenesis. Some other data demonstrating purinergic activation of mTOR, PI3K/Akt and ERK, pathways implicated in cell growth, survival, migration and proliferation, support this hypothesis [47]. However, contrary to numerous in vitro and in vivo data showing that nucleotides increase SMC proliferation, similar data for ECs are missing. Our own (unpublished) data indicate that depending on the EC type and experimental conditions, nucleotides can stimulate, inhibit or have no effect on EC proliferation.

A role for purinergic signaling in EC apoptosis, postulated to contribute to vascular injury or the regulation of angiogenesis, also awaits additional study, as not much data are available on this subject. Adenosine, but not ATP or ADP, has been shown to induce apoptosis in pulmonary artery ECs by increasing intracellular S-adenosylhomocysteine levels and impairing methylation of some proteins or nucleic acids [33, 98]. In another study suggesting that ATP and ADP induce EC apoptosis, very high levels of ATP were used (10 mM) to induce cell death, whereas 100 μM and 1 mM ATP had no effect on EC apoptosis [119]. Moreover, in this report, high concentrations of ATP and ADP activated NFκB, whereas in ECs this pathway exerts anti-apoptotic effects [52]. Again, the role of nucleotides in EC apoptosis has to be further investigated.

Our data indicate that extracellular nucleotides induce rapid and transient activation of eNOS, AMPK, and mTOR/p70S6k and a delayed increase in ATPi. However, some of these data are not consistent, *i.e.*, activation of AMPK should turn-off the mTOR pathway, which is energy-dependent.

EC activation and damage are associated with blood coagulation, inflammatory responses (e.g., up-regulation of adhesion molecules), endothelial dysfunction (e.g., reduced NO generation), and atherosclerotic lesion development (e.g., neointima formation). Hyperglycemia/diabetes mellitus and hypertension are conditions that negatively affect vascular function. Therefore, nucleotide-induced activation of eNOS and AMPK may exert protective effects in some vascular-related diseases. This hypothesis needs further validation.

Although the multiple functions of eNOS/NO in ECs are well established, and we showed that extracellular nucleotides activate eNOS, outcomes associated with NO generation induced by extracellular nucleotides should be confirmed. We expect to verify that P2 receptor signaling via NO may protect ECs from high glucose-induced apoptosis, attenuate cytokine-mediated up-regulation of pro-inflammatory proteins or reduce oxidative stress.

Much less is known about AMPK function in ECs. Contrary to other reports, we have not confirmed AMPK participation in eNOS activation. AMPK regulates not only cellular energy metabolism, but can protect the endothelium from various stresses, including high levels of glucose and fatty acids, such as observed in obesity and type 2 diabetes [31]. AMPK also modulates signaling pathways that control cell growth, differentiation, and survival and therefore affects angiogenesis [82]. Other studies indicate that AMPK activated under hypoxia is involved in up-regulation of hypoxia induced factor-1 (HIF-1), a transcription factor that is critical for hypoxic induction of physiologically important genes, as well as secretion of VEGF and stimulation of glucose uptake [70]. Other evidence suggests that AMPK inhibits the activation of NFκB in HUVECs induced by both palmitate and TNFα [21]. Further studies on the physiological and pathophysiological roles of AMPK in endothelium are necessary. Again, whether extracellular nucleotides regulate EC metabolism or angiogenesis via AMPK needs to be verified.

We have also shown that not only nucleotides, but also adenosine activates AMPK and increases levels of ATPi. Interestingly, these effects of adenosine are not mediated by P1 receptors but are related to intracellular effects of adenosine after uptake via adenosine transporters.

Finally, from a translational point of view, activation of eNOS and AMPK seems to be relevant to endothelial dysfunction linked to metabolic syndrome and diabetes. We propose that P2 receptor-mediated signaling "rescues" eNOS activation and NO generation under high glucose conditions, even when the PI3K-Akt pathway is disabled, whereas VEGF-induced activation of eNOS is significantly attenuated by high glucose concentrations. This is a novel observation suggesting that P2 receptors may represent viable targets for pharmacological intervention to regulate eNOS activity and hence protect the endothelium from the deleterious effects of high levels of glucose.

We expect that in a few years many of the unknown functions of P2 receptors in the endothelium will have been elucidated.

Acknowledgments Published and unpublished data that were presented and discussed in this chapter, were obtained by Cleide da Silva, Katarzyna Koziak, Robert Jarzyna, Anke Specht, Anne Spristersbach, Eva Csizmadia, Maryse Picher, Veronica Klepeis, Vickery Trinkaus-Randall, Jan Krzysztof Blusztajn, Agnes Kittel, Laurie Erb, Gary Weisman, Marcia Wink, Olaf Guckelberger, Simon Robson, Barbara Wegiel and Christiane Ferran, whom I would like to thank.

This work was supported in part by the National Institutes of Health (HL66167) and the Juvenile Diabetes Research Foundation (5-2007-736).

References

1. Abedi H, Zachary I. (1997) Vascular endothelial growth factor stimulates tyrosine phosphorylation and recruitment to new focal adhesions of focal adhesion kinase and paxillin in endothelial cells. J Biol Chem 272:15442–51.
2. Ahmad S, Ahmad A, Ghosh M, Leslie CC, White CW. (2004) Extracellular ATP-mediated signaling for survival in hyperoxia-induced oxidative stress. J Biol Chem 279: 16317–25.

3. Ahmad S, Ahmad A, White CW. (2006) Purinergic signaling and kinase activation for survival in pulmonary oxidative stress and disease. Free Radic Biol Med 41:29–40.
4. Albert JL, Boyle JP, Roberts JA, Challiss RA, Gubby SE, Boarder MR. (1997) Regulation of brain capillary endothelial cells by P2Y receptors coupled to Ca^{2+}, phospholipase C and mitogen-activated protein kinase. Br J Pharmacol 122:935–41.
5. Anter E, Thomas SR, Schulz E, Shapira OM, Vita JA, Keaney JF, Jr. (2004) Activation of endothelial nitric-oxide synthase by the p38 MAPK in response to black tea polyphenols. J Biol Chem 279:46637–43.
6. Atkinson L, Milligan CJ, Buckley NJ, Deuchars J. (2002) An ATP-gated ion channel at the cell nucleus. Nature 420:42.
7. Avraham HK, Lee TH, Koh Y, Kim TA, Jiang S, Sussman M, Samarel AM, Avraham S. (2003) Vascular endothelial growth factor regulates focal adhesion assembly in human brain microvascular endothelial cells through activation of the focal adhesion kinase and related adhesion focal tyrosine kinase. J Biol Chem 278:36661–8.
8. Bagchi S, Liao Z, Gonzalez FA, Chorna NE, Seye CI, Weisman GA, Erb L. (2005) The P2Y2 nucleotide receptor interacts with alphav integrins to activate Go and induce cell migration. J Biol Chem 280:39050–7.
9. Ballerini P, Di Iorio P, Ciccarelli R, Nargi E, D′Alimonte I, Traversa U, Rathbone MP, Caciagli F. (2002) Glial cells express multiple ATP binding cassette proteins which are involved in ATP release. Neuroreport 13:1789–92.
10. Bar I, Guns PJ, Metallo J, Cammarata D, Wilkin F, Boeynams JM, Bult H, Robaye B. (2008) Knockout mice reveal a role for P2Y6 receptor in macrophages, endothelial cells, and vascular smooth muscle cells. Mol Pharmacol 74:777–84.
11. Bodin P, Burnstock G. (2001) Evidence that release of adenosine triphosphate from endothelial cells during increased shear stress is vesicular. J Cardiovasc Pharmacol 38:900–8.
12. Bodor ET, Waldo GL, Blaesius R, Harden TK. (2004) Delineation of ligand binding and receptor signaling activities of purified P2Y receptors reconstituted with heterotrimeric G proteins. Purinergic Signal 1:43–9.
13. Brake AJ, Wagenbach MJ, Julius D. (1994) New structural motif for ligand-gated ion channels defined by an ionotropic ATP receptor. Nature 371:519–23.
14. Brown CA, Patel V, Wilkinson G, Boarder MR. (1996) P2 purinoceptor-stimulated conversion of arginine to citrulline in bovine endothelial cells is reduced by inhibition of protein kinase C. Biochem Pharmacol 52:1849–54.
15. Burnstock G. (2002) Purinergic signaling and vascular cell proliferation and death. Arterioscler Thromb Vasc Biol 22:364–73.
16. Burnstock G. (2006) Purinergic signalling. Br J Pharmacol 147(Suppl 1):S172–81.
17. Burnstock G. (2007) Purine and pyrimidine receptors. Cell Mol Life Sci 64:1471–83.
18. Burnstock G, Kennedy C. (1985) Is there a basis for distinguishing two types of P2-purinoceptor? Gen Pharmacol 16:433–40.
19. Buvinic S, Briones R, Huidobro-Toro JP. (2002) P2Y(1) and P2Y(2) receptors are coupled to the NO/cGMP pathway to vasodilate the rat arterial mesenteric bed. Br J Pharmacol 136:847–56.
20. Buvinic S, Poblete MI, Donoso MV, Delpiano AM, Briones R, Miranda R, Huidobro-Toro JP. (2006) P2Y1 and P2Y2 receptor distribution varies along the human placental vascular tree: role of nucleotides in vascular tone regulation. J Physiol 573:427–43.
21. Cacicedo JM, Yagihashi N, Keaney JF, Jr., Ruderman NB, Ido Y. (2004) AMPK inhibits fatty acid-induced increases in NF-kappaB transactivation in cultured human umbilical vein endothelial cells. Biochem Biophys Res Commun 324:1204–9.
22. Chiao CW, Tostes RC, Webb RC. (2008) P2X7 receptor activation amplifies lipopolysaccharide-induced vascular hyporeactivity via interleukin-1 beta release. J Pharmacol Exp Ther 326:864–70.
23. Chu S, Bohlen HG. (2004) High concentration of glucose inhibits glomerular endothelial eNOS through a PKC mechanism. Am J Physiol Renal Physiol 287:F384–92.

24. Communi D, Raspe E, Pirotton S, Boeynaems JM. (1995) Coexpression of P2Y and P2U receptors on aortic endothelial cells. Comparison of cell localization and signaling pathways. Circ Res 76:191–8.
25. Connelly L, Palacios-Callender M, Ameixa C, Moncada S, Hobbs AJ. (2001) Biphasic regulation of NF-kappa B activity underlies the pro and anti-inflammatory actions of nitric oxide. J Immunol 166:3873–81.
26. Corton JM, Gillespie JG, Hardie DG. (1994) Role of the AMP-activated protein kinase in the cellular stress response. Curr Biol 4:315–24.
27. Cotran RS, Pober JS. (1989) Effects of cytokines on vascular endothelium: their role in vascular and immune injury. Kidney Int 35:969–75.
28. da Silva CG, Jarzyna R, Specht A, Kaczmarek E. (2006) Extracellular nucleotides and adenosine independently activate AMP-activated protein kinase in endothelial cells: involvement of P2 receptors and adenosine transporters. Circ Res 98:e39–e47.
29. da Silva CG, Specht A, Wegiel B, Ferran C, Kaczmarek E. (2009) Mechanism of purinergic activation of endothelial nitric oxide synthase in endothelial cells. Circulation 119:871–9.
30. Dagher Z, Ruderman N, Tornheim K, Ido Y. (1999) The effect of AMP-activated protein kinase and its activator AICAR on the metabolism of human umbilical vein endothelial cells. Biochem Biophys Res Commun 265:112–5.
31. Dagher Z, Ruderman N, Tornheim K, Ido Y. (2001) Acute regulation of fatty acid oxidation and amp-activated protein kinase in human umbilical vein endothelial cells. Circ Res 88:1276–82.
32. Daniel T, Carling D. (2002) Expression and regulation of the AMP-activated protein kinase-SNF1 (sucrose non-fermenting 1) kinase complexes in yeast and mammalian cells: studies using chimaeric catalytic subunits. Biochem J 365:629–38.
33. Dawicki DD, Chatterjee D, Wyche J, Rounds S. (1997) Extracellular ATP and adenosine cause apoptosis of pulmonary artery endothelial cells. Am J Physiol 273:L485–94.
34. De Caterina R, Libby P, Peng HB, Thannickal VJ, Rajavashisth TB, Gimbrone MA, Jr., Shin WS, Liao JK. (1995) Nitric oxide decreases cytokine-induced endothelial activation. Nitric oxide selectively reduces endothelial expression of adhesion molecules and proinflammatory cytokines. J Clin Invest 96:60–8.
35. Di T, Sullivan JA, Magness RR, Zhang L, Bird IM. (2001) Pregnancy-specific enhancement of agonist-stimulated ERK-1/2 signaling in uterine artery endothelial cells increases Ca(2+) sensitivity of endothelial nitric oxide synthase as well as cytosolic phospholipase A(2). Endocrinology 142:3014–26.
36. Dimmeler S, Fleming I, Fisslthaler B, Hermann C, Busse R, Zeiher AM. (1999) Activation of nitric oxide synthase in endothelial cells by Akt-dependent phosphorylation. Nature 399:601–5.
37. Dolmetsch RE, Xu K, Lewis RS. (1998) Calcium oscillations increase the efficiency and specificity of gene expression [see comments]. Nature 392:933–6.
38. Du XL, Edelstein D, Dimmeler S, Ju Q, Sui C, Brownlee M. (2001) Hyperglycemia inhibits endothelial nitric oxide synthase activity by posttranslational modification at the Akt site. J Clin Invest 108:1341–8.
39. Dubyak GR, el-Moatassim C. (1993) Signal transduction via P2-purinergic receptors for extracellular ATP and other nucleotides. Am J Physiol 265:C577–C606.
40. Dudzinski DM, Michel T. (2007) Life history of eNOS: partners and pathways. Cardiovasc Res 75:247–60.
41. Dulak J, Jozkowicz A, Dembinska-Kiec A, Guevara I, Zdzienicka A, Zmudzinska-Grochot D, Florek I, Wojtowicz A, Szuba A, Cooke JP. (2000) Nitric oxide induces the synthesis of vascular endothelial growth factor by rat vascular smooth muscle cells. Arterioscler Thromb Vasc Biol 20:659–66.
42. el-Remessy AB, Bartoli M, Platt DH, Fulton D, Caldwell RB. (2005) Oxidative stress inactivates VEGF survival signaling in retinal endothelial cells via PI 3-kinase tyrosine nitration. J Cell Sci 118:243–52.

43. Erb L, Liu J, Ockerhausen J, Kong Q, Garrad RC, Griffin K, Neal C, Krugh B, Santiago-Perez LI, Gonzalez FA, Gresham HD, Turner JT, Weisman GA. (2001) An RGD sequence in the P2Y(2) receptor interacts with alpha(V)beta(3) integrins and is required for G(o)-mediated signal transduction. J Cell Biol 153:491–501.
44. Federici M, Menghini R, Mauriello A, Hribal ML, Ferrelli F, Lauro D, Sbraccia P, Spagnoli LG, Sesti G, Lauro R. (2002) Insulin-dependent activation of endothelial nitric oxide synthase is impaired by O-linked glycosylation modification of signaling proteins in human coronary endothelial cells. Circulation 106:466–72.
45. Fulton D, Gratton JP, Sessa WC. (2001) Post-translational control of endothelial nitric oxide synthase: why isn't calcium/calmodulin enough? J Pharmacol Exp Ther 299:818–24.
46. Gerasimovskaya EV, Ahmad S, White CW, Jones PL, Carpenter TC, Stenmark KR. (2002) Extracellular ATP is an autocrine/paracrine regulator of hypoxia-induced adventitial fibroblast growth. Signaling through extracellular signal-regulated kinase-1/2 and the Egr-1 transcription factor. J Biol Chem 277:44638–50.
47. Gerasimovskaya EV, Woodward HN, Tucker DA, Stenmark KR. (2008) Extracellular ATP is a pro-angiogenic factor for pulmonary artery vasa vasorum endothelial cells. Angiogenesis 11:169–82.
48. Gomes P, Srinivas SP, Van Driessche W, Vereecke J, Himpens B. (2005) ATP release through connexin hemichannels in corneal endothelial cells. Invest Ophthalmol Vis Sci 46: 1208–18.
49. Gomes P, Srinivas SP, Vereecke J, Himpens B. (2006) Gap junctional intercellular communication in bovine corneal endothelial cells. Exp Eye Res 83:1225–37.
50. Gordon EL, Pearson JD, Slakey LL. (1986) The hydrolysis of extracellular adenine nucleotides by cultured endothelial cells from pig aorta. Feed-forward inhibition of adenosine production at the cell surface. J Biol Chem 261:15496–507.
51. Graham A, McLees A, Kennedy C, Gould GW, Plevin R. (1996) Stimulation by the nucleotides, ATP and UTP of mitogen-activated protein kinase in EAhy 926 endothelial cells. Br J Pharmacol 117:1341–7.
52. Grimaldo S, Tian F, Li LY. (2009) Sensitization of endothelial cells to VEGI-induced apoptosis by inhibiting the NF-kappaB pathway. Apoptosis 14:788–95.
53. Guns PJ, Korda A, Crauwels HM, Van Assche T, Robaye B, Boeynaems JM, Bult H. (2005) Pharmacological characterization of nucleotide P2Y receptors on endothelial cells of the mouse aorta. Br J Pharmacol 146:288–95.
54. Hardie DG, Salt IP, Davies SP. (2000) Analysis of the role of the AMP-activated protein kinase in the response to cellular stress. Methods Mol Biol 99:63–74.
55. Harrington LS, Evans RJ, Wray J, Norling L, Swales KE, Vial C, Ali F, Carrier MJ, Mitchell JA. (2007) Purinergic 2X1 receptors mediate endothelial dependent vasodilation to ATP. Mol Pharmacol 72:1132–6.
56. Harrington LS, Mitchell JA. (2004) Novel role for P2X receptor activation in endothelium-dependent vasodilation. Br J Pharmacol 143:611–7.
57. Harrington LS, Mitchell JA. (2005) P2X1 receptors and the endothelium. Mem Inst Oswaldo Cruz 100(Suppl 1):111–2.
58. Hawley SA, Davison M, Woods A, Davies SP, Beri RK, Carling D, Hardie DG. (1996) Characterization of the AMP-activated protein kinase from rat liver and identification of threonine 172 as the major site at which it phosphorylates AMP-activated protein kinase. J Biol Chem 271:27879–87.
59. Hechler B, Freund M, Ravanat C, Magnenat S, Cazenave JP, Gachet C. (2008) Reduced atherosclerotic lesions in P2Y1/apolipoprotein E double-knockout mice: the contribution of non-hematopoietic-derived P2Y1 receptors. Circulation 118:754–63.
60. Hida A, Kawakami A, Miyashita T, Yamasaki S, Nakashima K, Tanaka F, Izumi Y, Tamai M, Huang M, Ida H, Nakamura H, Origuchi T, Ueki Y, Eguchi K. (2004) Nitric oxide acts on the mitochondria and protects human endothelial cells from apoptosis. J Lab Clin Med 144:148–55.

61. Ido Y, Carling D, Ruderman N. (2002) Hyperglycemia-induced apoptosis in human umbilical vein endothelial cells: inhibition by the AMP-activated protein kinase activation. Diabetes 51:159–67.
62. Kaczmarek E, Erb L, Koziak K, Jarzyna R, Wink MR, Guckelberger O, Blusztajn JK, Trinkaus-Randall V, Weisman GA, Robson SC. (2005) Modulation of endothelial cell migration by extracellular nucleotides: involvement of focal adhesion kinase and phosphatidylinositol 3-kinase-mediated pathways. Thromb Haemost 93:735–42.
63. Kaczmarek E, Koziak K, Sevigny J, Siegel JB, Anrather J, Beaudoin AR, Bach FH, Robson SC. (1996) Identification and characterization of CD39/vascular ATP diphosphohydrolase. J Biol Chem 271:33116–22.
64. Kalinowski L, Dobrucki LW, Szczepanska-Konkel M, Jankowski M, Martyniec L, Angielski S, Malinski T. (2003) Third-generation beta-blockers stimulate nitric oxide release from endothelial cells through ATP efflux: a novel mechanism for antihypertensive action. Circulation 107:2747–52.
65. Knight GE, Oliver-Redgate R, Burnstock G. (2003) Unusual absence of endothelium-dependent or -independent vasodilatation to purines or pyrimidines in the rat renal artery. Kidney Int 64:1389–97.
66. Koszalka P, Ozuyaman B, Huo Y, Zernecke A, Flogel U, Braun N, Buchheiser A, Decking UK, Smith ML, Sevigny J, Gear A, Weber AA, Molojavyi A, Ding Z, Weber C, Ley K, Zimmermann H, Godecke A, Schrader J. (2004) Targeted disruption of cd73/ecto-5′-nucleotidase alters thromboregulation and augments vascular inflammatory response. Circ Res 95:814–21.
67. Kronenwett R, Graf T, Nedbal W, Weber M, Steidl U, Rohr UP, Mohler T, Haas R. (2002) Inhibition of angiogenesis in vitro by alphav integrin-directed antisense oligonucleotides. Cancer Gene Ther 9:587–96.
68. Laird SM, Graham A, Paul A, Gould GW, Kennedy C, Plevin R. (1998) Tumour necrosis factor stimulates stress-activated protein kinases and the inhibition of DNA synthesis in cultures of bovine aortic endothelial cells. Cell Signal 10:473–80.
69. Lasso de la Vega MC, Terradez P, Obrador E, Navarro J, Pellicer JA, Estrela JM. (1994) Inhibition of cancer growth and selective glutathione depletion in Ehrlich tumour cells in vivo by extracellular ATP. Biochem J 298 (Pt 1):99–105.
70. Lee M, Hwang JT, Lee HJ, Jung SN, Kang I, Chi SG, Kim SS, Ha J. (2003) AMP-activated protein kinase activity is critical for hypoxia-inducible factor-1 transcriptional activity and its target gene expression under hypoxic conditions in DU145 cells. J Biol Chem 278: 39653–61.
71. Li W-H, Llopis J, Whitney M, Zlokarnik G, Tsien RY. (1998) Cell-permeant caged InsP3 ester shows that Ca^{2+} spike frequency can optimize gene expression [see comments]. Nature 392:936–41.
72. Liao Z, Seye CI, Weisman GA, Erb L. (2007) The P2Y2 nucleotide receptor requires interaction with alpha v integrins to access and activate G12. J Cell Sci 120: 1654–62.
73. Liu HT, Toychiev AH, Takahashi N, Sabirov RZ, Okada Y. (2008) Maxi-anion channel as a candidate pathway for osmosensitive ATP release from mouse astrocytes in primary culture. Cell Res 18:558–65.
74. Loscalzo J, Welch G. (1995) Nitric oxide and its role in the cardiovascular system. Prog Cardiovasc Dis 38:87–104.
75. Mathie RT, Ralevic V, Alexander B, Burnstock G. (1991) Nitric oxide is the mediator of ATP-induced dilatation of the rabbit hepatic arterial vascular bed. Br J Pharmacol 103: 1602–6.
76. Matsubara M, Hayashi N, Jing T, Titani K. (2003) Regulation of endothelial nitric oxide synthase by protein kinase C. J Biochem (Tokyo) 133:773–81.
77. Menzies J, Paul A, Kennedy C. (2003) P2X7 subunit-like immunoreactivity in the nucleus of visceral smooth muscle cells of the guinea pig. Auton Neurosci 106:103–9.

78. Miao LY, Tang JP, Esposito DP, Zhang JH. (2001) Age-related changes in P2 receptor mRNA of rat cerebral arteries. Exp Gerontol 37:67–79.
79. Montiel M, de la Blanca EP, Jimenez E. (2006) P2Y receptors activate MAPK/ERK through a pathway involving PI3K/PDK1/PKC-zeta in human vein endothelial cells. Cell Physiol Biochem 18:123–34.
80. Moser TL, Kenan DJ, Ashley TA, Roy JA, Goodman MD, Misra UK, Cheek DJ, Pizzo SV. (2001) Endothelial cell surface F1-F0 ATP synthase is active in ATP synthesis and is inhibited by angiostatin. Proc Natl Acad Sci USA 98:6656–61.
81. Motley ED, Eguchi K, Patterson MM, Palmer PD, Suzuki H, Eguchi S. (2007) Mechanism of endothelial nitric oxide synthase phosphorylation and activation by thrombin. Hypertension 49:577–83.
82. Nagata D, Mogi M, Walsh K. (2003) AMP-activated protein kinase (AMPK) signaling in endothelial cells is essential for angiogenesis in response to hypoxic stress. J Biol Chem 278:31000–6.
83. Naumann N, Siratska O, Gahr M, Rosen-Wolff A. (2005) P-glycoprotein expression increases ATP release in respiratory cystic fibrosis cells. J Cyst Fibros 4:157–68.
84. Nebe B, Bohn W, Sanftleben H, Rychly J. (1996) Induction of a physical linkage between integrins and the cytoskeleton depends on intracellular calcium in an epithelial cell line. Exp Cell Res 229:100–10.
85. Partovian C, Zhuang Z, Moodie K, Lin M, Ouchi N, Sessa WC, Walsh K, Simons M. (2005) PKCalpha activates eNOS and increases arterial blood flow in vivo. Circ Res 97:482–7.
86. Patel V, Brown C, Goodwin A, Wilkie N, Boarder MR. (1996) Phosphorylation and activation of p42 and p44 mitogen-activated protein kinase are required for the P2 purinoceptor stimulation of endothelial prostacyclin production. Biochem J 320 (Pt 1): 221–6.
87. Paul A, Torrie LJ, McLaren GJ, Kennedy C, Gould GW, Plevin R. (2000) P2Y receptor-mediated inhibition of tumor necrosis factor alpha -stimulated stress-activated protein kinase activity in EAhy926 endothelial cells. J Biol Chem 275:13243–9.
88. Peng HB, Libby P, Liao JK. (1995) Induction and stabilization of I kappa B alpha by nitric oxide mediates inhibition of NF-kappa B. J Biol Chem 270:14214–9.
89. Praetorius HA, Leipziger J. (2009) ATP release from non-excitable cells. Purinergic Signal 5:433–46.
90. Ralevic V, Burnstock G. (1991) Roles of P2-purinoceptors in the cardiovascular system. Circulation 84:1–14.
91. Ramirez AN, Kunze DL. (2002) P2X purinergic receptor channel expression and function in bovine aortic endothelium. Am J Physiol Heart Circ Physiol 282:H2106–16.
92. Randriamboavonjy V, Fleming I. (2005) Endothelial nitric oxide synthase (eNOS) in platelets: how is it regulated and what is it doing there? Pharmacol Rep 57(Suppl):59–65.
93. Ransford GA, Fregien N, Qiu F, Dahl G, Conner GE, Salathe M. (2009) Pannexin 1 Contributes to ATP Release in Airway Epithelia. Am J Respir Cell Mol Biol 41:523–34.
94. Ray FR, Huang W, Slater M, Barden JA. (2002) Purinergic receptor distribution in endothelial cells in blood vessels: a basis for selection of coronary artery grafts. Atherosclerosis 162:55–61.
95. Richardson A, Parsons T. (1996) A mechanism for regulation of the adhesion-associated proteintyrosine kinase pp125FAK [published erratum appears in Nature 1996 Jun 27;381(6585):810]. Nature 380:538–40.
96. Romanov RA, Rogachevskaja OA, Khokhlov AA, Kolesnikov SS. (2008) Voltage dependence of ATP secretion in mammalian taste cells. J Gen Physiol 132:731–44.
97. Romer LH, McLean N, Turner CE, Burridge K. (1994) Tyrosine kinase activity, cytoskeletal organization, and motility in human vascular endothelial cells. Mol Biol Cell 5:349–61.
98. Rounds S, Yee WL, Dawicki DD, Harrington E, Parks N, Cutaia MV. (1998) Mechanism of extracellular ATP- and adenosine-induced apoptosis of cultured pulmonary artery endothelial cells. Am J Physiol 275:L379–88.

99. Rumjahn SM, Baldwin KA, Buxton IL. (2007) P2y receptor-mediated angiogenesis via vascular endothelial growth factor receptor 2 signaling. Proc West Pharmacol Soc 50:58–60.
100. Rumjahn SM, Yokdang N, Baldwin KA, Thai J, Buxton IL. (2009) Purinergic regulation of vascular endothelial growth factor signaling in angiogenesis. Br J Cancer 100:1465–70.
101. Schenk U, Westendorf AM, Radaelli E, Casati A, Ferro M, Fumagalli M, Verderio C, Buer J, Scanziani E, Grassi F. (2008) Purinergic control of T cell activation by ATP released through pannexin-1 hemichannels. Sci Signal 1:ra6.
102. Schneider JC, El Kebir D, Chereau C, Lanone S, Huang XL, De Buys Roessingh AS, Mercier JC, Dall'Ava-Santucci J, Dinh-Xuan AT. (2003) Involvement of Ca^{2+}/calmodulin-dependent protein kinase II in endothelial NO production and endothelium-dependent relaxation. Am J Physiol Heart Circ Physiol 284:H2311–9.
103. Schock SC, Leblanc D, Hakim AM, Thompson CS. (2008) ATP release by way of connexin 36 hemichannels mediates ischemic tolerance in vitro. Biochem Biophys Res Commun 368:138–44.
104. Sessa WC. (2004) eNOS at a glance. J Cell Sci 117:2427–9.
105. Sessa WC. (2005) Regulation of endothelial derived nitric oxide in health and disease. Mem Inst Oswaldo Cruz 100(Suppl 1):15–8.
106. Seye CI, Kong Q, Yu N, Gonzalez FA, Erb L, Weisman GA. (2007) P2 receptors in atherosclerosis and postangioplasty restenosis. Purinergic Signal 3:153–62.
107. Seye CI, Yu N, Gonzalez FA, Erb L, Weisman GA. (2004) The P2Y2 nucleotide receptor mediates vascular cell adhesion molecule-1 expression through interaction with VEGF receptor-2 (KDR/Flk-1). J Biol Chem 279:35679–86.
108. Shattil SJ, Haimovich B, Cunningham M, Lipfert L, Parsons JT, Ginsberg MH, Brugge JS. (1994) Tyrosine phosphorylation of pp125FAK in platelets requires coordinated signaling through integrin and agonist receptors. J Biol Chem 269:14738–45.
109. Shen J, DiCorleto PE. (2008) ADP stimulates human endothelial cell migration via P2Y1 nucleotide receptor-mediated mitogen-activated protein kinase pathways. Circ Res 102: 448–56.
110. Simon J, Filippov AK, Goransson S, Wong YH, Frelin C, Michel AD, Brown DA, Barnard EA. (2002) Characterization and channel coupling of the P2Y(12) nucleotide receptor of brain capillary endothelial cells. J Biol Chem 277:31390–400.
111. Soltoff SP. (1998) Related adhesion focal tyrosine kinase and the epidermal growth factor receptor mediate the stimulation of mitogen-activated protein kinase by the G-protein-coupled P2Y2 receptor. Phorbol ester or $[Ca^{2+}]i$ elevation can substitute for receptor activation. J Biol Chem 273:23110–7.
112. Soslau G, Silverberg M, Brodsky I, McCarty PP. (1997) ATP modification of serotonin-induced contraction of the rat pulmonary artery. Proc Soc Exp Biol Med 214:233–41.
113. Stein SC, Woods A, Jones NA, Davison MD, Carling D. (2000) The regulation of AMP-activated protein kinase by phosphorylation. Biochem J 345(Pt 3):437–43.
114. Sugiyama T, Kawamura H, Yamanishi S, Kobayashi M, Katsumura K, Puro DG. (2005) Regulation of P2X7-induced pore formation and cell death in pericyte-containing retinal microvessels. Am J Physiol Cell Physiol 288:C568–76.
115. Tanaka N, Kawasaki K, Nejime N, Kubota Y, Nakamura K, Kunitomo M, Takahashi K, Hashimoto M, Shinozuka K. (2004) P2Y receptor-mediated Ca(2+) signaling increases human vascular endothelial cell permeability. J Pharmacol Sci 95:174–80.
116. Thors B, Halldorsson H, Thorgeirsson G. (2004) Thrombin and histamine stimulate endothelial nitric-oxide synthase phosphorylation at Ser1177 via an AMPK mediated pathway independent of PI3K-Akt. FEBS Lett 573:175–80.
117. Van Daele P, Van Coevorden A, Roger PP, Boeynaems JM. (1992) Effects of adenine nucleotides on the proliferation of aortic endothelial cells. Circ Res 70:82–90.
118. Vincent AM, Russell JW, Low P, Feldman EL. (2004) Oxidative stress in the pathogenesis of diabetic neuropathy. Endocr Rev 25:612–28.

119. von Albertini M, Palmetshofer A, Kaczmarek E, Koziak K, Stroka D, Grey ST, Stuhlmeier KM, Robson SC. (1998) Extracellular ATP and ADP activate transcription factor NF-kappa B and induce endothelial cell apoptosis. Biochem Biophys Res Commun 248:822–9.
120. Waldo GL, Harden TK. (2004) Agonist binding and Gq-stimulating activities of the purified human P2Y1 receptor. Mol Pharmacol 65:426–36.
121. Wang L, Karlsson L, Moses S, Hultgardh-Nilsson A, Andersson M, Borna C, Gudbjartsson T, Jern S, Erlinge D. (2002) P2 receptor expression profiles in human vascular smooth muscle and endothelial cells. J Cardiovasc Pharmacol 40:841–53.
122. Witters LA, Kemp BE, Means AR. (2006) Chutes and Ladders: the search for protein kinases that act on AMPK. Trends Biochem Sci 31:13–6.
123. Xiao D, Bird IM, Magness RR, Longo LD, Zhang L. (2001) Upregulation of eNOS in pregnant ovine uterine arteries by chronic hypoxia. Am J Physiol Heart Circ Physiol 280:H812–20.
124. Yamamoto K, Sokabe T, Matsumoto T, Yoshimura K, Shibata M, Ohura N, Fukuda T, Sato T, Sekine K, Kato S, Isshiki M, Fujita T, Kobayashi M, Kawamura K, Masuda H, Kamiya A, Ando J. (2006) Impaired flow-dependent control of vascular tone and remodeling in P2X4-deficient mice. Nat Med 12:133–7.
125. Yamamoto K, Sokabe T, Ohura N, Nakatsuka H, Kamiya A, Ando J. (2003) Endogenously released ATP mediates shear stress-induced Ca^{2+} influx into pulmonary artery endothelial cells. Am J Physiol Heart Circ Physiol 285:H793–803.
126. Yegutkin GG. (2008) Nucleotide- and nucleoside-converting ectoenzymes: Important modulators of purinergic signalling cascade. Biochim Biophys Acta 1783:673–94.
127. Yegutkin GG, Henttinen T, Jalkanen S. (2001) Extracellular ATP formation on vascular endothelial cells is mediated by ecto-nucleotide kinase activities via phosphotransfer reactions. Faseb J 15:251–60.
128. Zheng J, Wen Y, Chen DB, Bird IM, Magness RR. (2005) Angiotensin II elevates nitric oxide synthase 3 expression and nitric oxide production via a mitogen-activated protein kinase cascade in ovine fetoplacental artery endothelial cells. Biol Reprod 72:1421–8.
129. Zimmermann H. (2000) Extracellular metabolism of ATP and other nucleotides. Naunyn Schmiedebergs Arch Pharmacol 362:299–309.

Chapter 3
Extracellular Purines in Endothelial Cell Barrier Regulation

Nagavedi S. Umapathy, Evgeny A. Zemskov, Agnieszka Jezierska, Irina A. Kolosova, Rudolf Lucas, John D. Catravas, and Alexander D. Verin

Abstract The vascular endothelium is a semi-selective diffusion barrier that regulates a variety of functions including controlling of the passage of macromolecules and fluid between the blood and interstitial fluid. It is well known that loss of this barrier (permeability increase) results in tissue inflammation, the hall mark of inflammatory diseases such as acute lung injury (ALI) and a severe form of it, acute respiratory distress syndrome (ARDS). Apart from ventilation strategies, no standard treatment exists for ALI and ARDS, making the search for novel regulators of endothelial hyperpermeability and dysfunction important. Accumulating data suggest that extracellular purines are promising and physiologically relevant barrier-protective agents. Purines decrease transendothelial permeability by interacting with cell surface P1 and P2Y purinoceptors belonging to the superfamily of G-protein-coupled receptors (GPCR). Selective activation of endothelial purinoreceptors responsible for barrier protection might form a basis for the treatment of various disorders. The therapeutic potential of purinoreceptors is rapidly expanding field in pharmacology and some selective agonists became recently available. In this review, we demonstrate the comprehensive overview of the purinoceptors expression in the endothelium, their interaction with G-proteins and activation of various signal transduction pathways, which lead to an endothelial barrier enhancement and protection.

Keywords Purinoceptor · Vascular endothelium · G-Protein · Permeability · ATP · Barrier enhancement · VE-cadherin · MLC-phosphatase · GPCR · P2Y · Purines · Adenosine · Protein kinase A · LPS

A.D. Verin (✉)
Vascular Biology Center, Medical College of Georgia, Augusta, GA 30912, USA
e-mail: averin@mcg.edu

Nagavedi S. Umapathy and Evgeny A. Zemskov are equally contributed.
Conflict of Interest: The authors declare no conflicts of interest.

E. Gerasimovskaya, E. Kaczmarek (eds.), *Extracellular ATP and Adenosine as Regulators of Endothelial Cell Function*, DOI 10.1007/978-90-481-3435-9_3,

3.1 Introduction

The vascular endothelium is a semi-selective diffusion barrier between the plasma and interstitial fluid and is critical for normal vessel wall homeostasis. Endothelial permeability is known to be regulated by the balance between centripetal and centrifugal intracellular forces, provided by the contractile machinery and the elements opposing contraction, respectively. The latter include tethering complexes, responsible for cell-cell and cell-substrate contacts, and systems granting cell rigidity and preventing cell collapse, such as actin filaments, microtubules and intermediate filaments [28]. Some naturally occurring substances such as sphingosine-1-phosphate [27] and the second messenger cAMP [39] are known to enhance the endothelial cells (EC) barrier. Recently, much attention has been given to the therapeutic potential of purinergic agonists and antagonists for the treatment of cardiovascular and pulmonary diseases [4, 14, 75, 97]. Purines (ATP, ADP and adenosine) function as intercellular signaling molecules, which are released to extracellular compartments from different sources in the body and subsequently reach the target organs [15]. Accumulating experimental data suggest that ATP [54, 55] and other purines are promising physiologically relevant barrier-protective agents as they are readily present in the EC microenvironment in vivo, and they decrease transendothelial permeability in vitro. ATP can be released into the bloodstream from platelets [6] and red blood cells [7, 16] and its concentrations may temporarily exceed 100 μM in blood [19]. Furthermore, the endothelium is a source of ATP locally within the vascular bed, and ATP is released constitutively across the apical membrane of EC under basal conditions [91]. Enhanced release of ATP is observed from ECs in response to various stimuli, including hypotonic challenge [91], calcium agonists [91], shear stress [71], thrombin [71], ATP itself [11], and LPS [12]. Extracellular ATP can either signal directly [34, 107] or is rapidly degraded into adenosine leading to additional vascular nucleoside signaling [31–33, 59]. Purine action is mediated by cell surface P1 and P2Y purinergic receptors (purinoceptors) [76, 102]. Both P1 and P2Y purinoceptors belong to the superfamily of G-protein-coupled receptors (GPCR). GPCRs consist of seven transmembrane domains, three extracellular and three intracellular loops, extracellular N- and intracellular C-termini. The receptors are coupled to heterotrimeric G-proteins serving for them as guanidine exchange factors (GEFs). Heterotrimeric G-proteins, immediate targets for activated purinoceptors, exist as αβγ trimers. Gβγ is a non-dissociated dimer, whereas Gα-subunit is dissociated from the complex after GTP binding. There are 39 different G-protein subunits identified in mammalian cells: 21 α-subunits, 6 β-subunits and 12 γ-subunits. Such number of the subunits means a remarkable variety of the heterotrimers, although, likely, not all of them may be formed in vivo due to tissue-specific expression of some subunits and other factors [69].

Purinoceptors are activated upon extracellular adenosine (P1) and ATP/ADP/UTP/UDP-glucose (P2Y) stimulation. To date, twelve G-protein-coupled purinoceptors were identified in mammalian cells: four P1 receptors (A1, A2A, A2B, A3) and eight P2Y receptors (P2Y1, P2Y2, P2Y4, P2Y6, P2Y11, P2Y12, P2Y13, P2Y14) [90, 103] (Table 3.1). In EC of different origin, expressions

Table 3.1 The family of purinoceptors

Family	Receptor	Agonist	Transduction Mechanism
P1	A1	Adenosine	Gi, cAMP↓
	A2A	Adenosine	Gs, cAMP↑
	A2B	Adenosine	Gs, cAMP↑ Gq, Ca^{2+}, PKC
	A3	Adenosine	Gi, cAMP↓
P2	P2Y1	ADP	Gq /G11, Ca^{2+}, PKC
	P2Y2	ATP	Gq/G11, Ca^{2+}, PKC
	P2Y4	UTP	Gq/G11, Ca^{2+}, PKC
	P2Y6	UDP	Gq/G11, Ca^{2+}, PKC
	P2Y11	ATP	Gs, cAMP↑ Gq/G11, Ca^{2+}, PKC
	P2Y12	2-MesADP	Gi, cAMP↓
	P2Y13	ADP	Gi, cAMP↓
	P2Y14	UDP-glucose	Gi, cAMP↓

of A1, A2A, A2B, A3, P2Y1, P2Y2, P2Y4, P2Y6, and P2Y11 purinoceptors had been demonstrated [2, 77, 85, 103, 106]. Our recent study (data not shown) indicates the expression of P2Y13 in human pulmonary artery EC (HPAEC). Heterotrimeric G-proteins activated by P1 and P2Y receptors in EC belong to four functionally distinct subfamilies: Gαs, Gαq/11, Gαi and Gα12/13. Activation of these particular G-proteins determines a cell response upon agonist stimulation. In this review, we are mainly focused on the effects of purine-induced P1 and P2Y receptors on modulation of endothelial monolayer integrity, respective signaling pathways leading to an enhancement or to a loss of endothelial barrier function will be discussed. Several studies demonstrated that P2X receptors are abundant in EC [106, 108, 112]. However, the P2X specific agonist, AMP-CCP, was completely inactive on the HPAEC monolayer [54] suggesting that P2X receptors are unlikely to be involved in ATP-mediated EC barrier enhancement.

3.2 Purinoceptor-Mediated Regulation of Endothelial Barrier Function

3.2.1 Cell Signaling Pathways Activated Upon Purinoceptor Stimulation

Integrity of endothelium is determined by cell-cell and cell-matrix contacts, as well as by organization of cytoskeleton of EC. Signaling cascades are activated by ligand-bound purinoceptors. Dynamic changes in cytoskeleton organization and proteins link cytoskeletal structures to adherens junctions (AJ), tight junctions (TJ), and focal adhesion (FA) contacts. These are key targets of purinoceptor-mediated signaling pathways affecting endothelial barrier function. A major role in endothelial cell-cell contacts belong to the AJ and its transmembrane protein, vascular endothelial cadherin (VE-cadherin) involved in Ca^{2+}-dependent homotypic contacts with adjacent cells. Cytoplasmic domain of VE-cadherin is linked to the cortical actin ring via β/α-catenins which stabilizes AJ, as well as determines a mechanism for dynamic reorganization of cell-cell contacts. Actin-mediated disassembly/stabilization of AJ is determined mainly by phosphorylation/dephosphorylation equilibrium of actin-associated 20-kDa myosin light chain protein (MLC_{20}). Phosphorylation status of MLC_{20} plays an important role in cytoskeleton organization in EC and, therefore, is crucial for endothelium integrity [40, 101]. Phosphorylation of MLC_{20} by Ca^{2+}/calmodulin-dependent MLC kinase leads to an actin-myosin contraction and centripetal force-driven AJ results in a loss of endothelial integrity and hyperpermeability. In contrast, cell signaling pathways leading to deposphorylation of MLC_{20} by MLC phosphatase, protein phosphatase 1 (PP1), results in the formation of a thick cortical actin ring, cell relaxation and spreading. Agonist-mediated activation of purinoceptors on the surface of EC may enhance as well as decrease a barrier function of the endothelium (Fig. 3.1). Stimulation of A2A, A2B, or P2Y11 receptors coupled to Gαs proteins [21, 22, 51, 73] leads to direct interaction of dissociated Gαs with plasma membrane adenylyl cyclase and activation of the cAMP synthesis [111]. The cAMP is a classical activator of protein kinase A (PKA), although it may be involved in modulation of other pathways. As it was recently shown for 5′-N-ethylcarboxamidoadenosine (NECA)-stimulated A2B receptors in human umbilical vein EC (HUVEC), generation of cAMP may lead to alternative, PKA-independent Epac1/Rap1/B-Raf pathway resulting in ERK1/2 activation [35]. The cAMP-dependent activation of PKA has indispensible consequences as a potent positive regulator of endothelial monolayer integrity. Indeed, activated PKA shifts EC to relaxed status by prevention of MLC_{20} phosphorylation and, therefore, formation of stress fibers. Purinoceptors A1, A3, and P2Y14 are involved in Gi protein-mediated signaling [35, 51]. Activation of these receptors leads to an inhibition of adenylyl cyclase via its direct association with free Gαi [70]. Gi protein-derived free Gβγ dimers interact with PI3-kinase (PI3-K) or phospholipase Cβ (PLCβ) and initiate respective signaling pathways [44, 78, 90]. PI3-K activates PKB/Akt (via phosphatidyl-3,4,5-triphosphate (PIP_3) generation and activation of PDK1) [42] and ERK1/2 (via Ras/Raf-1/MEK1/2) [44]. PLC activation (and, therefore, generation of inositol 1,4,5-triphosphate (IP_3) and diacylglycerol (DAG)) may follow by Ca^{2+} influx due to stimulation of plasma membrane and endoplasmic reticulum Ca^{2+}-channels [78]. Elevation of $[Ca^{2+}]_i$ and DAG levels routinely follows by membrane translocation and activation of conventional (α, βI, βII, γ) and novel (δ, ϵ, η, θ) PKC isoforms [66]. PKCα isoform has been shown to activate the RhoA/Rho kinase (ROK) pathway by direct phosphorylation of the upstream effectors, RhoGDI and RhoGEF [52], which leads to phosphorylation of MLC_{20}, as well as to regulation of TJ disassembly via phosphorylation of p120 and β-catenin [82]. Adenosine-mediated activation of Gαi proteins has also been shown to promote an upregulation of p38 MAPK [44] and may possibly activate JNK by ROK-dependent phosphorylation [62]. The p38 MAPK can initiate stress fiber formation via phosphorylation of actin-capping protein hsp27 and its further dissociation from actin filaments [45, 86]. Another important event related

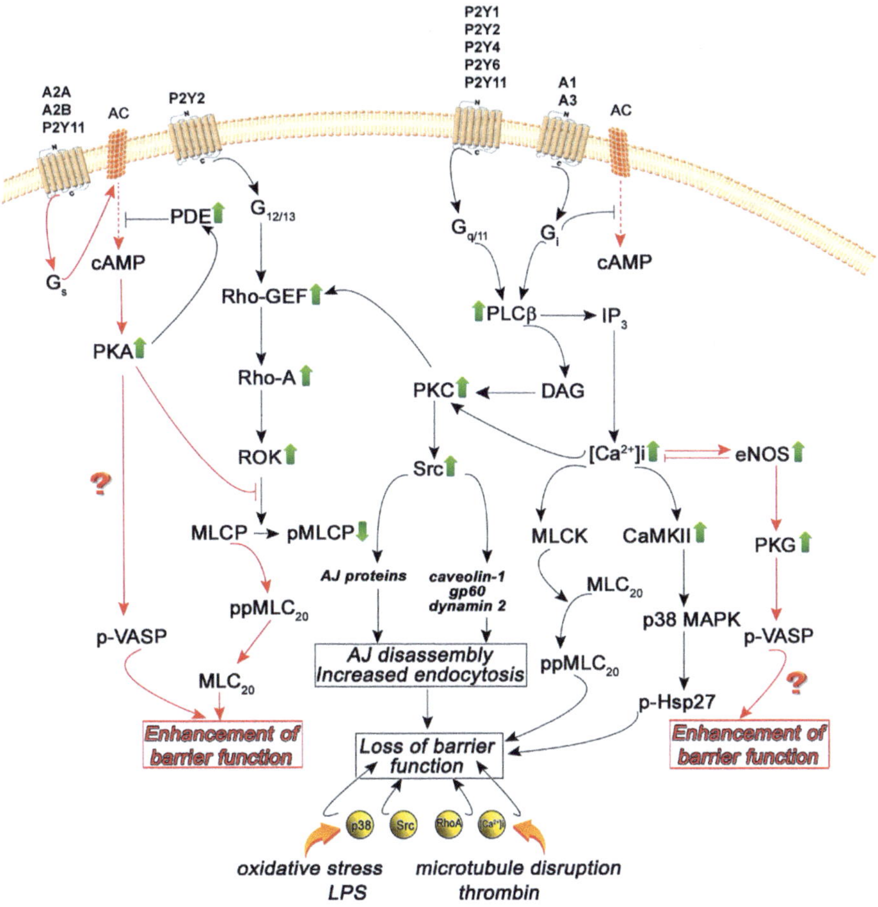

Fig. 3.1 Purinoceptor-activated cell signaling pathways, which affect barrier function of endothelial monolayers (see also explanations in the text). Pathways leading to enhancement of endothelial integrity are shown in red. Barrier-disruptive pathways are shown in *black*

to Gαi protein-mediated signaling is activation of Src kinase [61, 93]. In EC, Src kinase positively regulates caveolae-mediated transcytosis of albumin by phosphorylation of caveolin-1, gp60 and dynamin-2 [92, 96]. Phosphorylation of dynamin-2 can also stimulate clathrin- and caveolin-dependent endocytosis, and, therefore, enhance internalization of transmembrane proteins involved in cell-cell contacts. Activated Src may initiate transactivation of receptor tyrosine kinases (RTK) and RTK-mediated signaling.

Gq/11 protein-mediated signaling is activated by agonist stimulation of A2B, P2Y1, P2Y2, P2Y4, P2Y6 and P2Y11 receptors [21, 22, 36, 37, 65, 81, 105]. Dissociated Gαq or Gα11 subunits directly interact with PLCβ and positively modulate production of IP_3 and DAG [78]. This, in turn, leads to $[Ca^{2+}]_i$ influx and activation of PKC isoforms [73]. Extensive studies performed in EC have demonstrated Ca^{2+}-dependent activation of endothelial nitric oxide (NO) synthase (eNOS) via direct interaction with Ca^{2+}/calmodulin and/or via phosphorylation by Ca^{2+}/calmodulin-dependent protein kinase II (CaMKII)) [38, 87]. NO production leads to activation of guanylyl cyclase (GA), cGMP synthesis and stimulation of cGMP-dependent protein kinase (PKG) [53, 68]. While this pathway provides a negative feedback control of Ca^{2+} influx through down-regulation of endoplasmic reticulum (ER) IP_3-sensitive channels and plasma membrane Ca^{2+}-influx channels [25, 113], it can also increase Ca^{2+} uptake by ER via activation of ER Ca^{2+} ATPases [25]. Therefore, eNOS/GA/PKG cascade can down-regulate Ca^{2+}-mediated signaling leading to endothelial barrier dysfunction. In HUVEC, stimulatory phosphorylation of eNOS at Ser-1177 was shown to be adenosine-independent, but mediated by several nucleotides (ATP, UTP and ADP). Inhibitory analysis demonstrated an involvement of P2Y1, P2Y2, and possibly P2Y4 purinoceptors in the activation of eNOS via $[Ca^{2+}]_i$ elevation and DAG-dependent PKCδ [24]. Another protein target of activated PKG is vasodilator-stimulated phosphoprotein (VASP), a protein regulating actin polymerization and assembly [57]. Phosphorylated by PKG/PKA, VASP has been shown to localize to endothelial cell-cell junctions: in association with TJs and AJs [20, 79]. Although an entire role of the VASP phosphorylation in endothelial contraction/relaxation remains unclear, its modification correlates with enhancement of endothelial barrier function in purinoceptor agonist-stimulated cells [20, 54].

The P2Y2 receptor was also shown to activate Gα12 protein signaling. This activation requires an interaction of the receptor with αvβ3-integrin and can be inhibited either by αv-integrin antisense oligonucleotides or by point mutation in an integrin-binding sequence of the P2Y2 receptor [58]. Activation of Gα12 protein positively modulate Rho-guanine nucleotide exchange factor (p115Rho-GEF) via its interaction with Gα12 subunit [56] as well as via phosphorylation by activated PKCα [47]. This, in turn, promotes RhoA/ROK activation and phosphorylation of MLC_{20} and MLC phosphatase.

Elevation of cytosolic Ca^{2+} may be, apparently, attributed to any P1 or P2Y purinoceptor-mediated pathway in EC. Indeed, beside IP_3-mediated Ca^{2+} release stimulated by Gαq/11 and Gαi, Ca^{2+} influx from extracellular space may be stimulated by Gαs via cAMP-activated Ca^{2+}-channel, While $[Ca^{2+}]_i$ is essential for activation of eNOS and stimulation of endothelium-derived release of vasorelaxant, NO [29, 43], elevation of cytosolic Ca^{2+} is certainly a negative factor for endothelial integrity. Nevertheless, as was shown in purine-activated EC, Ca^{2+} influx is rather transient and does not interfere with barrier-enhancing effects [54, 67].

3.2.2 Adenosine-Activated Signaling and Endothelial Monolayer Integrity

Extracellular adenosine or ATP may have a positive, as well as a negative effect on the integrity of endothelium, depending on a prevalence of particular P1 and P2Y purinoceptors. Although extracellular ATP itself should have a positive effect on endothelial integrity (via specific P2Y receptors coupled to heterotrimeric Gαs and Gαq/11 proteins), a quick degradation of ATP by cell-surface ectonucleotidases to ADP and adenosine [80, 95] may be followed by activation of ADP-specific P2Y and adenosine-specific P1 receptors. Therefore, the resulting effect may reflect an interference of opposite signaling cascades. The response mediated by A1 and A3 receptors can definitively be a cause of the integrity loss. First, signaling pathways activated by these purinoceptors shift MLC_{20} to its phosphorylated form. Elevation of $[Ca^{2+}]_i$ activate Ca^{2+}/calmodulin-dependent phosphorylation of MLC_{20} along with RhoA/ROK-dependent inhibitory phosphorylation of regulatory myosin phosphatase target subunit (MYPT1) of MLC phosphatase, PP1, which, in turn, causes stress fiber formation with following polarization of EC. Second, an activation of Src-family kinases can increase clathrin- and caveolin-dependent endocytosis of adhesive proteins via phosphorylation of caveolin-1 and dynamin-2 [63], stimulate disassembly of AJs via tyrosine phosphorylation of p120-catenin and SHP-2 [3, 50, 99] and, therefore, may enhance a loss of cell-cell adhesion and formation of intercellular gaps.

On the other hand, signaling cascades activated by A2 receptors can decrease endothelial permeability by shifting MLC_{20} to its dephosphorylated form via PKA-dependent inhibition of RhoA/ROK pathway. As was recently demonstrated, PKA phosphorylation of RhoGDI at Ser-174 prevents RhoA activation by thrombin [74].

Activation of ERK1/2 is a common event in purinoceptor-mediated signaling. As has been shown in model experiments (Chinese hamster ovary cells stably transfected with human P1 receptors), all four adenosine receptors may activate ERK1/2 phosphorylation when stimulated by the non-P1 receptor agonist, NECA, although intensities of the activation were not equal. The most potent ERK1/2 activator was the A2B receptor, whereas the highest level of ERK1/2 phosphorylation was obtained in A1 and A3 receptor-transfected cells [88, 89]. However, the activation of ERK1/2 is, likely, cell/tissue-specific, and was not detected in some types of cells expressed endogenous P1 receptors upon agonist stimulation [60]. Although ERK1/2 may be responsible for Ca^{2+}-independent NO release by endothelium [110], its functions in EC may have a dual role, because of an involvement in barrier-protective and barrier-disruptive mechanisms. In cultured EC stimulated by extracellular ATP, ERK1/2 activity has been shown to be dispensable as an enhancement of endothelial barrier function under normal conditions [54], but it was important for ATP-mediated survival of hyperoxia-challenged EC [1].

3.2.3 Purinoceptor-Mediated Signaling as a Regulator of Endothelial Integrity at Various Pathological States

A number of agents which negatively affect endothelial integrity have been shown to activate mechanisms interfering with purinoceptor-mediated signaling. Activation of the RhoA/ROK signaling pathway followed by actin stress fiber formation and a loss of cell-cell contacts is a general cause of endothelial dysfunction upon various pathological stimuli. Thrombin, a protease produced from inactive prothrombin by proteolytic cleavage on the surface of injured endothelium, is an inflammation response mediator. Thrombin is responsible for specific cleavage and activation of protease-activated receptors (PARs), members of the GPCR superfamily [104]. In EC, PAR-1, which can couple to Gα12/13, Gαq/11 and Gαi heterotrimeric proteins, stimulates, in turn, PLCβ, PKCα and RhoA pathways and inhibits adenylyl cyclase [23, 41, 94]. These pathways can essentially lead to an activation of MLC kinase and inhibition of MLC phosphatase, stress fiber formation and disassembly of cell-cell junctions in EC [40, 84, 101]. Another agent compromising endothelial integrity is bacterial lipopolysaccharide (LPS), a component of gram-negative bacteria cell wall. LPS is a potent inflammatory agent involved in cell response in bacterial infections. Bovine pulmonary artery EC (BPAEC) exposed to LPS exhibited a concentration-dependent decrease in transendothelial electrical resistance, activation of Src, phosphorylation of the focal adhesion protein paxillin, and reduced expression of the AJ proteins, VE-cadherin and β-catenin [17]. LPS treatment, as well as stimulation of EC with TGF-β1, TNF-α and VEGF, increased endothelial permeability via phosphorylation of MLC_{20} and hsp27 and formation of stress fibers [5, 9, 64].

A number of microtubule (MT)-destabilizing agents were also shown to promote a loss of endothelial integrity via activation of RhoA/ROK pathway and MLC_{20} phosphorylation [10, 100]. In cultured HPAEC, RhoA activation was promoted by Rho-specific GEF-H1 released from disrupted MT upon nocodazole treatment [8]. Another MT-destabilizer, cytostatic anti-tumor agent, 2-methoxyestradiol, was also able to activate Rho kinase and p38 MAPK followed by hsp27 and MLC_{20} phosphorylation [13]. These studies done in our laboratory clearly demonstrated a link between MT disruption and following actin stress fiber formation. Moreover, the molecular mechanisms of endothelial dysfunction and a critical role of actin-mediated cytoskeleton rearrangement regulated by RhoA/ROK signaling suggest that purinoceptor-mediated pathways are potential therapeutical approaches to restore a barrier function of endothelium. Indeed, data obtained from our laboratory and by others using cellular and animal models of acute endothelial injuries have confirmed an ability of purinergic agonists to protect endothelial integrity from various disruptive stimuli. In particular, protective effects of ATP and NECA have been demonstrated in cellular and mouse models of LPS-induced endothelial injury [55] and our unpublished observation, Fig. 3.2), in thrombin-treated EC [83] and in cultured EC under oxidative stress conditions [1].

The protective function is mainly associated with PKA activation and decreased dephosphorylation of MLC_{20}. Under normal conditions, extracellular ATP added to the cultured HPAEC, enhanced the integrity rapidly increasing transendothelial electrical resistance [54]. Similar effect was also obtained upon stimulation by other P2Y purinoceptor agonists: ADP and 2MeSATP [54]. ATP-induced transendothelial electrical resistance increase was significantly attenuated by PKA inhibitors, H89 and KT5720A, and correlated with VASP phosphorylation, dephosphorylation of

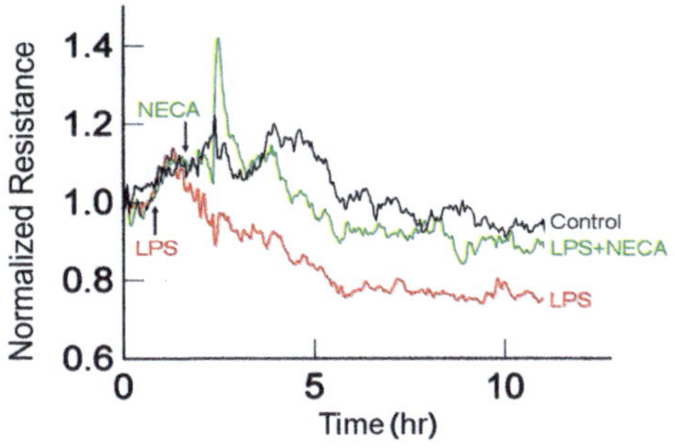

Fig. 3.2 P1 receptor agonist, NECA, attenuates the LPS-induced EC barrier permeability increase. HPAEC were challenged with LPS (100 ng/ml) for 1 h and then exposed to NECA (100 nM). Transendothelial electrical resistance (TER) of the cell monolayers was registered and presented after normalization

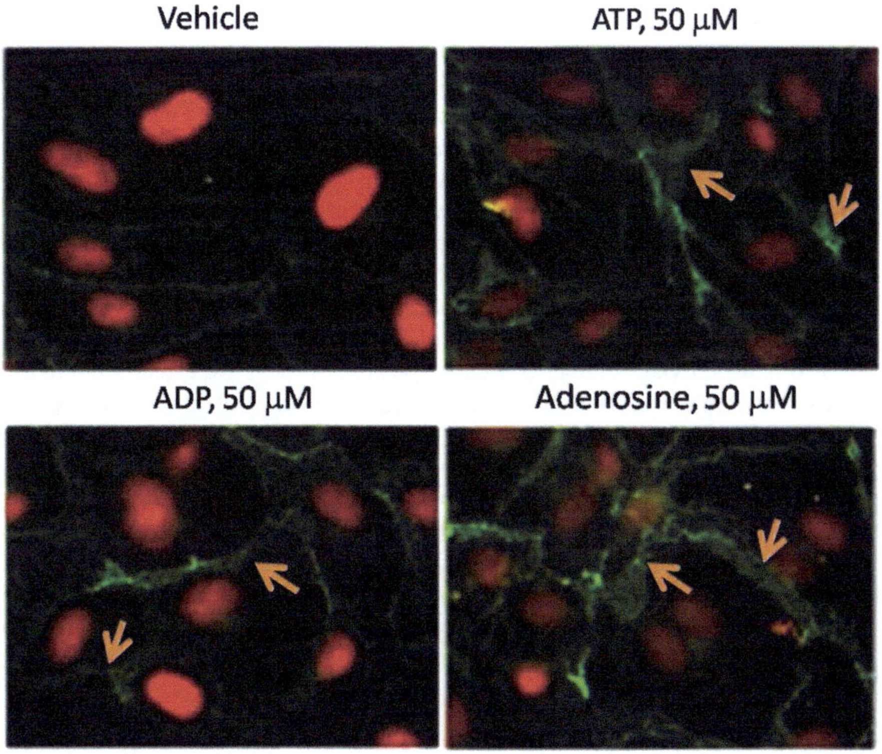

Fig. 3.3 Immunostaining for the VE-cadherin (adherent junctional protein) demonstrates that appreciably more VE-cadherin is recruited to the cell junctions (*arrows*) after treating HPAEC with purines

MLC_{20} and transient dephosphorylation of MLC phosphatase. Immunofluorescence staining of Purines-stimulated HPAEC revealed dramatic increase in VE-cadherin expression at the cellular periphery (Fig. 3.3) [54].

3.2.4 Down-Regulation (Desensitization) of Purinoceptor-Mediated Signaling

P1 and P2Y receptor-mediated signaling can be down-regulated at different levels of the signaling cascades. Activated PKA has inhibitory effect on some adenylyl cyclase isoforms (AC5 and AC6) via its direct phosphorylation. PKA phosphorylation of AC5 directly inhibits its activity, whereas phosphorylation of AC6 at Ser674 disrupts Gαs-binding [18, 49].

Moreover, PKA phosphorylation enhances activities of cAMP-specific phosphodiesterases (PDEs), particularly, PDE4 isoforms [26, 114], diminishing intracellular levels of cAMP more efficiently.

Activated GPCRs are phosphorylated by specific GPCR kinases (GRKs) docked to the receptors by membrane-bound free Gβγ dimers [72]. This phosphorylation serves as a signal for receptor internalization. Phosphorylated receptors are able to form complexes with adaptor proteins of clathrin-coated pits (such as β-arrestin) and are internalized by clathrin-dependent endocytosis followed by either recycling back to the cell surface or lysosomal degradation [46]. Interestingly, β-arrestin associated with purinoceptors may also serve as a scaffold for a signaling complex Raf-1/MEK1/2/ERK1/2 on the surface of endosomes [46]. In this case, activated ERK1/2 is not translocated to the cell nucleus but switched to phosphorylation of cytoplasmic protein substrates. Ubiquitous GRK2 is one of cytoplasmic ERK1/2 substrates. ERK1/2-dependent phosphorylation stimulates GRK2 activity and hence positively regulates purinoceptor internalization and desensitization [98]. ERK2 can also phosphorylate PDE4 with either positive or negative effect on activities of different PDE4 isoforms providing a fine modulation of intracellular cAMP levels [48].

A variety in presentation and expression levels of purinoceptors in EC of different origin determines a complexity of agonist-mediated cell response due to a possible cross-talk among several pathways with opposed effects on endothelial functions. Because of that, a usage of selective agonists/antagonists, either natural or synthetic, has been considered as a very attractive way for pharmacological manipulation of endothelial barrier function. Extensive research in this field already has provided several highly-specific P1 and P2Y receptor agonists and antagonists.

3.3 Barrier-Protective Potency of Receptor-Specific Purinergic Agonists

Selective activation of endothelial purinoreceptor(s) responsible for barrier protection might form a basis for the treatment of various disorders. The therapeutic potential of purinoreceptors is rapidly expanding field in pharmacology and some selective agonists became recently available. One recent study demonstrates that the human bronchial epithelia express P2Y6 receptors on both apical and basolateral membranes and that the cAMP/PKA pathway regulates apical but not basolateral

P2Y6 receptor-coupled ion transport [109]. We speculate that extracellular purines released from the blood cells (for example, platelets), ECs or alveolar epithelial cells (basal side of EC) stimulate purinergic receptors based on their expression pattern (apical or basal) as schematically presented (Fig. 3.4). A recent study indicated the hetero-oligomerization between two metabotrofic purinoceptors, P2Y1 and P2Y11, co-expressed in HEK293 cells, promotes agonist-induced internalization of the P2Y11 receptor, which itself is unable to undergo endocytosis [30]. Moreover, the agonist profile for the co-expressed P2Y1 and P2Y11 was different from the agonist profile established for cells expressing the P2Y11 receptor only. The hetero-oligomerization of the P2Y1 and P2Y11 receptors modifies the functions of the P2Y11 receptor in response to extracellular nucleotides. More studies are needed to characterize the agonist profile of expressed purinoceptors, the expression profile of multiple purinoceptors on the apical or basal side of EC and the pathophysiological agonist concentrations that activate various purinoceptors. Obtained data will help to establish conditions for possible purine therapies.

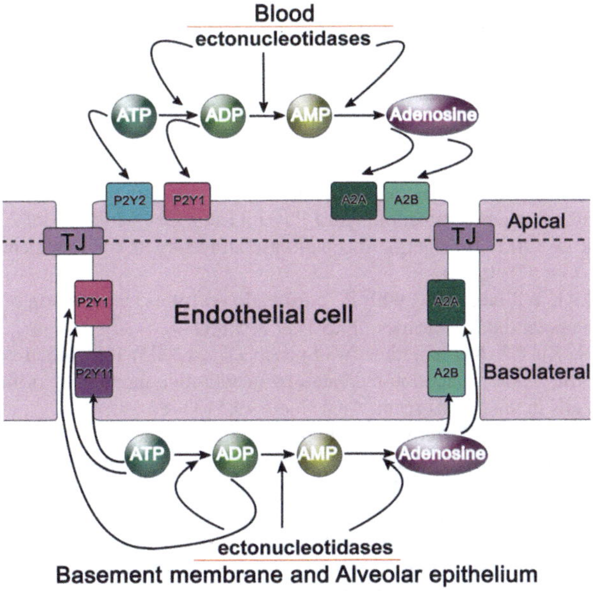

Fig. 3.4 Possible interaction between extracellular purines and purinoceptors in the endothelium. The apical and basolateral domains are separated by tight junctions

References

1. Ahmad S, Ahmad A, Ghosh M, Leslie CC, White CW. (2004) Extracellular ATP-mediated signaling for survival in hyperoxia-induced oxidative stress. J Biol Chem 279:16317–25.
2. Ahmad S, Ahmad A, White CW. (2006) Purinergic signaling and kinase activation for survival in pulmonary oxidative stress and disease. Free Radic Biol Med 41:29–40.

3. Anastasiadis PZ, Reynolds AB. (2000) The p120 catenin family: complex roles in adhesion, signaling and cancer. J Cell Sci 113 (Pt 8):1319–34.
4. Angiolillo DJ. (2007) ADP receptor antagonism: what's in the pipeline? Am J Cardiovasc Drugs 7:423–32.
5. Antonov A, Snead C, Gorshkov B, Antonova GN, Verin AD, Catravas JD. (2008) Heat shock protein 90 inhibitors protect and restore pulmonary endothelial barrier function. Am J Respir Cell Mol Biol 39:551–9.
6. Beigi R, Kobatake E, Aizawa M, Dubyak GR. (1999) Detection of local ATP release from activated platelets using cell surface-attached firefly luciferase. Am J Physiol 276:C267–78.
7. Bergfeld GR, Forrester T. (1992) Release of ATP from human erythrocytes in response to a brief period of hypoxia and hypercapnia. Cardiovasc Res 26:40–7.
8. Birukova AA, Adyshev D, Gorshkov B, Bokoch GM, Birukov KG, Verin AD. (2006) GEF-H1 is involved in agonist-induced human pulmonary endothelial barrier dysfunction. Am J Physiol Lung Cell Mol Physiol 290:L540–8.
9. Birukova AA, Birukov KG, Adyshev D, Usatyuk P, Natarajan V, Garcia JG, Verin AD. (2005) Involvement of microtubules and Rho pathway in TGF-beta1-induced lung vascular barrier dysfunction. J Cell Physiol 204:934–47.
10. Birukova AA, Smurova K, Birukov KG, Usatyuk P, Liu F, Kaibuchi K, Ricks-Cord A, Natarajan V, Alieva I, Garcia JG, Verin AD. (2004) Microtubule disassembly induces cytoskeletal remodeling and lung vascular barrier dysfunction: role of Rho-dependent mechanisms. J Cell Physiol 201:55–70.
11. Bodin P, Burnstock G. (1996) ATP-stimulated release of ATP by human endothelial cells. J Cardiovasc Pharmacol 27:872–5.
12. Bodin P, Burnstock G. (1998) Increased release of ATP from endothelial cells during acute inflammation. Inflamm Res 47:351–4.
13. Bogatcheva NV, Adyshev D, Mambetsariev B, Moldobaeva N, Verin AD. (2007) Involvement of microtubules, p38, and Rho kinases pathway in 2-methoxyestradiol-induced lung vascular barrier dysfunction. Am J Physiol Lung Cell Mol Physiol 292:L487–99.
14. Burnstock G. (2008) Purinergic receptors as future targets for treatment of functional GI disorders. Gut 57:1193–4.
15. Burnstock G, Williams M. (2000) P2 purinergic receptors: modulation of cell function and therapeutic potential. J Pharmacol Exp Ther 295:862–9.
16. Carroll JS, Ku CJ, Karunarathne W, Spence DM. (2007) Red blood cell stimulation of platelet nitric oxide production indicated by quantitative monitoring of the communication between cells in the bloodstream. Anal Chem 79:5133–8.
17. Chatterjee A, Snead C, Yetik-Anacak G, Antonova G, Zeng J, Catravas JD. (2008) Heat shock protein 90 inhibitors attenuate LPS-induced endothelial hyperpermeability. Am J Physiol Lung Cell Mol Physiol 294:L755–63.
18. Chen Y, Harry A, Li J, Smit MJ, Bai X, Magnusson R, Pieroni JP, Weng G, Iyengar R. (1997) Adenylyl cyclase 6 is selectively regulated by protein kinase A phosphorylation in a region involved in Galphas stimulation. Proc Natl Acad Sci USA 94:14100–4.
19. Coade SB, Pearson JD. (1989) Metabolism of adenine nucleotides in human blood. Circ Res 65:531–7.
20. Comerford KM, Lawrence DW, Synnestvedt K, Levi BP, Colgan SP. (2002) Role of vasodilator-stimulated phosphoprotein in PKA-induced changes in endothelial junctional permeability. FASEB J 16:583–5.
21. Communi D, Govaerts C, Parmentier M, Boeynaems JM. (1997) Cloning of a human purinergic P2Y receptor coupled to phospholipase C and adenylyl cyclase. J Biol Chem 272:31969–73.
22. Communi D, Robaye B, Boeynaems JM. (1999) Pharmacological characterization of the human P2Y11 receptor. Br J Pharmacol 128:1199–206.
23. Coughlin SR. (2000) Thrombin signalling and protease-activated receptors. Nature 407: 258–64.

24. da Silva CG, Specht A, Wegiel B, Ferran C, Kaczmarek E. (2009) Mechanism of purinergic activation of endothelial nitric oxide synthase in endothelial cells. Circulation 119:871–9.
25. Dedkova EN, Blatter LA. (2002) Nitric oxide inhibits capacitative Ca^{2+} entry and enhances endoplasmic reticulum Ca^{2+} uptake in bovine vascular endothelial cells. J Physiol 539:77–91.
26. Di Benedetto G, Zoccarato A, Lissandron V, Terrin A, Li X, Houslay MD, Baillie GS, Zaccolo M. (2008) Protein kinase A type I and type II define distinct intracellular signaling compartments. Circ Res 103:836–44.
27. Dudek SM, Camp SM, Chiang ET, Singleton PA, Usatyuk PV, Zhao Y, Natarajan V, Garcia JG. (2007) Pulmonary endothelial cell barrier enhancement by FTY720 does not require the S1P1 receptor. Cell Signal 19:1754–64.
28. Dudek SM, Garcia JG. (2001) Cytoskeletal regulation of pulmonary vascular permeability. J Appl Physiol 91:1487–500.
29. Duza T, Sarelius IH. (2003) Conducted dilations initiated by purines in arterioles are endothelium dependent and require endothelial Ca^{2+}. Am J Physiol Heart Circ Physiol 285:H26–37.
30. Ecke D, Hanck T, Tulapurkar ME, Schafer R, Kassack M, Stricker R, Reiser G. (2008) Hetero-oligomerization of the P2Y11 receptor with the P2Y1 receptor controls the internalization and ligand selectivity of the P2Y11 receptor. Biochem J 409:107–16.
31. Eckle T, Faigle M, Grenz A, Laucher S, Thompson LF, Eltzschig HK. (2008) A2B adenosine receptor dampens hypoxia-induced vascular leak. Blood 111:2024–35.
32. Eltzschig HK, Abdulla P, Hoffman E, Hamilton KE, Daniels D, Schonfeld C, Loffler M, Reyes G, Duszenko M, Karhausen J, Robinson A, Westerman KA, Coe IR, Colgan SP. (2005) HIF-1-dependent repression of equilibrative nucleoside transporter (ENT) in hypoxia. J Exp Med 202:1493–505.
33. Eltzschig HK, Faigle M, Knapp S, Karhausen J, Ibla J, Rosenberger P, Odegard KC, Laussen PC, Thompson LF, Colgan SP. (2006) Endothelial catabolism of extracellular adenosine during hypoxia: the role of surface adenosine deaminase and CD26. Blood 108:1602–10.
34. Erlinge D. (2004) Extracellular ATP: a central player in the regulation of vascular smooth muscle phenotype. Focus on "Dual role of PKA in phenotype modulation of vascular smooth muscle cells by extracellular ATP". Am J Physiol Cell Physiol 287: C260–2.
35. Fang Y, Olah ME. (2007) Cyclic AMP-dependent, protein kinase A-independent activation of extracellular signal-regulated kinase 1/2 following adenosine receptor stimulation in human umbilical vein endothelial cells: role of exchange protein activated by cAMP 1 (Epac1). J Pharmacol Exp Ther 322:1189–200.
36. Filippov AK, Fernandez-Fernandez JM, Marsh SJ, Simon J, Barnard EA, Brown DA. (2004) Activation and inhibition of neuronal G protein-gated inwardly rectifying K(+) channels by P2Y nucleotide receptors. Mol Pharmacol 66:468–77.
37. Filippov AK, Webb TE, Barnard EA, Brown DA. (1999) Dual coupling of heterologously-expressed rat P2Y6 nucleotide receptors to N-type Ca^{2+} and M-type K^{+} currents in rat sympathetic neurones. Br J Pharmacol 126:1009–17.
38. Fleming I, Fisslthaler B, Dimmeler S, Kemp BE, Busse R. (2001) Phosphorylation of Thr(495) regulates Ca(2+)/calmodulin-dependent endothelial nitric oxide synthase activity. Circ Res 88:E68–75.
39. Fukuhara S, Sakurai A, Sano H, Yamagishi A, Somekawa S, Takakura N, Saito Y, Kangawa K, Mochizuki N. (2005) Cyclic AMP potentiates vascular endothelial cadherin-mediated cell-cell contact to enhance endothelial barrier function through an Epac-Rap1 signaling pathway. Mol Cell Biol 25:136–46.
40. Garcia JG, Davis HW, Patterson CE. (1995) Regulation of endothelial cell gap formation and barrier dysfunction: role of myosin light chain phosphorylation. J Cell Physiol 163: 510–22.

41. Gavard J, Gutkind JS. (2008) Protein kinase C-related kinase and ROCK are required for thrombin-induced endothelial cell permeability downstream from Galpha12/13 and Galpha11/q. J Biol Chem 283:29888–96.
42. Germack R, Griffin M, Dickenson JM. (2004) Activation of protein kinase B by adenosine A1 and A3 receptors in newborn rat cardiomyocytes. J Mol Cell Cardiol 37:989–99.
43. Govers R, Rabelink TJ. (2001) Cellular regulation of endothelial nitric oxide synthase. Am J Physiol Renal Physiol 280:F193–206.
44. Hammarberg C, Fredholm BB, Schulte G. (2004) Adenosine A3 receptor-mediated regulation of p38 and extracellular-regulated kinase ERK1/2 via phosphatidylinositol-3′-kinase. Biochem Pharmacol 67:129–34.
45. Hirano S, Rees RS, Yancy SL, Welsh MJ, Remick DG, Yamada T, Hata J, Gilmont RR. (2004) Endothelial barrier dysfunction caused by LPS correlates with phosphorylation of HSP27 in vivo. Cell Biol Toxicol 20:1–14.
46. Hoffmann C, Ziegler N, Reiner S, Krasel C, Lohse MJ. (2008) Agonist-selective, receptor-specific interaction of human P2Y receptors with beta-arrestin-1 and -2. J Biol Chem 283:30933–41.
47. Holinstat M, Mehta D, Kozasa T, Minshall RD, Malik AB. (2003) Protein kinase Calpha-induced p115RhoGEF phosphorylation signals endothelial cytoskeletal rearrangement. J Biol Chem 278:28793–8.
48. Houslay MD, Baillie GS. (2003) The role of ERK2 docking and phosphorylation of PDE4 cAMP phosphodiesterase isoforms in mediating cross-talk between the cAMP and ERK signalling pathways. Biochem Soc Trans 31:1186–90.
49. Iwami G, Kawabe J, Ebina T, Cannon PJ, Homcy CJ, Ishikawa Y. (1995) Regulation of adenylyl cyclase by protein kinase A. J Biol Chem 270:12481–4.
50. Iyer S, Ferreri DM, DeCocco NC, Minnear FL, Vincent PA. (2004) VE-cadherin-p120 interaction is required for maintenance of endothelial barrier function. Am J Physiol Lung Cell Mol Physiol 286:L1143–53.
51. Klinger M, Freissmuth M, Nanoff C. (2002) Adenosine receptors: G protein-mediated signalling and the role of accessory proteins. Cell Signal 14:99–108.
52. Knezevic N, Roy A, Timblin B, Konstantoulaki M, Sharma T, Malik AB, Mehta D. (2007) GDI-1 phosphorylation switch at serine 96 induces RhoA activation and increased endothelial permeability. Mol Cell Biol 27:6323–33.
53. Koesling D, Mullershausen F, Lange A, Friebe A, Mergia E, Wagner C, Russwurm M. (2005) Negative feedback in NO/cGMP signalling. Biochem Soc Trans 33:1119–22.
54. Kolosova IA, Mirzapoiazova T, Adyshev D, Usatyuk P, Romer LH, Jacobson JR, Natarajan V, Pearse DB, Garcia JG, Verin AD. (2005) Signaling pathways involved in adenosine triphosphate-induced endothelial cell barrier enhancement. Circ Res 97: 115–24.
55. Kolosova IA, Mirzapoiazova T, Moreno-Vinasco L, Sammani S, Garcia JG, Verin AD. (2008) Protective effect of purinergic agonist ATPgammaS against acute lung injury. Am J Physiol Lung Cell Mol Physiol 294:L319–24.
56. Kozasa T, Jiang X, Hart MJ, Sternweis PM, Singer WD, Gilman AG, Bollag G, Sternweis PC. (1998) p115 RhoGEF, a GTPase activating protein for Galpha12 and Galpha13. Science 280:2109–11.
57. Krause M, Dent EW, Bear JE, Loureiro JJ, Gertler FB. (2003) Ena/VASP proteins: regulators of the actin cytoskeleton and cell migration. Annu Rev Cell Dev Biol 19:541–64.
58. Liao Z, Seye CI, Weisman GA, Erb L. (2007) The P2Y2 nucleotide receptor requires interaction with alpha v integrins to access and activate G12. J Cell Sci 120:1654–62.
59. Loffler M, Morote-Garcia JC, Eltzschig SA, Coe IR, Eltzschig HK. (2007) Physiological roles of vascular nucleoside transporters. Arterioscler Thromb Vasc Biol 27:1004–13.
60. Lynge J, Schulte G, Nordsborg N, Fredholm BB, Hellsten Y. (2003) Adenosine A 2B receptors modulate cAMP levels and induce CREB but not ERK1/2 and p38 phosphorylation in rat skeletal muscle cells. Biochem Biophys Res Commun 307:180–7.

61. Ma YC, Huang J, Ali S, Lowry W, Huang XY. (2000) Src tyrosine kinase is a novel direct effector of G proteins. Cell 102:635–46.
62. Marinissen MJ, Chiariello M, Tanos T, Bernard O, Narumiya S, Gutkind JS. (2004) The small GTP-binding protein RhoA regulates c-jun by a ROCK-JNK signaling axis. Mol Cell 14:29–41.
63. Minshall RD, Tiruppathi C, Vogel SM, Malik AB. (2002) Vesicle formation and trafficking in endothelial cells and regulation of endothelial barrier function. Histochem Cell Biol 117:105–12.
64. Mong PY, Wang Q. (2009) Activation of Rho kinase isoforms in lung endothelial cells during inflammation. J Immunol 182:2385–94.
65. Murthy KS, Makhlouf GM. (1998) Coexpression of ligand-gated P2X and G protein-coupled P2Y receptors in smooth muscle. Preferential activation of P2Y receptors coupled to phospholipase C (PLC)-beta1 via Galphaq/11 and to PLC-beta3 via Gbetagammai3. J Biol Chem 273:4695–704.
66. Newton R, Stevens DA, Hart LA, Lindsay M, Adcock IM, Barnes PJ. (1997) Superinduction of COX-2 mRNA by cycloheximide and interleukin-1beta involves increased transcription and correlates with increased NF-kappaB and JNK activation. FEBS Lett 418:135–8.
67. Noll T, Holschermann H, Koprek K, Gunduz D, Haberbosch W, Tillmanns H, Piper HM. (1999) ATP reduces macromolecule permeability of endothelial monolayers despite increasing [Ca^{2+}]i. Am J Physiol 276:H1892–901.
68. Olanrewaju HA, Mustafa SJ. (2000) Adenosine A(2A) and A(2B) receptors mediated nitric oxide production in coronary artery endothelial cells. Gen Pharmacol 35:171–7.
69. Oldham WM, Hamm HE. (2008) Heterotrimeric G protein activation by G-protein-coupled receptors. Nat Rev Mol Cell Biol 9:60–71.
70. Palmer TM, Gettys TW, Stiles GL. (1995) Differential interaction with and regulation of multiple G-proteins by the rat A3 adenosine receptor. J Biol Chem 270:16895–902.
71. Pearson JD, Gordon JL. (1979) Vascular endothelial and smooth muscle cells in culture selectively release adenine nucleotides. Nature 281:384–6.
72. Pitcher JA, Hall RA, Daaka Y, Zhang J, Ferguson SS, Hester S, Miller S, Caron MG, Lefkowitz RJ, Barak LS. (1998) The G protein-coupled receptor kinase 2 is a microtubule-associated protein kinase that phosphorylates tubulin. J Biol Chem 273:12316–24.
73. Qi AD, Kennedy C, Harden TK, Nicholas RA. (2001) Differential coupling of the human P2Y(11) receptor to phospholipase C and adenylyl cyclase. Br J Pharmacol 132:318–26.
74. Qiao J, Holian O, Lee BS, Huang F, Zhang J, Lum H. (2008) Phosphorylation of GTP dissociation inhibitor by PKA negatively regulates RhoA. Am J Physiol Cell Physiol 295:C1161–8.
75. Raju NC, Eikelboom JW, Hirsh J. (2008) Platelet ADP-receptor antagonists for cardiovascular disease: past, present and future. Nat Clin Pract Cardiovasc Med 5:766–80.
76. Ralevic V. (2009) Purines as neurotransmitters and neuromodulators in blood vessels. Curr Vasc Pharmacol 7:3–14.
77. Ray CJ, Marshall JM. (2006) The cellular mechanisms by which adenosine evokes release of nitric oxide from rat aortic endothelium. J Physiol 570:85–96.
78. Rebecchi MJ, Pentyala SN. (2000) Structure, function, and control of phosphoinositide-specific phospholipase C. Physiol Rev 80:1291–335.
79. Reinhard M, Jarchau T, Walter U. (2001) Actin-based motility: stop and go with Ena/VASP proteins. Trends Biochem Sci 26:243–9.
80. Robson SC, Wu Y, Sun X, Knosalla C, Dwyer K, Enjyoji K. (2005) Ectonucleotidases of CD39 family modulate vascular inflammation and thrombosis in transplantation. Semin Thromb Hemost 31:217–33.
81. Ryzhov S, Zaynagetdinov R, Goldstein AE, Matafonov A, Biaggioni I, Feoktistov I. (2009) Differential role of the carboxy-terminus of the A(2B) adenosine receptor in stimulation of adenylate cyclase, phospholipase Cbeta, and interleukin-8. Purinergic Signal 5: 289–98.

82. Sandoval R, Malik AB, Minshall RD, Kouklis P, Ellis CA, Tiruppathi C. (2001) Ca(2+) signalling and PKCalpha activate increased endothelial permeability by disassembly of VE-cadherin junctions. J Physiol 533:433–45.
83. Satpathy M, Gallagher P, Jin Y, Srinivas SP. (2005) Extracellular ATP opposes thrombin-induced myosin light chain phosphorylation and loss of barrier integrity in corneal endothelial cells. Exp Eye Res 81:183–92.
84. Satpathy M, Gallagher P, Lizotte-Waniewski M, Srinivas SP. (2004) Thrombin-induced phosphorylation of the regulatory light chain of myosin II in cultured bovine corneal endothelial cells. Exp Eye Res 79:477–86.
85. Schaddelee MP, Voorwinden HL, van Tilburg EW, Pateman TJ, Ijzerman AP, Danhof M, de Boer AG. (2003) Functional role of adenosine receptor subtypes in the regulation of blood-brain barrier permeability: possible implications for the design of synthetic adenosine derivatives. Eur J Pharm Sci 19:13–22.
86. Schneider GB, Hamano H, Cooper LF. (1998) In vivo evaluation of hsp27 as an inhibitor of actin polymerization: hsp27 limits actin stress fiber and focal adhesion formation after heat shock. J Cell Physiol 177:575–84.
87. Schneider JC, El Kebir D, Chereau C, Lanone S, Huang XL, De Buys Roessingh AS, Mercier JC, Dall'Ava-Santucci J, Dinh-Xuan AT. (2003) Involvement of Ca^{2+}/calmodulin-dependent protein kinase II in endothelial NO production and endothelium-dependent relaxation. Am J Physiol Heart Circ Physiol 284:H2311–9.
88. Schulte G, Fredholm BB. (2000) Human adenosine A(1), A(2A), A(2B), and A(3) receptors expressed in Chinese hamster ovary cells all mediate the phosphorylation of extracellular-regulated kinase 1/2. Mol Pharmacol 58:477–82.
89. Schulte G, Fredholm BB. (2002) Signaling pathway from the human adenosine A(3) receptor expressed in Chinese hamster ovary cells to the extracellular signal-regulated kinase 1/2. Mol Pharmacol 62:1137–46.
90. Schulte G, Fredholm BB. (2003) Signalling from adenosine receptors to mitogen-activated protein kinases. Cell Signal 15:813–27.
91. Schwiebert LM, Rice WC, Kudlow BA, Taylor AL, Schwiebert EM. (2002) Extracellular ATP signaling and P2X nucleotide receptors in monolayers of primary human vascular endothelial cells. Am J Physiol Cell Physiol 282:C289–301.
92. Shajahan AN, Timblin BK, Sandoval R, Tiruppathi C, Malik AB, Minshall RD. (2004) Role of Src-induced dynamin-2 phosphorylation in caveolae-mediated endocytosis in endothelial cells. J Biol Chem 279:20392–400.
93. Shajahan AN, Tiruppathi C, Smrcka AV, Malik AB, Minshall RD. (2004) Gbetagamma activation of Src induces caveolae-mediated endocytosis in endothelial cells. J Biol Chem 279:48055–62.
94. Singh I, Knezevic N, Ahmmed GU, Kini V, Malik AB, Mehta D. (2007) Galphaq-TRPC6-mediated Ca^{2+} entry induces RhoA activation and resultant endothelial cell shape change in response to thrombin. J Biol Chem 282:7833–43.
95. Thompson LF, Eltzschig HK, Ibla JC, Van De Wiele CJ, Resta R, Morote-Garcia JC, Colgan SP. (2004) Crucial role for ecto-5′-nucleotidase (CD73) in vascular leakage during hypoxia. J Exp Med 200:1395–405.
96. Tiruppathi C, Song W, Bergenfeldt M, Sass P, Malik AB. (1997) Gp60 activation mediates albumin transcytosis in endothelial cells by tyrosine kinase-dependent pathway. J Biol Chem 272:25968–75.
97. Trevethick MA, Mantell SJ, Stuart EF, Barnard A, Wright KN, Yeadon M. (2008) Treating lung inflammation with agonists of the adenosine A2A receptor: promises, problems and potential solutions. Br J Pharmacol 155:463–74.
98. Trincavelli ML, Tuscano D, Marroni M, Klotz KN, Lucacchini A, Martini C. (2002) Involvement of mitogen protein kinase cascade in agonist-mediated human A(3) adenosine receptor regulation. Biochim Biophys Acta 1591:55–62.

99. Ukropec JA, Hollinger MK, Salva SM, Woolkalis MJ. (2000) SHP2 association with VE-cadherin complexes in human endothelial cells is regulated by thrombin. J Biol Chem 275:5983–6.
100. Verin AD, Birukova A, Wang P, Liu F, Becker P, Birukov K, Garcia JG. (2001) Microtubule disassembly increases endothelial cell barrier dysfunction: role of MLC phosphorylation. Am J Physiol Lung Cell Mol Physiol 281:L565–74.
101. Verin AD, Patterson CE, Day MA, Garcia JG. (1995) Regulation of endothelial cell gap formation and barrier function by myosin-associated phosphatase activities. Am J Physiol 269:L99–108.
102. Verkhrasky A, Krishtal OA, Burnstock G. (2009) Purinoceptors on neuroglia. Mol Neurobiol 39:190–208.
103. Volonte C, Amadio S, D'Ambrosi N, Colpi M, Burnstock G. (2006) P2 receptor web: complexity and fine-tuning. Pharmacol Ther 112:264–80.
104. Vu TK, Hung DT, Wheaton VI, Coughlin SR. (1991) Molecular cloning of a functional thrombin receptor reveals a novel proteolytic mechanism of receptor activation. Cell 64:1057–68.
105. Waldo GL, Harden TK. (2004) Agonist binding and Gq-stimulating activities of the purified human P2Y1 receptor. Mol Pharmacol 65:426–36.
106. Wang L, Karlsson L, Moses S, Hultgardh-Nilsson A, Andersson M, Borna C, Gudbjartsson T, Jern S, Erlinge D. (2002) P2 receptor expression profiles in human vascular smooth muscle and endothelial cells. J Cardiovasc Pharmacol 40:841–53.
107. Weissmuller T, Eltzschig HK, Colgan SP. (2005) Dynamic purine signaling and metabolism during neutrophil-endothelial interactions. Purinergic Signal 1:229–39.
108. Wilson HL, Varcoe RW, Stokes L, Holland KL, Francis SE, Dower SK, Surprenant A, Crossman DC. (2007) P2X receptor characterization and IL-1/IL-1Ra release from human endothelial cells. Br J Pharmacol 151:115–27.
109. Wong AM, Chow AW, Au SC, Wong CC, Ko WH. (2008) Apical vs. Basolateral P2Y6 Receptor-mediated Cl^- Secretion in Immortalized Bronchial Epithelia. Am J Respir Cell Mol Biol.
110. Wyatt AW, Steinert JR, Wheeler-Jones CP, Morgan AJ, Sugden D, Pearson JD, Sobrevia L, Mann GE. (2002) Early activation of the p42/p44MAPK pathway mediates adenosine-induced nitric oxide production in human endothelial cells: a novel calcium-insensitive mechanism. FASEB J 16:1584–94.
111. Yaar R, Jones MR, Chen JF, Ravid K. (2005) Animal models for the study of adenosine receptor function. J Cell Physiol 202:9–20.
112. Yamamoto K, Korenaga R, Kamiya A, Qi Z, Sokabe M, Ando J. (2000) P2X(4) receptors mediate ATP-induced calcium influx in human vascular endothelial cells. Am J Physiol Heart Circ Physiol 279:H285–92.
113. Yao X, Huang Y. (2003) From nitric oxide to endothelial cytosolic Ca^{2+}: a negative feedback control. Trends Pharmacol Sci 24:263–6.
114. Zhu B, Kelly J, Vemavarapu L, Thompson WJ, Strada SJ. (2004) Activation and induction of cyclic AMP phosphodiesterase (PDE4) in rat pulmonary microvascular endothelial cells. Biochem Pharmacol 68:479–91.

Chapter 4
The P2Y$_2$ Nucleotide Receptor in Vascular Inflammation and Angiogenesis

Cheikh I. Seye and Gary A. Weisman

Abstract There is compelling evidence for increased neovascularization of blood vessel walls at sites of intimal hyperplasia in models of arterial stenting, angioplasty, and venous bypass graft failure. We have shown that activation of the P2Y$_2$ nucleotide receptor (P2Y$_2$R) in vascular endothelial cells in vitro induces VCAM-1 (vascular cell adhesion molecule-1) expression and promotes the adherence of monocytes. VEGF, a pleiotropic factor that regulates endothelial cell survival, proliferation, and migration through its interaction with two receptor tyrosine kinases, Flt-1 or VEGF receptor-1 and Flk-1/KDR or VEGF receptor-2 also stimulates the expression of VCAM-1 in endothelial cells. Interestingly, our studies indicate that P2Y$_2$R activation promotes the VEGF-independent activation of Flk-1/KDR via Src binding to Src homology (SH3) binding domains in the C-terminal tail of the P2Y$_2$R leading to the up-regulation of VCAM-1. The P2Y$_2$R also contains an arginine-glycine-aspartate domain that mediates interactions with $\alpha_v\beta_3/\beta_5$ integrins that enable nucleotides to increase cell migration, an important event in inflammation and tumor angiogenesis. Taken together, these findings indicate that P2Y$_2$Rs can stimulate multiple signaling pathways to promote a variety of pathological responses underlying chronic inflammation, angiogenesis in tumors and atherosclerosis.

Keywords Nucleotide · P2Y$_2$ receptor · Endothelial cell · Inflammation · Atherosclerosis · Adhesion · Migration · Angiogenesis · Proliferation · Artery · Smooth muscle cell · G protein · Monocyte · Tyrosine kinase · Calcium · Microvessels

---

C.I. Seye (✉)
Department of Cellular and Integrative Physiology, Indiana University School of Medicine, 635 Barnhill Drive, MS 332, Indianapolis, Indiana 46202
e-mail: cseye@iupui.edu

E. Gerasimovskaya, E. Kaczmarek (eds.), *Extracellular ATP and Adenosine as Regulators of Endothelial Cell Function*, DOI 10.1007/978-90-481-3435-9_4,

4.1 Role of $P2Y_2$ Receptors in Vascular Inflammation

4.1.1 Introduction

Extracellular nucleotides cause a wide range of cellular responses and appear to play a role in the regulation of many vascular functions [63]. Vascular cells release nucleotides in response to ischemia, hypoxia, and chemical or mechanical stress [2, 10, 15, 58]. Other sources of extracellular nucleotides include aggregating platelets, degranulating macrophages, and excitatory neurons [12, 61]. Release of nucleotides has been proposed to occur by exocytosis of ATP/UTP-containing vesicles, facilitated diffusion by putative ABC transporters, cytoplasmic leakage, and electrodiffusional movements through ATP/nucleotide channels [92]. It is becoming apparent that extracellular nucleotides can promote the development of a variety of pathologies including disorders of the immune system and neurodegenerative and vascular diseases [88]. Indeed, ATP or UTP induces proliferation and migration of vascular smooth muscle cells, two processes involved in the development of intimal lesions found in atherosclerosis and post-angioplasty restenosis. The biological effects of extracellular nucleotides are mediated through activation of P1 and P2 purinergic receptors. P2 receptors are subdivided into two distinct categories, the metabotropic G protein-coupled (P2Y) receptors and ionotropic ligand-gated channel (P2X) receptors [57]. Vascular cells have been shown to express metabotropic P2Y and ionotropic P2X receptors [63].

Vascular diseases such as atherosclerosis and post-angioplasty restenosis are initiated by vascular injury. Previous reports demonstrate that large amounts of nucleotides are released into the extracellular milieu in response to vascular stress conditions, including ischemia/oxidative stress, flow, and mechanical stretch [32, 60]. In addition, high concentrations of extracellular ATP and UTP generated in vivo by platelet aggregation or cell damage that occurs during transluminal angioplasty likely increase the concentration of nucleotides in the vessel wall. Although the metabolism of extracellular nucleotides by ecto-nucleotidases plays a role in the regulation of purinergic signaling, recent studies indicate that chronic hypoxic exposure of pulmonary artery endothelial cells can decrease the rate of ecto-nucleotidase activity [28]. In addition, it has been reported that endothelial cell activation can decrease ATP-diphosphohydrolase activity [66], suggesting that pathological conditions affecting blood vessels contribute to local elevations in nucleotide concentrations in the vessel wall.

4.1.2 $P2Y_2$ Receptor Up-Regulation and Endothelial Expression of Adhesion Molecules: Role in Endothelial Inflammation

$P2Y_2R$ up-regulation occurs in various models of vascular injury including balloon denudation of the rat aorta [72], placement of a silicone collar around a rabbit carotid artery [73] and stenting of the pig coronary artery [77]. In the collared rabbit carotid artery, we showed that activation of the $P2Y_2R$ by UTP stimulates intimal hyperplasia and induces monocyte/macrophage infiltration into the vessel wall [73], suggesting a potential role for this receptor in monocyte recruitment by vascular endothelium. We have demonstrated that exogenous ATP or UTP stimulates vascular cell adhesion molecule-1 (VCAM-1) expression via $P2Y_2$ receptor activation in human coronary artery endothelial cells (HCAECs) [75]. Our studies also indicate that the $P2Y_2R$-mediated increase in VCAM-1 expression promotes the adherence of U937 monocytes and Jurkat lymphocytes to HCAECs and human submandibular gland cells, respectively [6, 75]. These data, for the first time, link P2Y purinergic receptor signaling to mechanisms controlling the expression of a cell adhesion molecule involved in monocyte/lymphocyte recruitment. In these studies, the relative equipotency and equiefficacy of UTP and ATP for induction of VCAM-1 expression strongly suggests the involvement of $P2Y_2$ receptors. Nonetheless, the ability of antisense, but not sense, $P2Y_2R$ oligonucleotides to inhibit UTP-induced VCAM-1 expression [75] unambiguously demonstrates the role of $P2Y_2Rs$. Furthermore, UTP increased the expression of VCAM-1 in human 1321N1 astrocytoma cells expressing recombinant $P2Y_2Rs$, but not in control 1321N1 cells [75] that lack endogenous P2 receptors [74]. Moreover, the inability of 2MeS-ATP and UDP to induce VCAM-1 expression in HCAECs [75] suggests that activation of $P2Y_1$, $P2Y_4$, and $P2Y_6$ receptors is not involved.

Stimulation of the $P2Y_2R$ in endothelial cells is known to activate a variety of signaling molecules including the mitogen- and stress-activated protein kinases ERK1/2, p38, and c-Jun NH_2-terminal kinase, the small GTPase RhoA, and the release of the second messenger Ca^{2+} from intracellular stores into the cytoplasm [11, 14, 70]. UTP-induced VCAM-1 expression in HCAECs was prevented by the calcium chelator BAPTA or by depletion of intracellular calcium stores with thapsigargin [75], indicating a role for $P2Y_2R$-mediated increases in $[Ca^{2+}]_i$ in VCAM-1 expression. UTP-induced VCAM-1 expression also was suppressed by inhibitors of p38 and Rho kinase, but not by inhibition of ERK1/2 [75], suggesting that p38 and RhoA activation are involved in the increased expression of VCAM-1 mediated by $P2Y_2Rs$.

Vascular endothelial growth factor (VEGF) induces VCAM-1, *i*ntra*c*ellular *a*dhesion *m*olecule-*1* (ICAM-1), and E-selectin expression in endothelial cells through a pathway involving the MEK/ERK-independent activation of NF-kB [44]. Activation of the $P2Y_2R$ has been reported to increase prostacyclin and nitric oxide release from endothelial cells [51, 76]. Studies have shown that nitric oxide decreases the level of VCAM-1 protein expression in endothelial cells [17]. Although nitric oxide does not play a role in VEGF-induced adhesion molecule expression [44], it could potentially modulate the effects of UTP on VCAM-1 expression in endothelial cells. Further studies are needed to identify the factors involved in nucleotide-induced VCAM-1 expression both at the transcriptional and translational levels.

Consistent with earlier reports showing that ATP and UTP stimulate cell-cell adhesion in a monocyte/macrophage lineage and enhance neutrophil adherence to endothelial cell monolayers [58, 85], our data demonstrate that ATP and UTP stimulate U937 monocyte and Jurkat lymphocyte adherence to HCAECs and human submandibular gland cells, respectively [6, 75]. The $P2Y_2R$ contains the integrin binding domain arginine-glycine-aspartic acid (RGD) in its first extracellular loop [21]. The RGD motif has been shown to be the core recognition sequence for many integrins including $\alpha_v\beta_3$ and $\alpha_v\beta_5$ [39]. Recent studies have indicated that the RGD sequence in the $P2Y_2R$ mediates its interaction with $\alpha_v\beta_3/\beta_5$ integrins, and is required for G_o- and $G_{12/13}$-mediated signal transduction [5, 21, 47]. It has been reported previously that the $\alpha_v\beta_3$ integrin on monocytes mediates their adherence to endothelial and epithelial cells, an early event in the acute inflammatory response [8]. Thus, protein complex formation between $\alpha_v\beta_3/\beta_5$ integrins and $P2Y_2Rs$ also may be involved in the regulation of monocyte binding to endothelial cells mediated by $P2Y_2Rs$.

Our studies also have focused on the mechanisms whereby $P2Y_2R$ activation increases the expression of VCAM-1 [74]. VEGF, a pleiotropic factor that regulates multiple biological phenomena, including endothelial cell survival, proliferation, and migration, also stimulates endothelial cell expression of adhesion molecules, including VCAM-1, ICAM-1, and selectins [44]. These multiple responses are mediated through the interaction of VEGF with two receptor tyrosine kinases, Flt-1 or VEGF receptor-1 and Flk-1/KDR or VEGF receptor-2 [44]. It is widely accepted that VEGFR-2 mediates the growth and migration of endothelial cells, the permeability of blood cells to ions and small molecules, and the expression of VCAM-1 in human umbilical vein endothelial cells [91]. It is increasingly apparent that VEGFR-2 serves as a point of convergence for multiple signals arising from diverse stimuli. VEGFR-2 transactivation follows activation of G protein-coupled receptors, including the endothelial differentiation gene product and the bradykinin receptor [81, 83]. We hypothesized that transactivation of VEGFR-2 may be involved in $P2Y_2R$-mediated VCAM-1 expression in endothelial cells.

We demonstrated for the first time that activation of the $P2Y_2R$ in human coronary artery endothelial cells induces the rapid tyrosine phosphorylation of VEGFR-2 and the activation of Vav2, a Rho guanine nucleotide exchange factor, leading to increased expression of VCAM-1 [74]. Two lines of evidence demonstrate the involvement of VEGFR-2 in this process. First, SU1498, a specific VEGFR-2 tyrosine kinase inhibitor, diminished UTP-induced VCAM-1 expression [74]. Second, selective inhibition of VEGFR-2 expression by siRNA inhibited UTP-induced VCAM-1 expression [74]. Ligand-independent activation of receptor tyrosine kinases is involved in many biological responses initiated by G protein-coupled receptors. It has been reported that activation of the endothelial differentiation gene product in response to sphingosine 1-phosphate stimulates endothelial nitric oxide synthase through activation of Flk-1/KDR [81]. EGF receptor transactivation by G protein-coupled receptor ligands may involve activation of matrix metalloproteinases, release of membrane heparin-bound EGF, and binding of heparin to the EGF receptor [83]. At the structural level, we have identified two SH3-binding sites (PXXP motifs) in the intracellular C terminus of the $P2Y_2R$ that enable $P2Y_2R$ interaction with Src and the regulation of EGF and PDGF receptor activity [50]. We demonstrated that PP2, the Src kinase inhibitor, decreased

P2Y$_2$R-mediated VEGFR-2 tyrosine phosphorylation, consistent with the involvement of Src in EGF and PDGF receptor transactivation by P2Y$_2$Rs [50]. We also found that the P2Y$_2$R and VEGFR-2 co-localize in cells upon UTP stimulation and demonstrated for the first time a direct interaction between these receptors in a UTP-dependent manner [74]. Deletion or mutation of the SH3-binding domains prevented UTP-induced expression of VCAM-1 [74]. Because expression of dominant negative Src inhibited the P2Y$_2$R/VEGFR-2 interaction, we concluded that Src binding to the SH3-binding domains in the P2Y$_2$R mediates both the UTP-induced P2Y$_2$R and VEGFR-2 interaction and the transactivation of VEGFR-2 by UTP that regulates VCAM-1 expression.

Since leukocyte infiltration and migration are key processes involved in atherosclerosis, these findings suggest that P2Y$_2$ receptors represent a novel target for reducing arterial inflammation associated with cardiovascular disease. Recent advances in the field of cell migration have identified many cellular events linking a motility signal to remodeling of the actin cytoskeleton that enables a cell to move [54]. Briefly, these events include receptor-mediated activation of Rac GTPase that, in turn, activates WASP protein binding to allow a complex of proteins called Arp2/3 to nucleate actin and facilitate formation of lamellipodia or branched actin filaments at the leading edge of a cell. A capping protein terminates actin polymerization while another protein (cofilin), a downstream target of RhoA GTPase, begins severing actin filaments so that the cell can continue to move. We found that actin stress fiber formation induced by UTP occurred in 1321N1 cells expressing the RGD-containing wild type P2Y$_2$R but not in cells expressing an RGE mutant P2Y$_2$R [47]. Furthermore, stress fiber formation and cytoskeletal rearrangements mediated by P2Y$_2$R/$\alpha_V\beta_{3/5}$ interactions required the activation of $G_{i/o}$ and G_{12} proteins, and not the activation of phospholipase C, intracellular calcium mobilization or protein kinase C (PKC) that occurs concomitantly due to the coupling of the P2Y$_2$R to G_q protein [88]. Our data indicate that mutation of the RGD domain in the P2Y$_2$R to RGE (arginine-glycine-glutamate), which does not bind integrins, inhibits RhoA GTPase and cofilin activities [5, 47], consistent with the involvement of $\alpha_V\beta_{3/5}$ integrins in these responses. Figure 4.1 summarizes the signaling pathways that are coupled to P2Y$_2$ receptor activation.

4.1.3 Potential Role of Nucleotides and P2Y$_2$ Receptors in Neuroinflammation

ATP and UTP induce cytokine-like effects in astrocytes that promote neuroinflammatory responses [55, 56, 87]. These nucleotides also induce cell-cell adhesion in the monocyte/macrophage lineage and neutrophil adherence to endothelial cell monolayers [58, 85], supporting the notion that released ATP and UTP lead to endothelial cell activation by autocrine/paracrine mechanisms. It also has been found that ATP and UTP induce cell migration [14, 31, 41, 62] and chemotaxis of microglial cells [38] and primary rat cortical astrocytes [87].

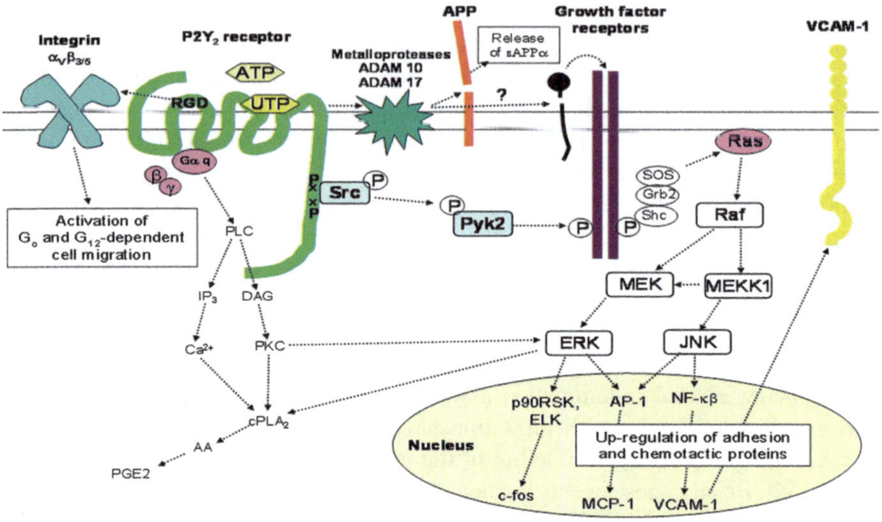

Fig. 4.1 Protein:protein interactions and signaling pathways linked to the P2Y$_2$ receptor. Activation of the G$_q$-coupled P2Y$_2$ receptor (P2Y$_2$R) stimulates phospholipase C-dependent formation of the second messengers IP$_3$ and diacylglycerol, which stimulate release of calcium from intracellular stores and activation of protein kinase C, respectively. The P2Y$_2$R also contains an RGD integrin-binding domain that interacts with integrins ($\alpha_V\beta_3$ and $\alpha_V\beta_5$) and enables activation by nucleotides of G$_o$ and G$_{12}$ proteins. The RGD domain is necessary for P2Y$_2$R-mediated chemotaxis, actin stress fiber formation, and coupling to signaling pathways involving RhoA and Rac GTPases. Two consensus SH3 binding sites (PXXP) in the C-terminal tail of the P2Y$_2$ receptor bind directly to Src upon receptor activation and mediate Src-dependent co-localization and transactivation of growth factor receptors that are required for up-regulation of cell adhesion/chemotactic proteins (e.g., VCAM-1). The P2Y$_2$R also can stimulate metalloprotease activity leading to the cleavage of membrane associated proteins. Abbreviations used: AA, arachidonic acid; ADAM, a disintegrin and metalloprotease family; ADAM10, the Kuz enzyme; ADAM 17 or TACE, tumor necrosis factor-α converting enzyme; AP-1, activator protein-1; APP, amyloid precursor protein; ATP, adenosine 5′-triphosphate; Ca^{2+}, calcium ion; c-fos, the cellular counterpart of oncogene v-*fos;* cPLA$_2$, cytosolic phospholipase A2; DAG, diacylglycerol; ELK, nuclear transcription factor E-26-like protein; ERK, extracellular-signal regulated kinase; Grb2, growth factor receptor binding protein 2; IP$_3$, inositol-1,4,5-trisphosphate; JNK, Jun kinase; MCP-1, monocyte chemoattractant protein-1; MEK, MAP/ERK kinase; MEKK1, MAP kinase kinase kinase 1; NF-kB, nuclear factor kB; p90RSK, 90 kDa ribosomal S6 kinase; PGE2, prostaglandin E2; PKC, protein kinase C; Pyk2, Proline-rich tyrosine kinase; PLC, phospholipase C; PXXP, Src homology 3 binding sites; Raf, rapidly accelerated fibrosarcoma; Ras, superfamily of small GTPases; RGD, arginine-glycine-aspartic acid; sAPPα, a soluble, non-amyloidogenic N-terminal fragment of APP; Shc, an adaptor protein; SOS, Ras-specific nucleotide exchange factor Son of sevenless; Src, a tyrosine kinase; UTP, uridine 5′-triphosphate; VCAM-1, vascular cell adhesion molecule-1

Recent studies suggest that inflammation associated with atherosclerotic lesion development in brain microvessels can contribute to Alzheimer's disease [3, 84], and may be a factor in the failure to eliminate amyloidogenic Aβ peptide from aging human brain [89]. Moreover, Aβ accumulation promotes cerebrovascular inflammation [84], most likely due to increased adherence of blood monocytes to cytokine-activated microvascular endothelium that facilitates pro-inflammatory monocyte migration across the blood-brain barrier. Indeed, increased monocyte migration across human brain endothelium is augmented by Aβ, and monocyte adherence and transendothelial migration are inhibited by antibodies to the putative Aβ receptor for advanced glycated endproducts (RAGE) or the platelet endothelial cell adhesion molecule (PECAM-1) [29, 30]. Under normal conditions, the vascular endothelium maintains blood fluidity and provides a barrier separating blood cells and plasma factors from highly reactive elements in the underlying tissue [53, 67]. Quiescent endothelial cells produce substances to maintain vascular homeostasis [24, 35, 48, 53]. However, injury or stress in surrounding tissue can shift endothelial cells from a quiescent to a pro-inflammatory status, largely due to the increased release of cytokines. Besides Aβ, other factors that can induce a pro-inflammatory status in endothelium include obesity, hypertension, diabetes, cigarette smoking, high serum LDL or homocysteine levels, chronic viral or bacterial infections, endotoxaemia, and an unfavorable genetic background [18]. Thus, stress-induced injury to the vasculature and underlying tissue would be expected to contribute similarly to the development of chronic inflammation under pathological conditions, including Alzheimer's disease, atherosclerosis and cancer. Interestingly, recent results indicate that cytokine-induced $P2Y_2R$ up-regulation and activation in primary rat cortical neurons promotes metalloprotease-dependent non-amyloidogenic peptide release (see Fig. 4.1), rather than Aβ formation [45], suggesting that modulation of $P2Y_2R$ activity in Alzheimer's disease may be a useful therapeutic approach for limiting chronic inflammatory and neurodegenerative responses.

4.1.4 Development of Transgenic Rats Overexpressing the $P2Y_2$ Receptor: A Model to Study Inflammatory Diseases

Recently, we developed a transgenic (Tg) rat model overexpressing the $P2Y_2R$ [1]. We demonstrated that incorporation of the $P2Y_2R$ transgene into a lentiviral vector enabled stable $P2Y_2R$ mRNA expression through 6 generations and the expressed $P2Y_2R$ was functionally active, as indicated by increased $P2Y_2R$ agonist-induced intracellular calcium mobilization in single smooth muscle cells (SMCs) from $P2Y_2R$ overexpressing Tg rats, as compared to wild type (WT) rats. In addition, there was a significant increase in the percentage of responsive aortic SMCs from $P2Y_2R$ overexpressing Tg rats, versus WT rats. We found that $P2Y_2R$ overexpressing Tg rats express increased levels of $P2Y_2R$ mRNA in major organs and tissues, including aorta, brain, heart, kidney, liver, lung, lacrimal gland and leg muscle. Comparisons of WT and $P2Y_2R$ overexpressing Tg rats showed that expression levels of the $P2Y_2R$ transgene in tissues did not change between the third and sixth generations. The endogenous $P2Y_2R$ in WT rats is expressed in liver, lung and muscle and to a lesser extent in heart, kidney, and brain, but was undetectable in lacrimal and salivary glands. Expression of the $P2Y_2R$ transgene in lacrimal and salivary glands of $P2Y_2R$ overexpressing Tg rats was elevated as compared to WT rats, but was lower than in the other tissues analyzed. Overall, these results indicate that functional $P2Y_2R$ expression is significantly higher in $P2Y_2R$ overexpressing Tg rat tissue, as compared to WT controls.

In the $P2Y_2R$ overexpressing rat, $P2Y_2R$-mediated increases in $[Ca^{2+}]_i$ in aortic SMCs were 18 times greater than in WT rat aortic SMCs [1]. Similarly, $P2Y_2R$ mRNA expression in aorta was 27 times greater in $P2Y_2R$ overexpressing Tg rats as compared to WT rats, indicating a close correlation between increased $P2Y_2R$ mRNA expression and functional activity, which suggests that these Tg rats will be ideal for investigations on the role of $P2Y_2R$ overexpression in vascular lesion development.

4.2 $P2Y_2$ Receptors in Angiogenesis

Vascular endothelial cells are involved in a variety of physiological processes, including vasoregulation, repair of the vascular intima, blood clotting, and development of new blood vessels (i.e., angiogenesis). Vascular tone and blood flow can be regulated by release from endothelium of compounds such as nitric oxide (NO), prostacyclin (PGI_2), ATP, and endothelin, whereas endothelial cell proliferation and migration are important processes required for the continual repair of established blood vessels [33]. During the formation of new blood vessels, endothelial cells release enzymes that degrade the basement membrane, enabling some endothelial cells to migrate through the vascular wall and proliferate to create capillaries that extend perpendicular to the original blood vessel [25]. Angiogenesis occurs in many physiological and pathophysiological conditions, including development, wound healing, ovarian and menstrual cycling, rheumatoid arthritis, and tumor growth.

4.2.1 $P2Y_2$ Receptors and Neovascularization of Atherosclerotic Plaques

Neovascularization in atherosclerotic plaques was first noted by Koster [46] more than 100 years ago. Barger also documented the existence of neovascularization in atherosclerosis using cineangiography at autopsy [7]. Neovascularization has been postulated to play a role in atherosclerosis by providing growth factors and cytokines to regions of plaque growth [42]. The clinical significance of neovascularization in atherosclerosis is unknown. Accumulating histopathological data associate plaque angiogenesis with rapid progression of unstable angina [82]. These studies confirm that intimal neovascularization is a ubiquitous feature of atherosclerosis, correlating with both histological grade and disease symptoms [82]. Vasa vasorum-derived microvessels do not extend to the intima of normal arteries, penetrating only the adventitia and outer media [27]. Diffusion of oxygen and other nutrients is limited to 100 μm from the lumen of the blood vessel, which in normal arteries is adequate to nourish the inner media and intimal layers. As vessel wall thickness increases in the setting of vascular disease, proliferation of the vasa vasorum and intimal neovascularization is observed.

The recruitment of pericytes and vascular smooth muscle cells (VSMCs) to endothelial tubes plays a key role in angiogenesis. These mural cells stabilize nascent endothelial tubes and therefore are essential for the formation of functional vessels [4].

Many growth factors have been suggested to play pivotal roles in different aspects of neovascularization. For instance, platelet-derived growth factor (PDGF) induces differentiation of mesenchymal cells into pericytes and VSMCs and stimulates the migration of these cells to the newly formed vessels [34, 36, 49, 80]. However, little is known about the contribution of extracellular nucleotides and their receptors to neovascularization.

4.2.2 Chemotactic Effects Mediated by $P2Y_2$ Receptors

In cultured endothelial cells (CECs) from guinea pig cardiac vasculature, both UTP and vascular endothelial growth factor (VEGF) are significant mitogenic and chemotactic factors [69]. In chick chorioallantoic membranes (CAMs), UTP and VEGF produced statistically significant increases in CAM vascularity [69]. Given the lack of selective ligands for receptors that bind UTP, these authors were unable to definitively identify the receptors mediating mitogenic or chemotactic responses to UTP in cultured endothelial cells or the receptor(s) that are responding to UTP by promoting angiogenesis in chick CAMs. However, the existence in CECs of two receptors capable of responding to UTP ($P2Y_2R$ and a UTP-selective receptor) has been previously reported [90]. Recent studies indicate that the extracellular nucleotides ATP, ADP, UTP and UDP serve as directional cues for the migration of rat aortic SMCs [14]. Nucleotide-induced migration of SMCs is the consequence of both chemotaxis and chemokinesis and may result from the activation of one or more P2 nucleotide receptor subtypes. The ability of UTP at submicromolar levels to stimulate migration of SMCs supports the hypothesis that this response could have physiological consequences and is mediated by $P2Y_2$ receptor activation without excluding participation of other P2Y receptor subtypes. The increased migration of SMCs in response to extracellular nucleotides could be related to increases in extracellular matrix (ECM) protein expression. Indeed, previous studies have shown that UTP induces osteopontin expression in rat and rabbit aortic SMCs [26, 73]. Increased expression of osteopontin, an arginine-glycine-aspartate (RGD)-containing ECM protein, is associated with the activation of rat arterial SMCs in vitro and in vivo [26, 52, 73]. The increase in osteopontin expression plays a key role in UTP-induced migration of rat aortic SMCs, since a monoclonal antibody against osteopontin fully abolished UTP-induced migration [62], whereas an antibody against vitronectin, another ECM protein also involved in migration of human SMCs [19], had no effect on the migration of rat aortic SMCs [62]. UTP induces increases in osteopontin mRNA expression by increasing both osteopontin mRNA stabilization and *osteopontin* promoter activity [64]. Recent studies have shown that activation of an AP-1 binding site located 76 bp upstream of the transcription start site in the rat *osteopontin* promoter is involved in UTP-induced osteopontin expression [65].

The ability of extracellular nucleotides to act as chemoattractants for rat arterial SMCs in a concentration range potentially found in vessels under pathological conditions [16] and the findings of previous studies demonstrating the mitogenic activity of extracellular nucleotides for these cells [22, 23, 52, 86] suggest that nucleotides released from mechanically-stretched or damaged vascular cells during angioplasty therapy for occluded arteries may promote angiogenesis.

4.2.3 $P2Y_2$ Receptors in Tumor Angiogenesis

Purinergic regulation of angiogenesis by human breast carcinoma-secreted nucleoside diphosphate kinase (NDPK) has been recently described [68]. These authors showed that MDA-MB-435S human breast cancer cell-secreted NDPK-B supports tumor formation by modulating the local ATP concentration thereby activating endothelial cell P2Y receptor-mediated angiogenesis. Their model suggests an NDPK-dependent recycling of ADP back to ATP in both arteries and capillaries. NDPK in the arteries may elevate local ATP concentrations to produce P2YR-mediated vasodilation and inhibit ADP-dependent platelet aggregation, mechanisms advantageous to the transit of cancer cells to secondary sites. NDPK secretion in capillaries may elevate local nucleotide concentrations and produce P2Y nucleotide receptor-mediated angiogenesis. We have previously shown that the $P2Y_2R$ can mediate transactivation of VEGFR-2 [74] and directly link nucleotide receptor activation to an established tumor angiogenesis signaling pathway (VEGF-VEGFR signaling). P2Y receptor potentiation of VEGFR-2 signaling, therefore, may be essential for manifestation of the angiogenic properties of nucleotides such as ATP. P2Y and VEGF receptors are colocalized to caveolar domains in the cell membrane, which could propagate a substantial pro-angiogenic signal in response to small amounts of receptor agonist [43].

Jackson et al. have reported that disordered purinergic signaling inhibits pathological angiogenesis in Cd39/Entpd1-null mice [40]. CD39/ecto-nucleoside triphosphate diphosphohydrolase-type-1 (ENTPD1) is the dominant vascular ectonucleotidase that catalyzes the phosphohydrolysis of extracellular nucleotides in the blood and extracellular space. This ecto-enzymatic process modulates endothelial cell, leukocyte, and platelet purinergic receptor-mediated responses to extracellular nucleotides. These authors reported that deletion of Cd39/Entpd1 results in abrogation of angiogenesis, causing decreased growth of implanted tumors and inhibition of pulmonary metastases. These changes were associated with decreased activation of focal adhesion kinase (FAK) and extracellular signal-regulated kinase-1 and -2 in endothelial cells. The authors also reported markedly abnormal angiogenesis in Cd39-null mice characterized by poor recruitment of surrounding pericytes and SMCs to new blood vessels. Because embryonic vascular development is normal in Cd39-null mice [20], the data suggest that purinergic signaling plays differential roles in physiological blood vessel development as opposed to pathological angiogenesis. Indeed, in vitro analyses indicated that cell migration in Cd39-null mice is likely to be secondary to abnormal adhesion and vitronectin receptor function in Cd39-null endothelial cells. Several members of the integrin family have been implicated in angiogenesis, although the largest body of data has linked the vitronectin receptor $\alpha_v\beta_3$ with the promotion of neovascularization [37]. The $P2Y_2R$ contains an integrin-binding RGD domain in its first extracellular loop that is required for $P2Y_2R$ interactions with $\alpha_v\beta_3$ integrin [21]. Mutation of the RGD sequence to RGE decreased the co-localization of $\alpha_v\beta_3$ integrin with the $P2Y_2R$ by 10-fold and greatly impaired UTP-induced phosphorylation of FAK [21]. Direct association between the $P2Y_2R$ and α_v integrins is necessary for UTP-induced endothelial cell migration [41], implying an important link between purinergic signaling, integrin function, and pro-angiogenic cellular behaviors. As discussed above, $P2Y_2R$ activation increases osteopontin expression in vascular SMCs. Osteopontin (OPN), one of the cytokines produced by various tumor cells, is suggested to be involved in angiogenesis by up-regulating endothelial cell migration in coordination with vascular endothelial growth factor (VEGF). OPN is secreted from various cancer cell lines and is postulated to promote malignant transformation [71]. When ras-transformed fibroblasts were transfected with OPN antisense RNA, their tumorigenic and malignant growth were suppressed significantly [9]. OPN also is overexpressed in human cancers [13] and serum levels of OPN are substantially elevated in patients with metastatic cancer [78]. Thus, OPN is postulated to be related with cancer progression.

Takahashi et al. generated a stable transfectant of murine neuroblastoma C1300 cells that constitutively secretes high levels of murine OPN [79]. To demonstrate the effect of OPN on tumor-induced angiogenesis in vivo, Millipore chambers containing OPN-transfected or control cells were implanted in the dorsal air sac of mice. The OPN-transfected cells significantly induced neovascularization in comparison to the control cells in mice, thus providing direct evidence for OPN involvement in tumor angiogenesis.

Our data link P2Y purinergic receptor signaling to mechanisms controlling the expression of a cell adhesion molecule involved in monocyte/lymphocyte recruitment. Therefore, up-regulation of P2Y receptors may be a potential diagnostic indicator for the early stages of atherosclerosis. We also demonstrated for the first time that the $P2Y_2R$ can mediate activation of VEGFR-2 and directly link nucleotide receptor activation to an established tumor angiogenesis signaling pathway. Thus, particular effort must be made to understand the consequences of nucleotide release from cells in the cardiovascular system and the subsequent effects of P2 nucleotide receptor activation in blood vessels, which may reveal novel therapeutic strategies for atherosclerosis, restenosis after angioplasty and angiogenesis.

References

1. Agca C, Seye C, Kashuba Benson CM, Rikka S, Chan AW, Weisman GA, Agca Y. (2009) Development of a Novel Transgenic Rat Overexpressing the P2Y$_2$ Nucleotide Receptor Using a Lentiviral Vector. J Vasc Res 46:447–58.
2. Ahmed SM, Rzigalinski BA, Willoughby KA, Sitterding HA, Ellis EF. (2000) Stretch-induced injury alters mitochondrial membrane potential and cellular ATP in cultured astrocytes and neurons. J Neurochem 74:1951–60.
3. Aliev G, Seyidova D, Neal ML, Shi J, Lamb BT, Siedlak SL, Vinters HV, Head E, Perry G, Lamanna JC, Friedland RP, Cotman CW. (2002) Atherosclerotic lesions and mitochondria DNA deletions in brain microvessels as a central target for the development of human AD and AD-like pathology in aged transgenic mice. Ann N Y Acad Sci 977:45–64.
4. Armulik A, Abramsson A, Betsholtz C. (2005) Endothelial/pericyte interactions. Circ Res 97:512–23.
5. Bagchi S, Liao Z, Gonzalez FA, Chorna NE, Seye CI, Weisman GA, Erb L. (2005) The P2Y$_2$ nucleotide receptor interacts with α_v integrins to activate G$_o$ and induce cell migration. J Biol Chem 280:39050–7.
6. Baker OJ, Camden JM, Rome DE, Seye CI, Weisman GA. (2008) P2Y$_2$ nucleotide receptor activation up-regulates vascular cell adhesion molecule-1 [corrected] expression and enhances lymphocyte adherence to a human submandibular gland cell line. Mol Immunol 45: 65–75.
7. Barger AC, Beeuwkes R, 3rd, Lainey LL, Silverman KJ. (1984) Hypothesis: vasa vasorum and neovascularization of human coronary arteries. A possible role in the pathophysiology of atherosclerosis. N Engl J Med 310:175–7.
8. Beekhuizen H, van Furth R. (1993) Monocyte adherence to human vascular endothelium. J Leukoc Biol 54:363–78.
9. Behrend EI, Craig AM, Wilson SM, Denhardt DT, Chambers AF. (1994) Reduced malignancy of ras-transformed NIH 3T3 cells expressing antisense osteopontin RNA. Cancer Res 54: 832–7.
10. Bergfeld GR, Forrester T. (1992) Release of ATP from human erythrocytes in response to a brief period of hypoxia and hypercapnia. Cardiovasc Res 26:40–7.
11. Boarder MR, Hourani SM. (1998) The regulation of vascular function by P2 receptors: multiple sites and multiple receptors. Trends Pharmacol Sci 19:99–107.
12. Bodin P, Burnstock G. (2001) Purinergic signalling: ATP release. Neurochem Res 26: 959–69.
13. Brown LF, Papadopoulos-Sergiou A, Berse B, Manseau EJ, Tognazzi K, Perruzzi CA, Dvorak HF, Senger DR. (1994) Osteopontin expression and distribution in human carcinomas. Am J Pathol 145:610–23.
14. Chaulet H, Desgranges C, Renault MA, Dupuch F, Ezan G, Peiretti F, Loirand G, Pacaud P, Gadeau AP. (2001) Extracellular nucleotides induce arterial smooth muscle cell migration via osteopontin. Circ Res 89:772–8.
15. Ciccarelli R, Di Iorio P, Giuliani P, D′Alimonte I, Ballerini P, Caciagli F, Rathbone MP. (1999) Rat cultured astrocytes release guanine-based purines in basal conditions and after hypoxia/hypoglycemia. Glia 25:93–8.
16. Communi D, Janssens R, Suarez-Huerta N, Robaye B, Boeynaems JM. (2000) Advances in signalling by extracellular nucleotides. the role and transduction mechanisms of P2Y receptors. Cell Signal 12:351–60.
17. De Caterina R, Libby P, Peng HB, Thannickal VJ, Rajavashisth TB, Gimbrone MA, Jr., Shin WS, Liao JK. (1995) Nitric oxide decreases cytokine-induced endothelial activation. Nitric oxide selectively reduces endothelial expression of adhesion molecules and proinflammatory cytokines. J Clin Invest 96:60–8.
18. Di Virgilio F, Solini A. (2002) P2 receptors: new potential players in atherosclerosis. Br J Pharmacol 135:831–42.

19. Dufourcq P, Louis H, Moreau C, Daret D, Boisseau MR, Lamaziere JM, Bonnet J. (1998) Vitronectin expression and interaction with receptors in smooth muscle cells from human atheromatous plaque. Arterioscler Thromb Vasc Biol 18:168–76.
20. Enjyoji K, Sevigny J, Lin Y, Frenette PS, Christie PD, Esch JS, 2nd, Imai M, Edelberg JM, Rayburn H, Lech M, Beeler DL, Csizmadia E, Wagner DD, Robson SC, Rosenberg RD. (1999) Targeted disruption of cd39/ATP diphosphohydrolase results in disordered hemostasis and thromboregulation. Nat Med 5:1010–7.
21. Erb L, Liu J, Ockerhausen J, Kong Q, Garrad RC, Griffin K, Neal C, Krugh B, Santiago-Perez LI, Gonzalez FA, Gresham HD, Turner JT, Weisman GA. (2001) An RGD sequence in the $P2Y_2$ receptor interacts with $\alpha_V\beta_3$ integrins and is required for G_o-mediated signal transduction. J Cell Biol 153:491–501.
22. Erlinge D, Yoo H, Edvinsson L, Reis DJ, Wahlestedt C. (1993) Mitogenic effects of ATP on vascular smooth muscle cells vs. other growth factors and sympathetic cotransmitters. Am J Physiol 265:H1089–97.
23. Erlinge D, You J, Wahlestedt C, Edvinsson L. (1995) Characterisation of an ATP receptor mediating mitogenesis in vascular smooth muscle cells. Eur J Pharmacol 289:135–49.
24. Esmon CT, Taylor FB, Jr., Snow TR. (1991) Inflammation and coagulation: linked processes potentially regulated through a common pathway mediated by protein C. Thromb Haemost 66:160–5.
25. Folkman J. (1985) Toward an understanding of angiogenesis: search and discovery. Perspect Biol Med 29:10–36.
26. Gadeau AP, Campan M, Millet D, Candresse T, Desgranges C. (1993) Osteopontin over-expression is associated with arterial smooth muscle cell proliferation in vitro. Arterioscler Thromb 13:120–5.
27. Geiringer E. (1951) Intimal vascularization and atherosclerosis. J Pathol Bacteriol 63:201–11.
28. Gerasimovskaya EV, Ahmad S, White CW, Jones PL, Carpenter TC, Stenmark KR. (2002) Extracellular ATP is an autocrine/paracrine regulator of hypoxia-induced adventitial fibroblast growth. Signaling through extracellular signal-regulated kinase-1/2 and the Egr-1 transcription factor. J Biol Chem 277:44638–50.
29. Giri R, Selvaraj S, Miller CA, Hofman F, Yan SD, Stern D, Zlokovic BV, Kalra VK. (2002) Effect of endothelial cell polarity on beta-amyloid-induced migration of monocytes across normal and AD endothelium. Am J Physiol Cell Physiol 283:C895–904.
30. Giri R, Shen Y, Stins M, Du Yan S, Schmidt AM, Stern D, Kim KS, Zlokovic B, Kalra VK. (2000) Beta-amyloid-induced migration of monocytes across human brain endothelial cells involves RAGE and PECAM-1. Am J Physiol Cell Physiol 279:C1772–81.
31. Goepfert C, Sundberg C, Sevigny J, Enjyoji K, Hoshi T, Csizmadia E, Robson S. (2001) Disordered cellular migration and angiogenesis in cd39-null mice. Circulation 104: 3109–15.
32. Grierson JP, Meldolesi J. (1995) Shear stress-induced $[Ca^{2+}]i$ transients and oscillations in mouse fibroblasts are mediated by endogenously released ATP. J Biol Chem 270: 4451–6.
33. Hayden MR, Tyagi SC. (1998) Arterial vascular remodeling: the endothelial cell's central role. Mol Med 95:213–7.
34. Hellstrom M, Kalen M, Lindahl P, Abramsson A, Betsholtz C. (1999) Role of PDGF-B and PDGFR-beta in recruitment of vascular smooth muscle cells and pericytes during embryonic blood vessel formation in the mouse. Development 126:3047–55.
35. Hirschi KK, D'Amore PA. (1996) Pericytes in the microvasculature. Cardiovasc Res 32: 687–98.
36. Hirschi KK, Rohovsky SA, D'Amore PA. (1998) PDGF, TGF-beta, and heterotypic cell-cell interactions mediate endothelial cell-induced recruitment of 10T1/2 cells and their differentiation to a smooth muscle fate. J Cell Biol 141:805–14.
37. Hodivala-Dilke KM, Reynolds AR, Reynolds LE. (2003) Integrins in angiogenesis: multitalented molecules in a balancing act. Cell Tissue Res 314:131–44.

38. Honda S, Sasaki Y, Ohsawa K, Imai Y, Nakamura Y, Inoue K, Kohsaka S. (2001) Extracellular ATP or ADP induce chemotaxis of cultured microglia through Gi/o-coupled P2Y receptors. J Neurosci 21:1975–82.
39. Hynes RO. (1992) Integrins: versatility, modulation, and signaling in cell adhesion. Cell 69:11–25.
40. Jackson SW, Hoshi T, Wu Y, Sun X, Enjyoji K, Cszimadia E, Sundberg C, Robson SC. (2007) Disordered purinergic signaling inhibits pathological angiogenesis in cd39/Entpd1-null mice. Am J Pathol 171:1395–404.
41. Kaczmarek E, Erb L, Koziak K, Jarzyna R, Wink MR, Guckelberger O, Blusztajn JK, Trinkaus-Randall V, Weisman GA, Robson SC. (2005) Modulation of endothelial cell migration by extracellular nucleotides: involvement of focal adhesion kinase and phosphatidylinositol 3-kinase-mediated pathways. Thromb Haemost 93:735–42.
42. Kahlon R, Shapero J, Gotlieb AI. (1992) Angiogenesis in atherosclerosis. Can J Cardiol 8: 60–4.
43. Kaiser RA, Oxhorn BC, Andrews G, Buxton IL. (2002) Functional compartmentation of endothelial P2Y receptor signaling. Circ Res 91:292–9.
44. Kim I, Moon SO, Kim SH, Kim HJ, Koh YS, Koh GY. (2001) Vascular endothelial growth factor expression of intercellular adhesion molecule 1 (ICAM-1), vascular cell adhesion molecule 1 (VCAM-1), and E-selectin through nuclear factor-kappa B activation in endothelial cells. J Biol Chem 276:7614–20.
45. Kong Q, Peterson TS, Baker O, Stanley E, Camden J, Seye CI, Erb L, Simonyi A, Wood WG, Sun GY, Weisman GA. (2009) Interleukin-1β enhances nucleotide-induced and α-secretase-dependent amyloid precursor protein processing in rat primary cortical neurons via up-regulation of the $P2Y_2$ receptor. J Neurochem 109:1300–10.
46. Koster W. (1876) Endarteritis and arteritis. Berl Klin Wochenschr 13:454–55.
47. Liao Z, Seye CI, Weisman GA, Erb L. (2007) The $P2Y_2$ nucleotide receptor requires interaction with α_v integrins to access and activate G_{12}. J Cell Sci 120:1654–62.
48. Lindahl P, Hellstrom M, Kalen M, Betsholtz C. (1998) Endothelial-perivascular cell signaling in vascular development: lessons from knockout mice. Curr Opin Lipidol 9:407–11.
49. Lindahl P, Johansson BR, Leveen P, Betsholtz C. (1997) Pericyte loss and microaneurysm formation in PDGF-B-deficient mice. Science 277:242–5.
50. Liu J, Liao Z, Camden J, Griffin KD, Garrad RC, Santiago-Perez LI, Gonzalez FA, Seye CI, Weisman GA, Erb L. (2004) Src homology 3 binding sites in the $P2Y_2$ nucleotide receptor interact with Src and regulate activities of Src, proline-rich tyrosine kinase 2, and growth factor receptors. J Biol Chem 279:8212–8.
51. Lustig KD, Erb L, Landis DM, Hicks-Taylor CS, Zhang X, Sportiello MG, Weisman GA. (1992) Mechanisms by which extracellular ATP and UTP stimulate the release of prostacyclin from bovine pulmonary artery endothelial cells. Biochim Biophys Acta 1134:61–72.
52. Malam-Souley R, Campan M, Gadeau AP, Desgranges C. (1993) Exogenous ATP induces a limited cell cycle progression of arterial smooth muscle cells. Am J Physiol 264:C783–8.
53. Marcus AJ, Safier LB. (1993) Thromboregulation: multicellular modulation of platelet reactivity in hemostasis and thrombosis. FASEB J 7:516–22.
54. Marx J. (2003) Cell biology. How cells step out. Science 302:214–6.
55. Neary JT, Baker L, Jorgensen SL, Norenberg MD. (1994) Extracellular ATP induces stellation and increases glial fibrillary acidic protein content and DNA synthesis in primary astrocyte cultures. Acta Neuropathol 87:8–13.
56. Neary JT, Whittemore SR, Zhu Q, Norenberg MD. (1994) Synergistic activation of DNA synthesis in astrocytes by fibroblast growth factors and extracellular ATP. J Neurochem 63:490–4.
57. North RA, Surprenant A. (2000) Pharmacology of cloned P2X receptors. Annu Rev Pharmacol Toxicol 40:563–80.
58. Parker AL, Likar LL, Dawicki DD, Rounds S. (1996) Mechanism of ATP-induced leukocyte adherence to cultured pulmonary artery endothelial cells. Am J Physiol 270:L695–703.

59. Parr CE, Sullivan DM, Paradiso AM, Lazarowski ER, Burch LH, Olsen JC, Erb L, Weisman GA, Boucher RC, Turner JT. (1994) Cloning and expression of a human P_{2U} nucleotide receptor, a target for cystic fibrosis pharmacotherapy. Proc Natl Acad Sci USA 91:3275–9.
60. Pearson JD, Gordon JL. (1979) Vascular endothelial and smooth muscle cells in culture selectively release adenine nucleotides. Nature 281:384–6.
61. Pedersen S, Pedersen SF, Nilius B, Lambert IH, Hoffmann EK. (1999) Mechanical stress induces release of ATP from Ehrlich ascites tumor cells. Biochim Biophys Acta 1416:271–84.
62. Pillois X, Chaulet H, Belloc I, Dupuch F, Desgranges C, Gadeau AP. (2002) Nucleotide receptors involved in UTP-induced rat arterial smooth muscle cell migration. Circ Res 90:678–81.
63. Ralevic V, Burnstock G. (1991) Roles of P2-purinoceptors in the cardiovascular system. Circulation 84:1–14.
64. Renault MA, Jalvy S, Belloc I, Pasquet S, Sena S, Olive M, Desgranges C, Gadeau AP. (2003) AP-1 is involved in UTP-induced osteopontin expression in arterial smooth muscle cells. Circ Res 93:674–81.
65. Renault MA, Jalvy S, Potier M, Belloc I, Genot E, Dekker LV, Desgranges C, Gadeau AP. (2005) UTP induces osteopontin expression through a coordinate action of NFkappaB, activator protein-1, and upstream stimulatory factor in arterial smooth muscle cells. J Biol Chem 280:2708–13.
66. Robson SC, Kaczmarek E, Siegel JB, Candinas D, Koziak K, Millan M, Hancock WW, Bach FH. (1997) Loss of ATP diphosphohydrolase activity with endothelial cell activation. J Exp Med 185:153–63.
67. Ross R. (1995) Cell biology of atherosclerosis. Annu Rev Physiol 57:791–804.
68. Rumjahn SM, Javed MA, Wong N, Law WE, Buxton IL. (2007) Purinergic regulation of angiogenesis by human breast carcinoma-secreted nucleoside diphosphate kinase. Br J Cancer 97:1372–80.
69. Satterwhite CM, Farrelly AM, Bradley ME. (1999) Chemotactic, mitogenic, and angiogenic actions of UTP on vascular endothelial cells. Am J Physiol 276:H1091–7.
70. Sauzeau V, Le Jeune H, Cario-Toumaniantz C, Vaillant N, Gadeau AP, Desgranges C, Scalbert E, Chardin P, Pacaud P, Loirand G. (2000) $P2Y_1$, $P2Y_2$, $P2Y_4$, and $P2Y_6$ receptors are coupled to Rho and Rho kinase activation in vascular myocytes. Am J Physiol Heart Circ Physiol 278:H1751–61.
71. Senger DR, Perruzzi CA, Papadopoulos A. (1989) Elevated expression of secreted phosphoprotein I (osteopontin, 2ar) as a consequence of neoplastic transformation. Anticancer Res 9:1291–9.
72. Seye CI, Gadeau AP, Daret D, Dupuch F, Alzieu P, Capron L, Desgranges C. (1997) Overexpression of $P2Y_2$ purinoceptor in intimal lesions of the rat aorta. Arterioscler Thromb Vasc Biol 17:3602–10.
73. Seye CI, Kong Q, Erb L, Garrad RC, Krugh B, Wang M, Turner JT, Sturek M, Gonzalez FA, Weisman GA. (2002) Functional $P2Y_2$ nucleotide receptors mediate uridine 5′-triphosphate-induced intimal hyperplasia in collared rabbit carotid arteries. Circulation 106:2720–6.
74. Seye CI, Yu N, Gonzalez FA, Erb L, Weisman GA. (2004) The $P2Y_2$ nucleotide receptor mediates vascular cell adhesion molecule-1 expression through interaction with VEGF receptor-2 (KDR/Flk-1). J Biol Chem 279:35679–86.
75. Seye CI, Yu N, Jain R, Kong Q, Minor T, Newton J, Erb L, Gonzalez FA, Weisman GA. (2003) The $P2Y_2$ nucleotide receptor mediates UTP-induced vascular cell adhesion molecule-1 expression in coronary artery endothelial cells. J Biol Chem 278:24960–5.
76. Shen BQ, Lee DY, Zioncheck TF. (1999) Vascular endothelial growth factor governs endothelial nitric-oxide synthase expression via a KDR/Flk-1 receptor and a protein kinase C signaling pathway. J Biol Chem 274:33057–63.
77. Shen J, Seye CI, Wang M, Weisman GA, Wilden PA, Sturek M. (2004) Cloning, up-regulation, and mitogenic role of porcine $P2Y_2$ receptor in coronary artery smooth muscle cells. Mol Pharmacol 66:1265–74.

78. Singhal H, Bautista DS, Tonkin KS, O'Malley FP, Tuck AB, Chambers AF, Harris JF. (1997) Elevated plasma osteopontin in metastatic breast cancer associated with increased tumor burden and decreased survival. Clin Cancer Res 3:605–11.
79. Takahashi F, Akutagawa S, Fukumoto H, Tsukiyama S, Ohe Y, Takahashi K, Fukuchi Y, Saijo N, Nishio K. (2002) Osteopontin induces angiogenesis of murine neuroblastoma cells in mice. Int J Cancer 98:707–12.
80. Tallquist MD, French WJ, Soriano P. (2003) Additive effects of PDGF receptor beta signaling pathways in vascular smooth muscle cell development. PLoS Biol 1:288–99.
81. Tanimoto T, Jin ZG, Berk BC. (2002) Transactivation of vascular endothelial growth factor (VEGF) receptor Flk-1/KDR is involved in sphingosine 1-phosphate-stimulated phosphorylation of Akt and endothelial nitric-oxide synthase (eNOS). J Biol Chem 277:42997–3001.
82. Tenaglia AN, Peters KG, Sketch MH, Jr, Annex BH. (1998) Neovascularization in atherectomy specimens from patients with unstable angina: implications for pathogenesis of unstable angina. Am Heart J 135:10–4.
83. Thuringer D, Maulon L, Frelin C. (2002) Rapid transactivation of the vascular endothelial growth factor receptor KDR/Flk-1 by the bradykinin B2 receptor contributes to endothelial nitric-oxide synthase activation in cardiac capillary endothelial cells. J Biol Chem 277: 2028–32.
84. Townsend KP, Obregon D, Quadros A, Patel N, Volmar C, Paris D, Mullan M. (2002) Proinflammatory and vasoactive effects of Abeta in the cerebrovasculature. Ann N Y Acad Sci 977:65–76.
85. Ventura MA, Thomopoulos P. (1995) ADP and ATP activate distinct signaling pathways in human promonocytic U-937 cells differentiated with 1,25-dihydroxy-vitamin D3. Mol Pharmacol 47:104–14.
86. Wang DJ, Huang NN, Heppel LA. (1992) Extracellular ATP and ADP stimulate proliferation of porcine aortic smooth muscle cells. J Cell Physiol 153:221–33.
87. Wang M, Kong Q, Gonzalez FA, Sun G, Erb L, Seye C, Weisman GA. (2005) P2Y_2 nucleotide receptor interaction with α_v integrin mediates astrocyte migration. J Neurochem 95: 630–40.
88. Weisman GA, Yu N, Liao Z, Gonzalez F, Erb L, Seye CI. (2006) P2 receptors in health and disease. Biotechnol Genet Eng Rev 22:171–95.
89. Weller RO, Yow HY, Preston SD, Mazanti I, Nicoll JA. (2002) Cerebrovascular disease is a major factor in the failure of elimination of Abeta from the aging human brain: implications for therapy of Alzheimer's disease. Ann N Y Acad Sci 977:162–8.
90. Yang S, Buxton IL, Probert CB, Talbot JN, Bradley ME. (1996) Evidence for a discrete UTP receptor in cardiac endothelial cells. Br J Pharmacol 117:1572–8.
91. Zeng H, Sanyal S, Mukhopadhyay D. (2001) Tyrosine residues 951 and 1059 of vascular endothelial growth factor receptor-2 (KDR) are essential for vascular permeability factor/vascular endothelial growth factor-induced endothelium migration and proliferation, respectively. J Biol Chem 276:32714–9.
92. Zimmermann H, Braun N. (1996) Extracellular metabolism of nucleotides in the nervous system. J Auton Pharmacol 16:397–400.

Chapter 5
Role of Purine-Converting Ecto-Enzymes in Angiogenic Phenotype of Pulmonary Artery Adventitial Vasa Vasorum Endothelial Cells of Chronically Hypoxic Calves

Evgenia V. Gerasimovskaya, Kurt R. Stenmark, and Gennady G. Yegutkin

Abstract Our previous studies demonstrated that angiogenic expansion of the vasa vasorum network can be observed in the pulmonary artery adventitia of chronically hypoxic calves. Because extracellular ATP and ADP are important regulators of vascular cell function, we hypothesized that these nucleotides may contribute to vasa vasorum neovascularization induced by chronic hypoxia. Treatment of pulmonary artery adventitial vasa vasorum endothelial cells (VVEC) with exogenous ATP or non-hydrolysable nucleotide analogs dramatically increased DNA synthesis and migration. Similar mitogenic responses have been observed in lung microvascular endothelial cells (MVEC), but not in endothelial cell isolated from large blood vessels of the same animals, such as aorta (AOEC) and main pulmonary artery (MPAEC). By using thin-layer chromatography assay, we also found that cultured VVEC displayed substantially lower ecto-ADPase/NTPDase and ecto-5′-nucleotidase activities compared to MPAEC and MVEC. In addition, VVEC are characterized by higher activities of ATP-regenerating ecto-enzymes, adenylate kinase and nucleotide diphosphokinase, consistent with a potent mitogenic effect of exogenous nucleotides in VVEC versus MPA and MVEC. Together, these studies indicate that VVEC, isolated from the sites of active neovascularization, may represent a unique pro-angiogenic phenotype with an augmented reliance to extracellular ATP and ADP. Our studies also demonstrated important role of endothelial purine-converting ecto-enzymes in the control of angiogenesis via directional regulation of local nucleotide levels.

Keywords Angiogenesis · Vasa-vasorum · Vascular remodeling · Hypoxia · Pulmonary hypertension · Pulmonary circulation · Pulmonary artery · Aorta · Adventitia · Cell migration · Cell proliferation · Purinergic receptors · Intracellular signaling · Adenine nucleotides · Adenosine · ecto-NTPDase · ecto-5′-nucleotidase · Adenylate kinase · Nucleotide diphosphokinase

---

E.V. Gerasimovskaya (✉)
Department of Pediatrics Critical Care Medicine, University of Colorado Denver, Aurora, CO 80045, USA
e-mail: evgenia.gerasimovskaya@ucdenver.edu

E. Gerasimovskaya, E. Kaczmarek (eds.), *Extracellular ATP and Adenosine as Regulators of Endothelial Cell Function*, DOI 10.1007/978-90-481-3435-9_5,

5.1 Introduction

Extracellular adenine nucleotides have long been known as regulators of vascular tone, permeability, and homeostasis. However, the ability of extracellular nucleotides to control cell growth and proliferation remains poorly investigated. The variety of studies that have been performed on endothelial cells isolated from large systemic and pulmonary vessels that would presume the endothelial cells from these sources/tissues exhibit quiescent or differentiated phenotype with a limited angiogenic capacity. Keeping this in mind, we have developed an animal model system that allows investigation of phenotypical and functional properties of microvascular endothelial cells derived from the sites of active angiogenesis. These cells are vasa vasorum endothelial cells (VVEC) isolated from microvascular network that grow extensively within the pulmonary artery adventitial compartment of chronically hypoxic calves. This chapter summarizes our recent findings on the role of extracellular nucleotides as potent angiogenic activators of VVEC as well as on the role of purine-converting enzymes potentially contributing to angiogenic properties of VVEC under chronic hypoxic condition.

5.2 Hypoxia-Induced Vascular Remodeling and Vasa Vasorum Angiogenesis

Pathologic vascular remodeling plays a pivotal role in the progression of a variety of diseases and conditions where hypoxia, ischemia, or inflammation are prominent features [7, 13]. Using a neonatal model of pulmonary hypertension, we previously demonstrated that significant changes in the structure of pulmonary vessels accompany the hypertensive process and include marked thickening of both the media and adventitia with particularly striking fibroproliferative changes observed in adventitia of chronically hypoxic calves [17, 78]. Angiogenesis, a process of blood vessel expansion, can be observed in response to various stress conditions of microvascular beds such inflammation and fibrosis. In this context, it is important to note that the adventitial compartment of at least elastic and large muscular arteries, includes within its borders, a microcirculatory network, the vasa vasorum. The vasa vasorum network has traditionally been assigned a passive role in maintaining vessel integrity through the supply of oxygen and nutrients to the outer part of the vessel wall. However, recent studies have demonstrated that extensive neovascularization of the vasa vasorum may contribute to the progression of vascular diseases in the systemic circulation including atherosclerosis, restenosis, vasculitis type II diabetes and metabolic syndrome [34, 35, 43, 60, 71] Our group has demonstrated that marked neovascularization of the vasa vasorum of the pulmonary circulation has been observed in neonatal calves exposed to chronic hypoxia (Fig. 5.1). It has also been shown that angiogenesis of the pulmonary circulation can be observed in a form of plexiform lesions in the lung of patients with primary pulmonary hypertension, CREST and HIV infections [15, 56, 82]. In severe pulmonary hypertension

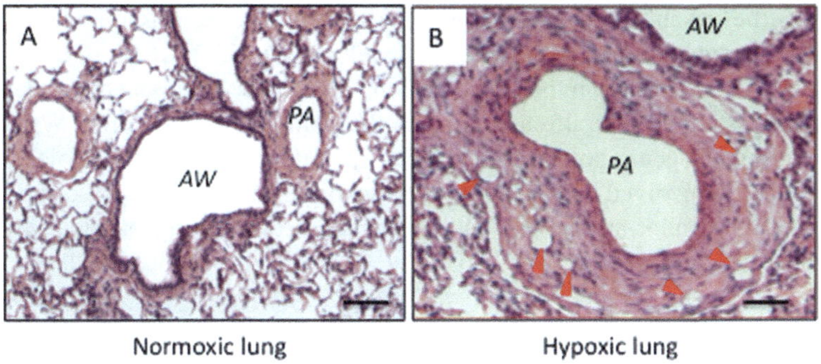

Fig. 5.1 Hypoxia induces vasa vasorum neovascularization in the PA adventitia of chronically hypoxic calves. H&E staining of lung sections of control and hypoxic and calves demonstrates that thickening of the pulmonary artery vessel wall is associated with an apparent increase in the density of vasa vasorum (*red arrows*) in vessels from hypoxic animals (**b**), compared with vessels from normoxic animals (**a**). *Scale bar* represents 100 μM; (AW = airways; PA = pulmonary artery)

excessively proliferating lung microvascular endothelial cells obliterate the vascular lumen and contribute to the disruption of the pulmonary blood flow [16]. Vasa vasorum endothelial cells observed in the adventitia of the large pulmonary arteries (PA) are though to be derived from bronchial circulation. Expansion of the bronchial vessels in the ischemic lung parenchyma and the pulmonary arteries has been demonstrated in patients with chronic thromboembolic disease that suggests a unique proliferative and invasive capacity of endothelial cells of the these systemic vessels [61]. Together, these observations imply that angiogenic vasa vasorum expansion may be a common feature of specific pulmonary and systemic vascular diseases and that hypoxic microenvironment may facilitate the angiogenic process [17, 26, 58, 60, 64].

5.3 Hypoxia-Induced Vasa Vasorum Neovascularization Represents an Angiogenesis of Systemic Vessels in the Pulmonary Circulation

In fact, the observed pulmonary artery adventitial thickening supported by an increased systemic blood flow (vasa vasorum), represents a unique angiogenesis model system. This involves a direct interaction between cells of pulmonary and systemic blood vessels, explicitly, angiogenesis of the systemic (bronchial) microcirculation in the pulmonary artery vascular wall. Moreover, we have demonstrated the recruitment of circulating progenitor and inflammatory cells to the PA adventitia of chronically hypoxic calves, suggesting that the expanding vasa vasorum appears to function as a port and conduit for entry and delivery of circulating inflammatory cells in the vessel wall [78]. Therefore, in many aspects hypoxia-induced adventitial inflammation and vasa vasorum neovascularization observed in neonatal calves, resembles the vascularization of tumors, as well as, the neovascularization of the vessel wall during atherosclerotic plague development [58, 60]. These observations raise questions as to how hypoxic conditions contribute to this neovascularization and the process of structural remodeling of the pulmonary circulation. While many factors with angiogenic potential have been identified, endogenous molecular factors specifically involved in hypoxia-induced expansion of the vasa vasorum, as well as the precise cellular and molecular mechanisms contributing to this process, are not completely understood.

5.4 Cultured PA Adventitial Vasa Vasorum Endothelial Cells is a Novel In Vitro Angiogenic Model System

The in vitro studies of angiogenic responses in microvascular endothelial cells are limited due to the difficulties of isolation and maintenance these cells in in vitro conditions. We have successfully established a culture of vasa vasorum endothelial cells (VVEC) from the adventitia of PA of neonatal calves, thereby presenting a physiologically relevant cell system to evaluate angiogenic effects of extracellular ATP and another nucleotides. To our knowledge, the use of this specific microvascular endothelial cell population has not been previously documented. VVEC cultures demonstrate two distinct advantages of using the hypoxic calf model: first, after two weeks under hypoxic conditions these animals develop pulmonary hypertension accompanied by remarkable pulmonary vascular remodeling, (similar to those observed in human) and second, large amounts of adventitial tissues are available, which allows large scale cell isolation. Using explant techniques followed by cloning ring differential trypsinization, VVECs can be successfully isolated from PA adventitial compartment. These cells express CD31, Flk-1, Tie-1, vWf and eNOS, as do pulmonary artery (Fig. 5.1) endothelial cells, cultured from the same vessel.

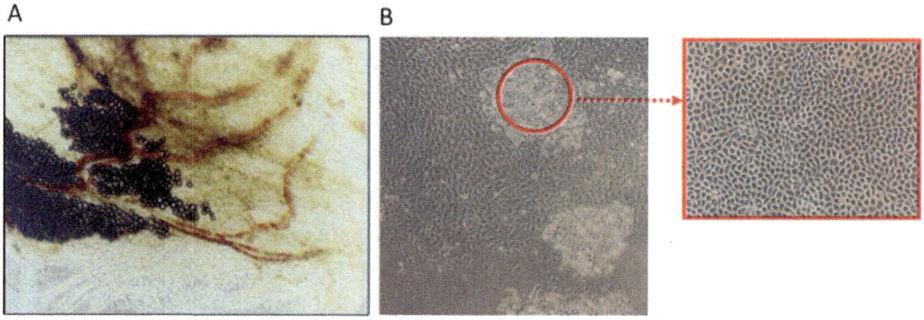

Fig. 5.2 Vasa vasorum endothelial cells (VVEC) can be isolated from the pulmonary artery adventitia and perpetuated in culture. (**a**) PA adventitial regions that develop massive neovascularization are used for VVEC isolation; (**b**) Co-culture of VVEC and adventitial fibroblasts. Pure endothelial cell culture can be obtained as a result of trypsinization by using cloning rings. VVEC exhibit a "cobblestone" morphology, a characteristics of endothelial phenotype

Table 5.1 Phenotypical characteristics of pulmonary artery adventitial VVEC

CD31/PECAM	+++
vWf	+
Flt-1	+
Flk-1	++
eNOS	+
Tie-2	++
Tie-1	+
Ephrin B2	+
CD133	++
CD34	+
c-kit	+
binding of lectin *Lycopersicon esculentum*	+++
uptake of acetylated low density lipoproteins (Dil-Ac-LDL)	+++

Data were obtained by immunofluorescence analysis: +++ indicates very high expression level; ++ indicates high expression level; + indicates variable expression level

In addition, VVEC express Flt-1, Tie-2, Ephrin B2, CD133 and c-kit. Our data also demonstrate the ability of VVEC to bind the lectin *Licopercsicon Esculentum*; and uptake acetylated low density lipoproteins labeled with 1,1′-dioctadecyl-3,3,3′,3′-tetramethylindo-carbocyanine perchlorate (DiI-Ac-LDL) (Table 5.1). Our data also demonstrates the ability of VVEC to form interconnecting networks on polymerized Matrigel and in co-cultures with adventitial fibroblasts suggesting an angiogenic capability of VVEC in vitro. Therefore, the cultured adventitial VVEC can be used as a physiologically relevant model cell system to study endothelial angiogenesis.

5.5 Extracellular Nucleotides as Important Regulators of Vascular Functions

Endogenous autocrine and paracrine soluble factors, released under hypoxic conditions from the cells of vasculature and blood circulation are important contributors to angiogenesis [27, 32, 50, 52, 74]. An autocrine mechanism for the regulation of vascular cell growth has been demonstrated for several vasoactive compounds, including TGF-β, VEGF, prostaglandin E2, and adrenomedulin [41, 44, 54, 93]. However, quite often the efforts to prevent pathologic angiogenesis and vascular remodeling by using tyrosine kinase or cytokine receptor antagonists have not resulted in a intended clinical outcome, suggesting that additional signaling mechanisms may contribute to the angiogenic process.

Extracellular purines and pyrimidines (ATP, ADP, AMP, adenosine, UTP and UDP) have been proposed as important regulators of vascular cell proliferation, migration, chemotaxis and secretory responses [2–4, 11, 19, 23, 24, 28, 72, 73, 83]. The effects of extracellular nucleotides are mediated through P2Y (metabotropic),

P2X (ionotropic) purinergic receptors, whereas the effects of adenosine are mediated through A1, A2A, A2B and A3 receptors [2, 11, 20]. The wide tissue distribution of purinergic receptors together with the expression of multiple receptor subtypes in vascular cells point to the physiological importance of purinergic signaling system for the regulation of vascular cell function. It is also becoming evident that alterations in the physiology of purinergic signaling may result in the development of a variety of neurogenerative, immune, and vascular pathologies [12, 19]. Considering this, by targeting purinergic receptors with specific ligands (either agonists or antagonists) would provide a useful tool for directional manipulation of pathway-specific intravascular responses associated with various pathologic conditions.

5.6 Sources and Potential Mechanisms of ATP Release into the Extracellular Milieu

Extracellular ATP concentrations are thought to be elevated in the local tissue microenvironment at various physiological and pathological conditions including hypoxia, inflammation, fluid shear stress, low osmolarity, thrombosis, and sympathetic stimulation (when ATP is co-released with noradrenalin), [8, 9, 10, 28, 39, 47]. It is also known that extracellular nucleotides are released in response to mechanical forces and other environmental stresses such as osmotic shock, acidosis, hyperoxia and hypoxia by cells of the vasculature, airways, and gut and play a key role in transducing cellular responses via activation of purinergic receptors in these organs [11, 50, 63]. Furthermore, along with transient release of nanomolar or sub-micromolar ATP concentrations in response to various stimuli, the cells can also constitutively release ATP at certain basal rates and maintain pericellular ATP levels in their vicinity within a high micromolar range [2, 91].

5.7 Implication of Extracellular ATP for Vascular Diseases

A number of studies support the idea that extracellular nucleotides could contribute to the development of vascular disease. Extracellular ATP has been implicated in the hyperplasia and hypertrophy of arterial walls in spontaneously hypertensive rats [66], the regulation of vascular permeability [40, 45, 62] and the control of proliferation and migration of vascular smooth muscle cells and hematopoietic stem cells [42, 51, 70, 72]. Purinergic antithrombotic drugs have been shown to reduce the risk of recurrent strokes and heart attacks [12]. Extracellular ATP and UTP have been demonstrated to stimulate DNA synthesis in vascular endothelial cells, smooth muscle cells (SMC), and adventitial fibroblasts [23, 28, 83]. In addition, extracellular nucleotides appear to directly affect migration of vascular as well as non-vascular cells including monocytes. Recent work demonstrates that in endothelial cells ATP- and UTP-induced stimulation of P2Y2 receptors is associated with co-activation of VEGF receptor-2, expression of Vascular Cell Adhesion Molecule-1 (VCAM-1), and monocyte recruitment, implying a link between purinergic signaling, angiogenesis and inflammatory responses [75, 76]. The product of ATP hydrolysis, adenosine, has also been shown to induce proliferative responses in various types of endothelial cells [24, 25, 30, 34]. Importantly, ATP acts synergistically with cytokines (e.g. platelet-derived growth factor, epidermal growth factor, and insulin-like growth factor) and integrins to stimulate vascular cell proliferation and migration, thereby supporting the physiological relevance of extracellular ATP under hypoxic and stress-related conditions [23, 28, 37, 42]. Recent findings also demonstrate a role of extracellular ATP and UTP in the stimulation and migration of human hematopoietic stem cells in vitro and in vivo [51, 70]. Together, these studies emphasize the angiogenic potency of extracellular adenine and pyrimidine nucleotides and their role in the regulation and modulation of vascular cell function during disease processes. However, despite the considerable progress in the understanding of the molecular and cellular mechanisms involved in the control of neovascularization as well as the continued interest in pro-angiogenic factors in various pathologic and physiologic states, including hypoxia and inflammation, the role of extracellular nucleotides in hypoxia-induced endothelial angiogenesis and the concomitant changes in microvascular phenotype is recently gaining appreciation.

5.8 Extracellular ATP and Pulmonary Artery Adventitia

Previously, we demonstrated that hypoxia stimulates release of ATP from adventitial fibroblasts and that this exogenous ATP acts in an autocrine manner to stimulate cell proliferation, a response critical in the vascular remodeling process observed in response to hypoxic conditions [28, 79]. These observations support the idea that in addition to nerve and circulating blood cells, vascular cells themselves appear to be potent sources of ATP and other adenine nucleotides [12, 29]. These observations prompted us to evaluate endothelium of newly formed adventitial vasa vasorum as a potential, yet unidentified, source of extracellular ATP in the pulmonary artery vessel wall. Indeed, our studies demonstrated that PA adventitial VVEC release ATP in response to hypoxia and low osmolarity, suggesting these cells may represent a potent source of extracellular ATP within the pulmonary artery vascular wall and that extracellular ATP could contribute to vasa vasorum neovascularization [79]. These observations support the notion that local purinergic signaling networks can be initiated by hypoxic stress, which in turn alter endothelial cell phenotype and function. In addition, since vascular cells express multiple purinergic receptors, extracellular nucleotides may play as an intercellular signaling molecule, providing cell-to-cell communication by an autocrine/paracrine mechanism.

5.9 Effects of Extracellular Nucleotides on Angiogenic Responses in Endothelial Cells

Although extracellular nucleotide-mediated signaling has been implicated in various physiological and pathologic conditions, there is limited information regarding the role of extracellular nucleotides in modulating endothelial cell proliferation. It has been postulated that endothelial cells from various vascular beds exhibit tremendous functional and morphological heterogeneity [5]. In particular, endothelial cells of large blood vessels such as pulmonary artery and aorta possess phenotypical characteristics different from those of endothelial cells of microvascular beds. In this regard, it was expected that the proliferative and angiogenic capacities of endothelial cells of large vessels and microvessels should also differ. To test this hypothesis, we measured the rate of DNA synthesis in endothelial cells isolated from aorta (AOEC), main pulmonary artery (MPAEC), lung microvessels (MVEC) and PA adventitial vasa vasorum (VVEC) (Fig. 5.3). It was found that extracellular ATP at concentrations of $10^{-6}-10^{-3}$ M, stimulated dramatic increases in DNA synthesis in VVEC and DPAEC (up to 12-fold), but has very little effect on AOEC and

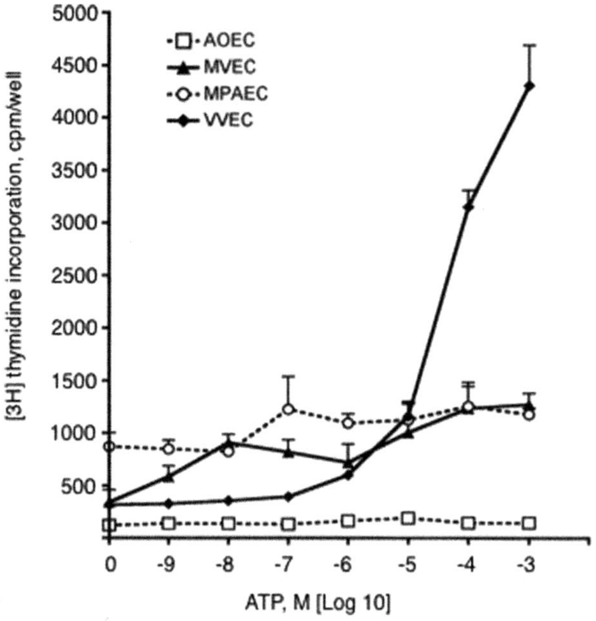

Fig. 5.3 Extracellular ATP exerts different proliferative responses in microvascular and macrovascular endothelial cells. Growth arrested (72 h, serum-free DMEM) vasa vasorum endothelial cells (VVEC), aorta endothelial cells (AOEC), main pulmonary artery endothelial cells (MPAEC) and lung microvascular endothelial cells (MVEC) were stimulated with extracellular ATP from 10^{-9} to 10^{-3} M in the presence of 0.125 μCi [^{3}H]-thymidine for 24 h, and incorporated radioactivity was determined in total cell lysates. The results are expressed as means ± S.E; ($n = 3$). Shown are the data representative for each endothelial cell type

MPAEC (1.37 and 1.32 fold increases respectively). In addition, under the same experimental conditions, extracellular ATP induced dramatic increases in DNA synthesis in pulmonary artery microvascular endothelial cells (3.76 fold) and vasa vasorum endothelial cells from aorta (6.16 fold, not shown). Therefore, our studies revealed that extracellular ATP is a more potent mitogen for microvascular endothelial (of both lung and systemic origin), than for endothelial cells of main pulmonary artery and aorta large vessels. Interestingly, microvascular endothelial cells of systemic origin (PA and aortic vasa vasorum) are slightly more "responsive" to ATP, than lung microvascular endothelial cells. The exceptional sensitivity of microvascular endothelial cells to stimulation with ATP and ADP is supported by other studies showing a mitogenic effect of extracellular ATP, UTP and 2MeSATP on brain capillary and corneal endothelial cells [6, 14] Previous studies on endothelial cells derived from large vessels also demonstrated that stimulation of purinergic receptors result in the increase of intracellular Ca^{2+}, and the release of nitric oxide, endothelium-dependent hyperpolarizing factor (EDHF), prostacyclin and t-PA [59, 86]. Similar to our observations, it was found that ATP and the other P2Y receptor agonists exerted only weak increases in DNA synthesis in bovine aortic endothelial cells [36, 83, 88], supporting the idea that differentiated endothelial cells possess a little proliferative capacity in response to extracellular nucleotides.

Since ATP hydrolysis products might be expected to contribute to DNA synthesis, we examined the effects of various adenine nucleotides, their non-hydrolyzable synthetic analogs, as well as UTP and UDP on DNA synthesis in VVEC, MPAEC, and AOEC. As shown in Fig. 5.4, ATP, ADP, MeSADP, significantly increased [methyl-^{3}H] thymidine incorporation in VVEC more than 5-fold. AMP, Ado, ATPγS, ADPβS, and MeSATP, also had significant effects on [methyl-^{3}H] thymidine incorporation with variable potency (2–5 fold) (Fig. 5.4a) BZ-ATP, and Ap_4A (not shown) were not effective in stimulating DNA synthesis. Surprisingly, no significant changes in DNA synthesis were observed in response to pyrimidine nucleotides, UTP and UDP, though a previous report demonstrated both of these pyrimidines stimulate guinea pig cardiac endothelial cell proliferation [72]. Importantly, not only are extracellular ATP and ADP potent mitogens for VVEC, but also extracellular adenosine and AMP (most likely, as a result of it's hydrolysis to adenosine by ecto-5′-nucleotidase/CD73). These observations suggest that stepwise hydrolysis of extracellular ATP to adenosine by ecto-NTDPase/CD39 and ecto-5′-NT/CD73 occurs in VVEC and that these enzymatic reactions can modulate the amplitude of ATP-induced mitogenic responses. Our data also demonstrates that all tested nucleotides and their non-hydrolysable synthetic analogs had significant effects on VVEC migration (Fig. 5.4b). The exaggerated mitogenic responses to extracellular nucleotides suggest that VVEC might be abundantly equipped with multiple purinergic receptors, coupled to mitogenic and angiogenic signaling pathways. Thus, purinergic receptors can also be considered as molecular determinants of such activated angiogenic phenotype of VVEC. The efficacy and the profile of extracellular nucleotides in increasing DNA synthesis and migration may suggest that the P2Y1 family of G protein-coupled purinergic receptors (P2Y1, P2Y11, P2Y12, P2Y13) is likely involved in mediating the effect of extracellular adenine

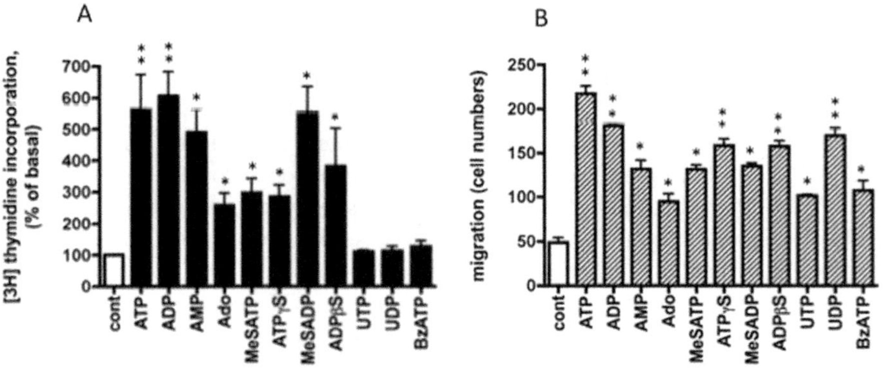

Fig. 5.4 The effects of purine and pyrimidine nucleotides and non-hydrolysable nucleotide analogs on VVEC DNA synthesis and migration. (**a**) Growth arrested VVEC (72 h, serum-free DMEM) were stimulated with indicated nucleotides (100 μM) in the presence of 0.125 μCi [methyl-^{3}H] thymidine for 24 h. Incorporated radioactivity was determined in total cell lysates. (**b**) Growth arrested VVEC (72 h, serum-free DMEM) were plated in permeable inserts (Costar, 8.0 μ pore size) in serum-free DMEM. The lower transwell compartment contained indicated nucleotides (100 μM). After 24 h, the amount of cells migrated through the filter was determined as described in [29]. Data represents means S.E. (as % of nonstimulated control) from three to five independent experiments, conducted on distinct VVEC populations; * $p<0.05$, ** $p<0.01$ vs. nonstimulated control

nucleotides in VVEC. Determining the purinergic receptor profile in VVEC will be an important step in further understanding the angiogenic characteristics of these cells.

5.10 Intracellular Signaling Pathways Associated with Extracellular ATP-Induced Vasa Vasorum Angiogenesis

Based on findings in other cell types possessing high proliferative potential, emphasis was placed on the Phosphoinositide 3-kinase (PI3K), mTOR (*m*ammalian *T*arget *o*f *R*apamycin), and ERK1/2. These pathways play key role in the regulation of metastatic cell growth, tumor progression and angiogenesis [31, 37, 84, 87]. However, there is a lack of evidence demonstrating the activation of these pathways in response to extracellular ATP, as well as evidence linking ATP-dependent activation of these pathways and angiogenic responses in endothelial cells. It was found that extracellular ATP mediates its angiogenic effects in VVEC through dramatic and prolonged activation of PI3K/mTOR and ERK1/2 pathways. The responses to extracellular ATP might be particularly important in the hypoxic and inflammed adventitial microenvironment where increased extracellular ATP level can be expected. In our recent studies we demonstrated that in contrast to VVEC, extracellular ATP had much lesser effect on mTOR, Akt, and ERK1/2 signaling in endothelial cells isolated from PA and aorta, along with a negligible angiogenic effects by ATP in these cells [29]. Together, these observations further support the idea that VVEC are distinct in their ability to proliferate in response to extracellular ATP. Therefore it can be speculated that VVEC isolated from the sites of hypoxic PA adventitia, may therefore have unique phenotypical characteristics, with a particular reliance on extracellular nucleotides as an environmental stimuli. In conjunction with other reports, our data suggest that there must be tremendous heterogeneity in endothelial phenotypes, with regard to extracellular nucleotide-induced functional responses and the signaling pathways through which these effects are mediated.

5.11 Purine-Converting Ecto-Enzymes and Their Role in the Regulation of Vascular Homeostasis

Subsequent to signal transduction, extracellular nucleotides need to be rapidly inactivated. The duration and magnitude of purinergic signaling in target cells is governed by a network of membrane-bound and soluble enzymes. General schemes of purine metabolism have included a role for the purine-inactivating enzymes of ecto-nucleoside triphosphate diphosphohydrolase (NTPDase) family, ecto-nucleotide pyrophosphatase/phosphodiesterase (E-NPP) family, ecto-5′-nucleotidase/CD73 and adenosine deaminase [91, 92]. The initial steps of ATP and ADP hydrolysis in the vasculature are primarily mediated by the enzymes known as ecto-ATPase, ATPDase, apyrase etc., which, according to the current nomenclature, should be termed as NTPDase family members and classified in order of discovery and classification [69]. Namely, eight different ENTPD genes encode members of the NTPDase protein family, with four of them (NTPDase1,2,3,8) being expressed as cell surface-located enzymes. NTPDases 5 and 6 exhibit intracellular localization and undergo secretion after heterologous expression, while NTPDases 4 and 7 are entirely intracellularly located, facing the lumen of cytoplasmic organelles. Subsequent breakdown of ATP/ADP-derived AMP into adenosine is mediated by another nucleotide-inactivating enzyme, ecto-5′-nucleotidase/CD73, which is anchored to the plasma membrane by glycosyl-phosphatidylinositol and expressed to a variable extent in vascular and other tissues [38, 69, 91, 92, 94].

In terms of the vascular system, endothelial E-NTPDase1/CD39 in conjunction with ecto-5′-nucleotidase/CD73, have been implicated in playing a critical role through the termination of prothrombotic and proinflammatory effects of circulating ATP and ADP and their conversion into adenosine. This keeps the haemostatic process tightly regulated by preventing excessive clot formation and vessel occlusion [53, 55, 69]. In addition, selective up-regulation of the CD39-CD73 axis on the surfaces of activated $CD4^{+}CD25^{+}Foxp3^{+}$ T-regulatory cells comprises an important constituent of their immunosuppressive machinery via sequential breakdown of lymphotoxic and proinflammatory ATP into anti-inflammatory adenosine, thus inhibiting T cell proliferation and secretion of cytokines [18, 33, 77]. Studies with mice lacking E-NTPDase-1/CD39 [57] or ecto-5′-nucleotidase/CD73

[80, 81] further confirmed the importance of coordinated nucleotide-inactivating cascade for proper immune and cardiovascular functions at various inflammatory, prothrombotic, and hypoxic states [48, 53, 65, 67].

Strikingly, along with the common view of extracellular ATP as a ligand for P2 receptors, substrate for ecto-nucleotidases and source of adenosine, this nucleotide was shown to concurrently serve as a phosphoryl donor for counteracting ecto-adenylate kinase and NDP kinase reactions [49, 92]. This backward ATP-regenerating pathway may represent a principally distinct route for the appearance of extracellular ATP and other purinergic agonists on the cell surface. However, the implication of these phosphorylating ecto-enzymes in regulation of purinergic signaling cascade is only beginning to be elucidated. Moreover, the identification of a complex mixture of soluble purine-converting enzymes freely circulating in the bloodstream adds another level of complexity to our understanding the regulatory mechanisms of purine homeostasis in the vasculature [90, 92]. Additionally, this may open up further research to assess the potential therapeutic and diagnostic applications of purinergic enzymes.

5.12 Potential Role of Purine-Converting Enzymes in Modulating Angiogenic Potential of VVEC

While studies from several laboratories clearly demonstrated the involvement of exogenous nucleotides in the regulation of vascular angiogenesis and haemostasis, there is a need to further elucidate a signalling role of individual nucleotides and nucleosides in these responses. In this context, the complexity is created by co-expression of different receptor subtypes on vascular cells, the respective redundancy of intracellular signalling pathways, as well as the ability of ATP and its metabolite adenosine to trigger divergent, often opposite, effects on the immune and mitogenic vascular functions. Moreover, direct quantification of extracellular ATP is always complicated firstly, due to high turnover sequence: "release-signaling-inactivation" and secondly, since physiologically relevant levels of this nucleotide in local tissue environment can substantially differ from bulk concentrations in the interstitial milieu [91]. Since several enzymatic activities are simultaneously involved in the active cycling between extracellular ATP and other purinergic agonists, we hypothesized that vascular remodelling and angiogenesis could be determined via regulated control of a unique purine-converting enzymatic network localized to the plasma membrane of vascular cells.

Despite the important regulatory role of extracellular nucleotides on angiogenesis, the contribution of ecto-enzymes to endothelial pro-angiogenic phenotype remains poorly understood. It was demonstrated that significant down-regulation of ecto-nucleotidase activities on vascular endothelium and other cell types during chronic hypoxia and oxidative stress was accompanied by triggering nucleotide-mediated signaling pathways, increased endothelial activation, and concomitant development of vascular disorders [28, 67, 68]. On the other hand, other investigators have shown endothelial nucleotide-inactivating ectoenzymes NTPDase1/CD39 and ecto-5′-nucleotidase/CD73 can be upregulated during acute hypoxia and inflammation. This increases the intravascular adenosine concentrations and dampens excessive inflammatory responses by affecting endothelial barrier function, adhesion and transmigration of lymphoid cells, and expression of other molecules involved in the adhesion cascade [22, 46, 85].

Specific purine-converting activities were quantified by thin-layer chromatography (TLC) using cultured endothelial cells as enzyme sources and saturating concentrations of ^{3}H-labeled and non-labelled nucleotides as appropriate substrates, as described previously [89]. In these studies, ATPase/NTPDase was evaluated as the rate of 500 μM [^{3}H]ATP dephosphorylation into ^{3}H-metabolites, that were quantified as pooled ADP, AMP and nucleoside fractions (Fig. 5.5a). Likewise, ADPase/NTPDase (Fig. 5.5b) and ecto-5′-nucleotidase (Fig. 5.5c) activities of the cells were determined by their ability to hydrolyse 500 μM [^{3}H]ADP and 300 μM [^{3}H]AMP, respectively. Backward adenylate kinase (Fig. 5.5d) and NDP kinase (Fig. 5.5e) activities were also determined by extent of ATP-induced phosphorylation of 500 μM [^{3}H]AMP or [^{3}H]ADP into high-energy ^{3}H-phosphoryls. Incubation times in these assays varied between 20 and 40 min, and were chosen so that the amount of metabolites did not exceed 10–15% of initially introduced substrate.

Consistent with our previous findings on human umbilical vein endothelial cells (HUVEC) [89], current studies using various cultured bovine endothelial cells also demonstrated the co-existence of extensive network of purine-converting endothelial ectoenzymes that regulate nucleotide levels via two counteracting, ATP-consuming and ATP-regenerating, pathways (Fig. 5.5f). Though, further comparative analysis of major exchange activities revealed substantial differences among the studied endothelial cells. Specifically, cultured VVEC hydrolyze various ^{3}H-labelled nucleotides with relative ATPase/ADPase/AMPase ratios of ∼4.4/9.3/1 and in addition, were able to sequentially phosphorylate exogenous AMP via ADP to ATP in the presence of γ-phosphate-donating ATP. Compared to VVEC, cultured MPAEC and especially MVEC (all isolated from chronically hypoxic calves) were characterized by substantially higher ATP- and ADP-inactivating capability, and at the same time, displayed relatively low activities of ecto-5′-nucleotidase as well as nucleotide-phosphorylating enzymes adenylate kinase and NDP kinase (Fig. 5.5a–e).

The lower rates of [^{3}H]ADP (Fig. 5.6b) hydrolysis in VVEC compared to MPAEC and MVEC is in agreement with previous reports demonstrating that decreases of ATP diphosphohydrolase activity/CD39 can be observed during oxidative stress-induced endothelial cell activation [1, 48, 67]. No significant hypoxia-dependent shifts were determined in studies with other purinergic enzymes, ecto-ATPase (Fig. 5.6a), adenylate kinase (Fig. 5.6d) and NDP kinase (Fig. 5.6e). Along with evidence for selectively diminished ADP- and AMP-inactivating capacity of the hypoxic VVEC, the unchanged activities of other ectoenzymes may additionally serve as suitable "internal controls" for comparable assay conditions among the endothelial cell types. Furthermore, it could be even more relevant to

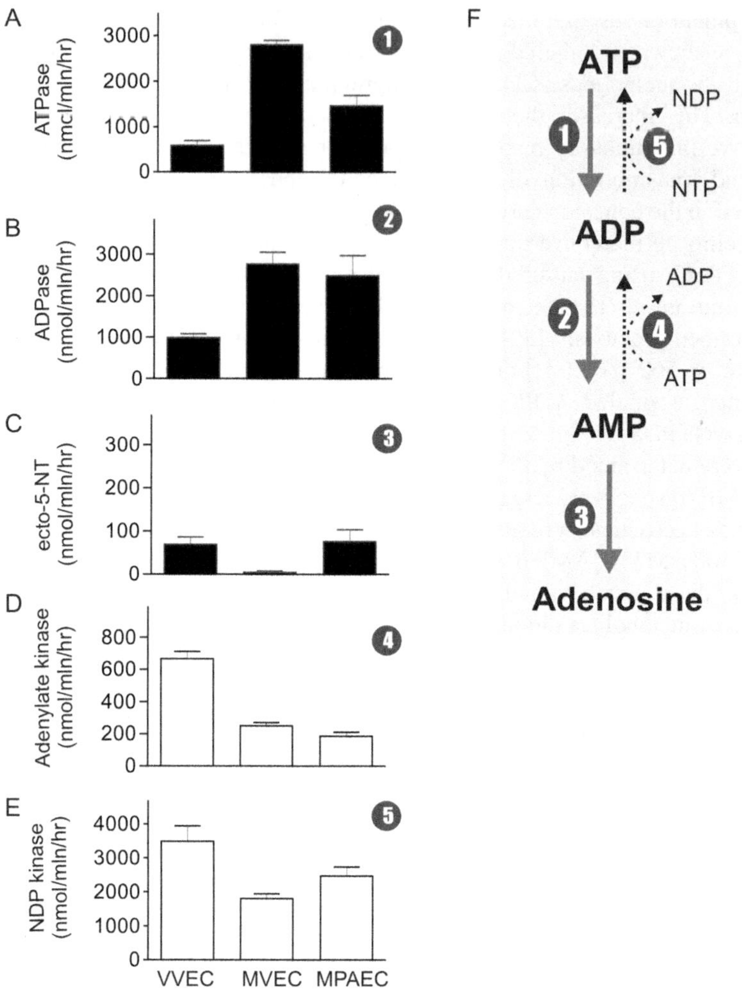

Fig. 5.5 Metabolism and inter-conversion of extracellular nucleotides on endothelial cell surfaces. Vascular endothelial cells were cultured overnight in 96-well plates followed by radio-TLC determination of ATPase (**a**), ADPase (**b**), ecto-5′-nucleotidase (**c**), adenylate kinase (**d**) and NDP kinase (**e**) activities. Enzymatic activities are represented as mean ± SEM (n=4−6). For experimental details, see [90]. (**f**) Scheme of major exchange pathways of extracellular nucleotides on endothelial cells. The elements of inactivating cascade comprise ATP/ADP-hydrolyzing enzyme NTPDase (1,2) and ecto-5′-nucleotidase (3), while the backward ATP-generating pathway is represented by adenylate kinase (4) and NDP kinase (5). (**g**) Comparative analysis of nucleotide catabolism in control and hypoxic VVEC. *Arrows* show the average enzyme activities relative to ecto-ATPase activity in control cells. *Numbers in shadowed circles* correspond to particular catalytic reaction within the larger network of purine-converting ectoenzymes, as specified in panel f

examine the differences in the ratio of key enzymes of nucleotide catabolism on endothelial cell surface. Comparative analysis of relative nucleotide-inactivating activities demonstrated the prevalence of ATPase and ADPase activities in MPAEC

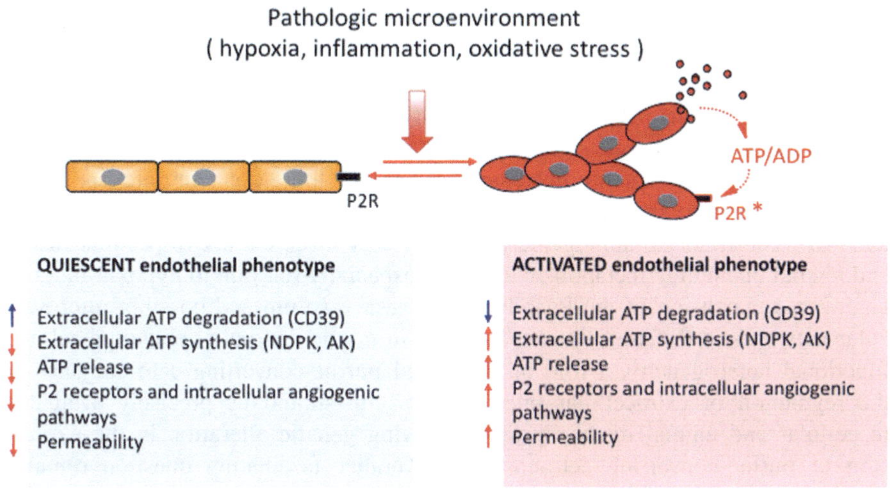

Fig. 5.6 Schematic view of the role of extracellular nucleotides in angiogenic phenotype of endothelial cells. Stress conditions such as hypoxia, inflammation and oxidative stress result in endogenous release of ATP (from both circulating cells and vascular wall) and in subsequent activation of endothelial cells towards pro-angiogenic phenotype characterized by decreased ATP/ADP-hydrolyzing and increased ATP-regenerating activities. Elevated levels of extracellular nucleotides (ATP and ADP) induce mitogenic, migratory and morphogenetic responses in EC. The effects of hypoxia on angiogenic activation of endothelial cells can be reversed by upregulation (down-regulation) of ecto-nicleotidase activities (ecto-ATPase/ADPase (CD39) and ecto-5′-nucleotidase (CD73) resulting in accumulation of extracellular adenosine

and MVEC over the activities in VVEC. However, ecto-5′-nucleotidase activity in MPAEC was almost equivalent, and in MVEC it was slightly lower than ecto-5′-nucleotidase activity observed in VVEC. In turn, VVEC exhibited significantly higher levels of AK and NDPK activities compared to MPAEC and MVEC. Interestingly, recent findings on the co- expression of counteracting ATP-regenerating ectoenzymes in the lymphoid cells suggest that regulation of purine metabolism might extend beyond the inactivating pathways. Specifically, along with low ecto-nucleotidase and high ADA activities, the lymphoid cells are characterized by relatively high ecto-AK and NDP kinase activities [90] and in addition, are capable of maintaining micromolar "ATP halo" in their immediate vicinity [91]. It is speculative that "ATP-regenerating" phenotype of hypoxic VVEC may resemble a phenotype of some circulating blood cells and could represent a novel mechanism of hypoxia-induced endothelial angiogenesis.

5.13 Conclusion and Perspectives

We demonstrated that extracellular nucleotides are both important angiogenic signaling molecules, and play an important physiological role in controlling angiogenic properties of vasa vasorum endothelial cells. Pathological conditions such as chronic hypoxia and inflammation can modulate extracellular nucleotide concentrations in the vessel wall through regulating the activity and expression levels of purine-converting ecto-enzymes, thereby providing their integration into purinergic signalling networks [21, 67, 68]. In view of the recent data on the important role of purinergic signaling in endothelial responses under various pathologic conditions, it remains necessary to explore the extent chronic and acute hypoxia might affect the extracellular nucleotide metabolism. Since exogenous ATP and ADP contribute to angiogenic activation of VVEC, the above peculiarities in endothelial nucleotide metabolism could be especially relevant to hypoxia-induced angiogenic expansion of the PA adventitial vasa vasorum and possibly microvascular networks in different tissues. Keeping in mind endothelial phenotypical and functional heterogeneity, a role of individual purine-converting ecto-enzymes in the regulation of extracellular nucleotide profile should be precisely evaluated in cellular and animal model systems allowing genetic alteration in the expression of purine-converting ecto-enzymes. Another fascinating question remains as to whether regulation of the expression and the activity of purine-converting enzymes under various pathological conditions occur in a coordinated manner, or they can be regulated independently in a cell-type and stimulus-dependent manner.

Acknowledgments This work was supported by National Heart, Lung, and Blood Institute Program Project Grant HL14985 (to K.R.S), the Finnish Academy of Sciences and the Sigrid Juselius Foundation (to G.Y.) National Heart, Lung, and Blood Institute R01 grant HL 086783 (to E.V.G.) and an American Heart Association grant 0665464Z (to E.V.G.). We also wish to thank Mr. Philip Weston for editorial help.

References

1. Aalto TK, Raivio KO. (1993) Metabolism of extracellular adenine nucleotides by human endothelial cells exposed to reactive oxygen metabolites. Am J Physiol 264:C282–6.
2. Abbracchio MP, Burnstock G. (1998) Purinergic signalling: pathophysiological roles. Jpn J Pharmacol 78:113–45.
3. Abbracchio MP, Burnstock G, Boeynaems JM, Barnard EA, Boyer JL, Kennedy C, Knight GE, Fumagalli M, Gachet C, Jacobson KA, Weisman GA. (2006) International Union of Pharmacology LVIII: update on the P2Y G protein-coupled nucleotide receptors: from molecular mechanisms and pathophysiology to therapy. Pharmacol Rev 58:281–341.
4. Agteresch HJ, Dagnelie PC, van den Berg JW, Wilson JH. (1999) Adenosine triphosphate: established and potential clinical applications. Drugs 58: 211–32.
5. Aird WC. (2007) Phenotypic heterogeneity of the endothelium: II. Representative vascular beds. Circ Res 100:174–90.
6. Albert JL, Boyle JP, Roberts JA, Challiss RA, Gubby SE, Boarder MR. (1997) Regulation of brain capillary endothelial cells by P2Y receptors coupled to Ca^{2+}, phospholipase C and mitogen-activated protein kinase. Br J Pharmacol 122:935–41.
7. Bicknell R, Harris, AL. (2004) Novel angiogenic signaling pathways and vascular targets. Annu Rev Pharmacol Toxicol 44:219–38.
8. Bodin P, Burnstock G. (1995) Nucleotide- and nucleoside-converting ectoenzymes: Important modulators of purinergic signalling cascade. Experientia 51:256–9.

9. Bodin P, Burnstock G . (2001) Evidence that release of adenosine triphosphate from endothelial cells during increased shear stress is vesicular. J Cardiovasc Pharmacol 38: 900–8.
10. Bours MJL, Swennen ELR, Di Virgilio F, Cronstein BN, Dagnelie PC. (2006) Adenosine 5′-triphosphate and adenosine as endogenous signaling molecules in immunity and inflammation. Pharmacol Therapeutics 112:358–404.
11. Burnstock G. (2002) Purinergic signaling and vascular cell proliferation and death. Arterioscler Thromb Vasc Biol 22:364–73.
12. Burnstock G. (2006) Pathophysiology and therapeutic potential of purinergic signaling. Pharmacol Rev 58:58–86.
13. Carmeliet P. (2003) Angiogenesis in health and disease. Nat Med 9:635–60.
14. Cha SH, Hahn TW, Sekine T, Lee KH, Endou H. (2000) Purinoceptor-mediated calcium mobilization and cellular proliferation in cultured bovine corneal endothelial cells. Jpn J Pharmacol 82:181–7.
15. Cool CD, Kennedy D, Voelkel NF, Tuder RM. (1997) Pathogenesis and evolution of plexiform lesions in pulmonary hypertension associated with scleroderma and human immunodeficiency virus infection. Hum Pathol 28:434–42.
16. Cool CD, Stewart JS, Werahera P, Miller GJ, Williams RL, Voelkel NF, Tuder RM. (1999) Three-dimensional reconstruction of pulmonary arteries in plexiform pulmonary hypertension using cell-specific markers. Evidence for a dynamic and heterogeneous process of pulmonary endothelial cell growth. Am J Pathol 155:411–9.
17. Davie NJ, Crossno JT, Jr., Frid MG, Hofmeister SE, Reeves JT, Hyde DM, Carpenter TC, Brunetti JA, McNiece IK, Stenmark KR. (2004) Hypoxia-induced pulmonary artery adventitial remodeling and neovascularization: contribution of progenitor cells. Am J Physiol Lung Cell Mol Physiol 286:L668–78.
18. Deaglio S, Dwyer KM, Gao W, Friedman D, Usheva A, Erat A, Chen JF, Enjyoji K, Linden J, 0 Oukka M, Kuchroo VK, Strom TB, Robson SC. (2007) Adenosine generation catalyzed by CD39 and CD73 expressed on regulatory T cells mediates immune suppression. J Exp Med 204:1257–65.
19. Di Virgilio F, Solini A. (2002) P2 receptors: new potential players in atherosclerosis. Br J Pharmacol 135:831–42.
20. Dubey RK, Gillespie DG, Jackson EK. (2002) A(2B) adenosine receptors stimulate growth of porcine and rat arterial endothelial cells. Hypertension 39:530–5.
21. Eltzschig HK, Ibla JC, Furuta GT, Leonard MO, Jacobson KA, Enjyoji K, Robson SC, Colgan SP.. (2003) Coordinated adenine nucleotide phosphohydrolysis and nucleoside signaling in posthypoxic endothelium: role of ectonucleotidases and adenosine A2B receptors. J Exp Med 198:783–96.
22. Eltzschig HK, Ibla JC, Furuta GT, Leonard MO, Jacobson KA, Enjyoji KS, Robson, C, Colgan, SP. (2003) Coordinated adenine nucleotidephosphohydrolysis and nucleoside signaling in posthypoxic endothelium: role of ectonucleotidases and adenosine A2B receptors, J Exp Med 198:783–96.
23. Erlinge D. (1998) Extracellular ATP: a growth factor for vascular smooth muscle cells. Gen Pharmacol 31:1–8.
24. Ethier MF, Chander V, and Dobson JG, Jr. (1993) Adenosine stimulates proliferation of human endothelial cells in culture. Am J Physiol 265:H131–8.
25. Feoktistov I, Ryzhov, S, Zhong H, Goldstein AE, Matafonov A, Zeng D, Biaggioni I. (2004) Hypoxia modulates adenosine receptors in human endothelial and smooth muscle cells toward an A2B angiogenic phenotype. Hypertension 44:649–54.
26. Fuchs S, Kornowski R, Leon MB, Epstein SE. (2001) Anti-angiogenesis: A new potential strategy to inhibit restenosis. Int J Cardiovasc Intervent 4:3–6.
27. Fukushi J, Ono M, Morikawa W, Iwamoto Y, Kuwano M. (2006) The activity of soluble VCAM-1 in angiogenesis stimulated by IL-4 and IL-13. J Immunol 165:2818–23.
28. Gerasimovskaya EV, Ahmad S, White CW, Jones PL, Carpenter TC, Stenmark KR. (2002) Extracellular ATP is an autocrine/paracrine regulator of hypoxia-induced adventitial fibroblast

growth. Signaling through extracellular signal-regulated kinase-1/2 and the Egr-1 transcription factor. J Biol Chem 277:44638–50.

29. Gerasimovskaya EV, Woodward H N, Tucker DA, Stenmark KR. (2008) Extracellular ATP is a pro-angiogenic factor for pulmonary artery vasa vasorum endothelial cells. Angiogenesis 11:169–82.
30. Grant MB, Davis MI, Caballero S, Feoktistov I, Biaggioni I, Belardinelli L.. (2001) Proliferation, migration, and ERK activation in human retinal endothelial cells through A(2B) adenosine receptor stimulation. Invest Ophthalmol Vis Sci 42: 2068–73.
31. Guba M, von Breitenbuch P, Steinbauer M, Koehl G, Flegel S, Hornung M, Bruns CJ, Zuelke C, Farkas S, Anthuber M, Jauch KW, Geissler EK. (2002) Rapamycin inhibits primary and metastatic tumor growth by antiangiogenesis: involvement of vascular endothelial growth factor. Nat Med 8:128–35.
32. Hartlapp I, Abe, R, Saeed RW, Peng T, Voelter, W, Bucala R, Metz, CN. (2001) Fibrocytes induce an angiogenic phenotype in cultured endothelial cells and promote angiogenesis in vivo. FASEB J 15:2215–24.
33. Haskó G, Linden J, Cronstein B, Pacher P. (2008) Adenosine receptors: therapeutic aspects for inflammatory and immune diseases. Nat Rev Drug Discov 7:759–70.
34. Hayden, MR, Tyagi SC. (2004) Vasa vasorum in plaque angiogenesis, metabolic syndrome, type 2 diabetes mellitus, and atheroscleropathy: a malignant transformation. Cardiovasc Diabetol 3:1–16.
35. Herrmann J, Best P J, Ritmann EL, Holmes DR, Lerman LO, Lerman A. (2002) Chronic endothelin receptor antagonism prevents coronary vasa vasorum neovascularization in experimental hypercholesterolemia. J Am Coll Cardiol 39:1555–61.
36. Hirakawa M, Oike M, Karashima Y, Ito Y. (2004) Sequential activation of RhoA and FAK/paxillin leads to ATP release and actin reorganization in human endothelium. J Physiol 558:479–88.
37. Huang N, Wang DJ, Heppel LA. (1989) Extracellular ATP is a mitogen for 3T3, 3T6, and A431 cells and acts synergistically with other growth factors. Proc Natl Acad Sci USA 86:7904–8.
38. Hunsucker SA, Mitchell BS, Spychala J. (2005) The 5′-nucleotidases as regulators of nucleotide and drug metabolism. Pharmacol Ther 107:1–30.
39. Idzko M, Hammad H, van Nimwegen M Kool M, Willart MAM, Muskens F, Hoogsteden HC, Luttmann W, Ferrari D, Di Virgilio F, Virchow JCM Jr, Lambrecht BN. (2007) Extracellular ATP triggers and maintains asthmatic airway inflammation by activating dendritic cells. Nat Med 13:913–919.
40. Jacobson JR, Dudek SM, Singleton PA, Kolosova IA, Verin AD, Garcia JGN. (2006) Endothelial cell barrier enhancement by ATP is mediated by the small GTPase Rac and cortactin Am J Physiol Lung Cell Mol Physiol 291:L289–95.
41. Jumblatt MM. (1994) Autocrine regulation of corneal endothelium by prostaglandin E2. Invest Ophthalmol Vis Sci 35:2783–90.
42. Kaczmarek E, Erb L, Koziak K, Jarzyna R, Wink MR, Guckelberger O, Blusztajn JK, Trinkaus-Randall V, Weisman GA, Robson SC. (2005) Modulation of endothelial cell migration by extracellular nucleotides: involvement of focal adhesion kinase and phosphatidylinositol 3-kinase-mediated pathways. Thromb Haemost 93:735–42.
43. Kaiser M, Younge B, Bjornsson J, Goronzy JJ, Weyand CM. (1999) Formation of new vasa vasorum in vasculitis. Production of angiogenic cytokines by multinucleated giant cells. Am J Pathol 155:765–74.
44. Kato H, Shichiri M, Marumo F, Hirata Y. (1997) Adrenomedullin as an autocrine/paracrine apoptosis survival factor for rat endothelial cells. Endocrinology 138: 2615–20.
45. Kolosova IA, Mirzapoiazova T, Adyshev D, Usatyuk P, Romer LH, Jacobson JR, Natarajan V, Pearse DB, Garcia JG, Verin AD. (2005) Signaling pathways involved in adenosine triphosphate-induced endothelial cell barrier enhancement. Circ Res 97:115–24.

46. Koszalka P, Ozuyaman, B, Huo Y, Zernecke A, Flogel U. Braun N, Buchheiser A, Decking UK, Smith ML, Sevigny J, Gear A, Weber AA, Molojavyi A, Ding Z, Weber C, Ley K, Zimmermann H, Godecke A, Schrader J. (2004) Targeted disruption of cd73/ecto-5′-nucleotidase alters thromboregulation and augments vascular inflammatory response. Circ Res 95:814–21.
47. Koyama T, Oike M, Ito Y. (2001) Involvement of Rho-kinase and tyrosine kinase in hypotonic stress-induced ATP release in bovine aortic endothelial cells J Physiol 532:759–69.
48. Krötz F, Sohn HY, Keller M, Gloe T, Bolz SS, Becker BF, Pohl U. (2002) Depolarization of endothelial cells enhances platelet aggregation through oxidative inactivation of endothelial NTPDase. Arterioscler Thromb Vasc Biol 22:2003–9.
49. Lazarowski ER, Homolya L, Boucher RC, Harden. (1997) TK Identification of an ecto-nucleoside diphosphokinase and its contribution to interconversion of P2 receptor agonists. J Biol Chem 272:20402–7.
50. Lazarowski, ER, Boucher RC, Harden TK. (2003) Mechanisms of release of nucleotides and integration of their action as P2X- and P2Y-receptor activating molecules. Mol Pharmacol 64:785–95.
51. Lemoli RM, Ferrari D, Fogli M, Rossi L, Pizzirani C, Forchap S, Chiozzi P, Vaselli D, Bertolini F, Foutz T, Aluigi M, Baccarani M, Di Virgilio F. (2004) Extracellular nucleotides are potent stimulators of human hematopoietic stem cells in vitro and in vivo. Blood 104:1662–70.
52. Liekens S, De Clercq E, Neyts, J. (2001) Angiogenesis: regulators and clinical applications. Biochem Pharmacol 61:253–70.
53. Marcus AJ, Broekman MJ, Drosopoulos JH, Islam N, Pinsky DJ, Sesti C, Levi R. (2003) Metabolic control of excessive extracellular nucleotide accumulation by CD39/ecto-nucleotidase-1: implications for ischemic vascular diseases. J Pharmacol Exp Ther 305: 9–16.
54. Mayer H, Bertram H. Lindenmaier W, Korff T. Weber, H, Weich H. (2005) Vascular endothelial growth factor. (VEGF-A) expression in human mesenchymal stem cells: autocrine and paracrine role on osteoblastic and endothelial differentiation. J Cell Biochem 95: 827–39.
55. Meghji P, Pearson JD, Slakey LL. (1995) Kinetics of extracellular ATP hydrolysis by microvascular endothelial cells from rat heart. Biochem J 308:725–31.
56. Meyrick B. (2001) The pathology of pulmonary artery hypertension. Clin Chest Med 22: 393–404, vii.
57. Mizumoto N, Kumamoto T, Robson SC, Sévigny J, Matsue H, Enjyoji K, Takashima A. (2002) CD39 is the dominant Langerhans cell-associated ecto-NTPDase: modulatory roles in inflammation and immune responsiveness. Nat Med 8:358–65.
58. Moreno PR, Fuster V. (2004) New aspects in the pathogenesis of diabetic atherothrombosis. J Am Coll Cardiol 44:2293–300.
59. Motte S, Pirroton S, Boeynaems JM. (1993) Heterogeneity of ATP receptors in aortic endothelial cells. Involvement of P2y and P2u receptors in inositol phosphate response. Circ Res 72:504–10.
60. Moulton KS, Vakili K, Zurakowski D, Soliman M, Butterfield C, Sylvin E, Lo KM, Gillies S, Javaherian K, Folkman J. (2003) Inhibition of plaque neovascularization reduces macrophage accumulation and progression of advanced atherosclerosis. Proc Natl Acad Sci USA 100:4736–41.
61. Mitzner W, Wagner EM. (2004) Vascular remodeling in the circulations of the lung. J Appl Physiol 97:1999–2004.
62. Noll T, Hölschermann H, Koprek K, Gündüz D, Haberbosch W, Tillmanns H, Piper HM. (1999) ATP reduces macromolecule permeability of endothelial monolayers despite increasing [Ca^{2+}]. Am J Physiol 276:H1892–1901.
63. Novak I. (2003) ATP as a signaling molecule: the exocrine focus. News Physiol Sci 18:12–7.
64. Numano F. (2000) Vasa vasoritis, vasculitis and atherosclerosis. Int J Cardiol 75(Suppl 1):S1–S8; discussion S17–S19.

65. Pinsky DJ, Broekman MJ, Peschon JJ, Stocking KL, Fujita T, Ramasamy R, Connolly ES, Jr, Huang J, Kiss S, Zhang Y, Choudhri TF, McTaggart RA, Liao H, Drosopoulos JH, Price VL, Marcus AJ, Maliszewski CR. (2002) Elucidation of the thromboregulatory role of CD39/ectoapyrase in the ischemic brain. J Clin Invest 109:1031–40.
66. Ralevic V, Burnstock G. (1998) Receptors for purines and pyrimidines. Pharmacol Rev 50:413–92.
67. Robson SC, Kaczmarek E, Siegel JB, Candinas D, Koziak K, Millan M, Hancock WW, Bach FH. (1997) Loss of ATP diphosphohydrolase activity with endothelial cell activation. J Exp Med 185:153–63.
68. Robson SC, Wu Y, Sun X, Knosalla C, Dwyer K, Enjyoji K. (2005) Ectonucleotidases of CD39 family modulate vascular inflammation and thrombosis in transplantation, Semin Thromb Hemost 31:217–33.
69. Robson SC, Sévigny J, Zimmermann H. (2006) The E-NTPDase family of ectonucleotidases: Structure function relationships and pathophysiological significance. Purinergic Signal 2:409–30.
70. Rossi L, Manfredini R, Bertolini F, Ferrari D, Fogli M, Zini R, Salati S, Salvestrini V, Gulinelli S, Adinolfi E, Ferrari S, Di Virgilio F, Baccarani M, Lemoli RM. (2007) The extracellular nucleotide UTP is a potent inducer of hematopoietic stem cell migration. Blood 109: 533–42.
71. Rupnick MA, Panigrahy D, Zhang CY, Dallabrida SM, Lowell BB, Langer R, Folkman MJ. (2002) Adipose tissue mass can be regulated through the vasculature. Proc Natl Acad Sci USA 99:10730–5.
72. Satterwhite CM, Farrelly AM, Bradley ME. (1999) Chemotactic, mitogenic, and angiogenic actions of UTP on vascular endothelial cells, Am J Physiol 276:H1091–7.
73. Schwiebert LM, Rice WC, Kudlow BA, Taylor AL, Schwiebert EM. (2002) Extracellular ATP signaling and P2X nucleotide receptors in monolayers of primary human vascular endothelial cells. Am J Physiol Cell Physiol 282:C289–C301.
74. Seghezzi G, Patel S, Ren CJ, Gualandris A, Pintucci G, Robbins ES, Shapiro RL, Galloway AC, Rifkin DB, Mignatti P. (1998) Fibroblast growth factor-2. (FGF-2) induces vascular endothelial growth factor. (VEGF) expression in the endothelial cells of forming capillaries: an autocrine mechanism contributing to angiogenesis. J Cell Biol 141:1659–73.
75. Seye CI, Yu N, Jain R, Kong Q, Minor T, Newton J, Erb L, Gonzalez FA, Weisman GA. (2003) The P2Y2 nucleotide receptor mediates UTP-induced vascular cell adhesion molecule-1 expression in coronary artery endothelial cells, J Biol Chem 278:24960–5.
76. Seye CI, Yu N, Gonzalez FA, Erb L, Weisman GA. (2004) The P2Y2 nucleotide receptor mediates vascular cell adhesion molecule-1 expression through interaction with VEGF receptor-2. (KDR/Flk-1). J Biol Chem 279:35679–86.
77. Sitkovsky MV. (2009) T regulatory cells: hypoxia-adenosinergic suppression and re-direction of the immune response. Trends Immunol 30:102–8.
78. Stenmark KR, Fagan KA, Frid MG. (2006) Hypoxia-induced pulmonary vascular remodeling: cellular and molecular mechanisms. Circ Res 99:675–91.
79. Stenmark KR, Gerasimovskaya E, Nemenoff RA, et al. (2002) Hypoxic activation of adventitial fibroblasts: role in vascular remodeling. Chest 122(6 Suppl):326S–34S.
80. Takedachi M, Qu D, Ebisuno Y, Oohara H, Joachims ML, McGee ST, Maeda E, McEver RP, Tanaka T, Miyasaka M, Murakami S, Krahn T, Blackburn MR, Thompson LF. (2008) CD73-generated adenosine restricts lymphocyte migration into draining lymph nodes. J Immunol 180:6288–96.
81. Thompson LF, Eltzschig HK, Ibla JC, Van De Wiele CJ, Resta R, Morote-Garcia JC, Colgan SP.. (2004) Crucial role for ecto-5′-nucleotidase. (CD73) in vascular leakage during hypoxia. J Exp Med 200:1395–405.
82. Tuder RM, Groves B, Badesch DB, Voelkel NF. (1994) Exuberant endothelial cell growth and elements of inflammation are present in plexiform lesions of pulmonary hypertension. Am L Pathol 144:275–85.

83. Van Daele P, Van Coevorden A, Roger PP, Boeynaems JM. (1992) Effects of adenine nucleotides on the proliferation of aortic endothelial cells. Circ Res 70:82–90.
84. Vanhaesebroeck B, Waterfield MD. (1999) Signaling by distinct classes of phosphoinositide 3-kinases. Exp Cell Res 253:239–54.
85. Van Linden A, Eltzschig HK. (2007) Role of pulmonary adenosine during hypoxia: extracellular generation, signaling and metabolism by surface adenosine deaminase/CD26. Expert Opin Biol Ther 7:1437–47.
86. Wang L, Karlsson L, Moses S, Hultgårdh-Nilsson A, Andersson M, Borna C, Gudbjartsson T, Jern S, Erlinge D. (2002) P2 receptor expression profiles in human vascular smooth muscle and endothelial cells. J Cardiovasc Pharmacol 40:841–53.
87. Wullschleger S, Loewith R, and Hall MN. (2006) TOR signaling in growth and metabolism. Cell 124:471–84.
88. Yamamoto K, Sokabe T, Ohura N, Nakatsuka H, Kamiya A, Ando J. (2003) Endogenously released ATP mediates shear stress-induced Ca^{2+} influx into pulmonary artery endothelial cells. Am J Physiol 285:H793–H803.
89. Yegutkin GG, Henttinen T, Jalkanen S. (2001) Extracellular ATP formation on vascular endothelial cells is mediated by ecto-nucleotide kinase activities via phosphotransfer reactions. FASEB J 15:251–60.
90. Yegutkin GG, Henttinen T, Samburski SS Spychala J, Jalkanen S. (2002) The evidence for two opposite, ATP-generating and ATP-consuming, extracellular pathways on endothelial and lymphoid cells. Biochem J 367:121–8.
91. Yegutkin GG, Mikhailov A, Samburski SS and Jalkanen S. (2006) The detection of micromolar pericellular ATP pool on lymphocyte surface by using lymphoid ecto-adenylate kinase as intrinsic ATP sensor. Mol Biol Cell 17:3378–85.
92. Yegutkin GG. (2008) Nucleotide- and nucleoside-converting ectoenzymes: Important modulators of purinergic signalling cascade. Biochim Biophys Acta 1783:673–94.
93. Yonekura H, Sakurai S, Liu X, Migita H, Wang H, Yamagishi S, Nomura M, Abedin MJ, Unoki H, Yamamoto Y, Yamamoto H. (1999) Placenta growth factor and vascular endothelial growth factor B and C expression in microvascular endothelial cells and pericytes: implication in autocrine and paracrine regulation of angiogenesis. J Biol Chem 274:35172–8.
94. Zimmermann H. (1992) 5′-Nucleotidase: molecular structure and functional aspects. Biochem J 285:345–65.

Chapter 6
Stimulation of Wound Revascularization by Adenosine Receptor Activation

M. Carmen Montesinos and María D. Valls

Abstract Adenosine is an endogenous mediator implicated in wound healing. The exact mechanisms and receptors involved are still under evaluation. We have observed that topical application of a selective adenosine A_{2A} receptor agonist promotes wound healing in experimental animals, both healthy and with impaired healing. Histological analysis revealed that adenosine promoted granulation tissue formation, with increased cellularity, matrix deposition, and vessel density. Adenosine may exert effects on many different cell types that finally result in promotion of angiogenesis and, furthermore, homing of endothelial precursor cells into sites of injury. We will review the effect of adenosine receptor activation on microvascular endothelial cell function and on wound revascularization.

Keywords Adenosine receptors · Wound healing · Angiogenesis · Vasculogenesis cell migration · Cell proliferation · Dermal microvascular endothelial cell · Endothelial precursor cells · Inflammation · Cytokines · Hypoxia · Hypoxia-inducible factor (HIF) · Vascular endothelial growth factor · Thrombospondin-1 · Murine air pouch model · Bone marrow transplantation · Fibrinolytic system · Plasminogen activation · Extracellular matrix · Annexin II

6.1 Introduction

Wound healing is a dynamic and complex process that involves a well coordinated, highly regulated series of events including inflammation, tissue formation, revascularization and tissue remodeling [13]. However, this orderly progression of

M.C. Montesinos (✉)
Department de Farmacologia, Universitat de València, Ave. Vicent Andrès Estellès s/n, 46100 Burjassot Valencia, Spain
Instituto de Reconocimiento Molecular y Desarrollo Tecnológico, Centro Mixto Universidad Politécnica de Valencia, Universidad de Valencia, Spain
Department of Medicine, New York University School of Medicine, 550 First Ave, New York, NY 10016, USA
e-mail: m.carmen.montesinos@uv.es

E. Gerasimovskaya, E. Kaczmarek (eds.), *Extracellular ATP and Adenosine as Regulators of Endothelial Cell Function*, DOI 10.1007/978-90-481-3435-9_6,

events is impaired in chronic wounds, including those due to diabetes. Loss of autonomic function and small fiber neuropathy is a characteristic feature of diabetic patients and can result in impaired neurogenic control of local microcirculatory blood flow and fluid homeostasis, diminished energy metabolism, oxygen delivery, and inflammatory responses. These altered processes could render diabetic patients more susceptible to tissue damage and infection [51]. Although many different interventions have been proposed to accelerate the healing process, very few have translated into positive clinical experience, and research is ongoing to provide better therapeutic alternatives to promote impaired wound healing.

The metabolic messenger adenosine is an endogenous regulator of inflammation that is generated as a result of ATP catabolism in ischemic or inflamed tissues. Adenosine mediates a wide variety of physiological effects through the activation of cell-surface receptors of which there are four subtypes, A_1, A_{2A}, A_{2B} and A_3, all of which belong to the superfamily of seven transmembrane-spanning G-protein coupled receptors [17, 22, 26]. All cell subtypes involved in wound healing, macrophages, epidermal cells, fibroblasts and microvascular endothelial cells, differentially express functional adenosine receptors, although the receptor expression patterns vary between cellular types. Adenosine A_{2A} and A_{2B} receptors, in particular, are expressed on all cell types involved in wound healing, while A_1 and A_3 receptors were not found in microvascular endothelial cells [6, 7, 14, 27]. Several studies indicate that adenosine agonists promote healing of cutaneous wounds through activation of surface receptors (reviewed in [62]). We and others have shown that selective agonists of the A_{2A} receptor subtype of adenosine promote tissue repair in mice, both healthy and with impaired healing, and one such agonist, Sonedenoson, is currently being evaluated as a prospective new therapy of diabetic foot ulcers. The examination of the mechanisms underlying this phenomenon showed that adenosine facilitates the formation of the granulation tissue and the blood vessels network that nourishes it [37, 38, 43, 63].

New vessel formation in the adult is the result of subtle and often complex interactions between regulatory and effector molecules and occurs through the cooperation of two distinct mechanisms: angiogenesis, the sprouting or partitioning of already existing vessels; and vasculogenesis, the differentiation and assembly of circulating bone marrow-derived endothelial cell precursors (angioblasts) [4]. During the sprouting phase, endothelial cells become activated and coordinately involved in the processes of proliferation, migration, and tube formation. Once a provisional vessel network is formed, the phase of resolution takes place by cessation of endothelial cell proliferation and migration, followed by recruitment and differentiation of smooth muscle cells and pericytes providing full functionality to the vessel [41]. This multistep process is tightly regulated by soluble growth factors and inhibitors, and a close interaction between adhesive extracellular matrix proteins and their integrin receptors.

Several vasoactive metabolites produced during tissue hypoxia have been implicated in the angiogenesis process [10]. Adenosine is also a vasodilating metabolite whose extracellular concentration dramatically increases in injured and oxygen deprived sites as consequence of ATP catabolism [52]. In addition, the pharmacological vasodilator dipyridamole that inhibits the activity of adenosine deaminase and adenosine re-uptake, elicits a neovascular response in rat cardiac and skeletal muscle, strongly supporting a role for adenosine in promoting angiogenesis [10]. We will review the mechanisms by which adenosine receptor activation affects the function of microvascular endothelial cells implicated in wound revascularization and will discuss the contribution of the adenosine receptors to blood vessel formation.

6.2 Adenosine Receptor Expression and Function on Dermal Microvascular Endothelial Cells

Most of our knowledge of the effects of adenosine receptors activation on endothelial cells: down-regulation of cytokine-induced adhesion molecules expression; increased vascular barrier; promotion of migration and proliferation; signal transduction mechanisms... comes from experiments carried out using endothelial cells from multiple origins, and mainly human umbilical vein (HUVEC). However, it should be taken in consideration that endothelial cells are exposed to particular physico-chemical and immunological condition depending on the vascular bed of origin, therefore phenotypical differences are expected [1, 49].

In most blood vessels, adenosine is a potent vasodilator via activation of A_{2A} and/or A_{2B} receptors, expressed in both the endothelium and smooth muscle. Conversely, adenosine is a potent vasoconstrictor in some branches of the pulmonary arteries and in renal afferent arterioles, through smooth muscle A_1 receptor activation [56]. Several expression profiling and functional studies have revealed that endothelial cells from multiple origins do not express A_1 and A_3 receptors [5, 14, 42]. Microvascular endothelial cells, in particular, play a central role in inflammation by recruiting circulating leukocytes to tissues, in part through cytokine-regulated expression of endothelial-leukocyte adhesion molecules that differ from large vessel endothelial cells [32, 45, 57].

The expression pattern of A_{2A} and A_{2B} receptors differs between endothelial cells from large vessels such as umbilical vein, which express higher levels of A_{2A} than A_{2B}, and dermal microvasculature, which preferentially express A_{2B}. This appears to have profound functional implications, as A_{2B}AR activation strongly induces the expression of angiogenic factors, such as interleukin-8 (IL-8), basic fibroblast growth factor (bFGF), and vascular endothelial growth factor (VEGF) via coupling to G(q), and possibly via G12/13, instead of the classical Gs pathway in a dermal microvascular endothelial cell line [14]. More importantly, the expression of both, A_{2A} and A_{2B}, subtypes in dermal microvascular endothelial cells is subject to dynamic regulation by pathophysiological conditions, such as hypoxia and cytokine exposure [12, 16, 42]. During episodes of hypoxia/ischemia, polymorphonuclear leukocytes are mobilized from the intravascular space to the interstitium, and such responses may contribute significantly to tissue damage during subsequent reperfusion. Acute increases in adenosine are important to counterbalance excessive inflammation or vascular leakage, and furthermore, to promote revascularization of the ischemic tissue [11, 56].

Hypoxia altered the expression of adenosine receptors by decreasing A_{2A} and increasing A_{2B} levels by a mechanism dependent on hypoxia-inducible factor (HIF) [16, 28]. Consistent with changes in receptor expression, adenosine acts in a cooperative fashion with hypoxia to stimulate VEGF and induce IL-8 secretion not stimulated by hypoxia alone [16, 47]. Further studies revealed that adenosine deaminase (ADA) and its complexing protein CD26 are coordinately induced by chronic hypoxia, effectively localizing ADA activity at the endothelial cell surface, as an innate metabolic adaptation to chronically elevated adenosine levels during hypoxia. In contrast, during acute hypoxia associated with vascular leakage and excessive inflammation, ADA inhibition may serve as therapeutic strategy [11].

Pro-inflammatory cytokines, such as IL-1 and TNF-α can increase both A_{2A} and A_{2B} adenosine receptor expression and function in primary dermal microvascular endothelial cells [42]. Interestingly, while A_{2B} receptor expression is also potentiated by IFN-γ in these cells, A_{2A} receptor levels are markedly decreased. Consequently, an A_{2A}-selective agonist stimulated cAMP accumulation in TNF-α treated, but not IFN-γ treated cells, whereas the non selective agonist NECA stimulated cAMP accumulation in cells treated with either cytokine. Thus, both receptors contribute to adenosine-induced cAMP accumulation in TNF-α stimulated cells, but only A_{2B} receptors caused cAMP elevation in IFN-γ treated cells. Another interesting observation in this study is that, under basal conditions, both receptors are poorly coupled to adenylyl cyclase, but they are more actively and functionally coupled following cytokine treatment (Fig. 6.1). This finding could be explained by cytokine-enhanced expression and/or recruitment of selective $G_{\beta 4}$ proteins (Fig. 6.2), or other signalling elements in the cell membrane, promoting receptor-G protein coupling [42]. In fact, $G_{\beta 4}$ subunit provides a high affinity agonist binding to the A_{2A} receptor-G protein complex that contains the subunit [40].

We further evaluated the contribution of adenosine A_{2A} receptors to VEGF secretion by dermal microvascular endothelial cells. The adenosine A_{2A} receptor agonist CGS21680 stimulated a marked increase in VEGF mRNA expression at a pharmacologically selective concentration (1 μM), which was reversed by an A_{2A} receptor antagonist. Moreover, pretreatment with TNF-α enhanced VEGF mRNA by the adenosine receptor agonist, whereas pretreatment with IFN-γ abrogated the CGS21680-mediated stimulation of VEGF mRNA (Fig. 6.3) [42]. These data further confirm a role of adenosine A_{2A} receptors in stimulating VEGF production in addition to the established role for A_{2B} receptors in these cells [14]. Therefore, it is likely then that at injured sites, where cytokines such as TNF-α are secreted by

Fig. 6.1 (Continued) (**e**) HMVECs with overnight cytokine pretreatment were stimulated for 15 min with the indicated concentrations of NECA. (**f**) ZM-241385 (10 μM) was added to the culture before NECA stimulation. Values are expressed as the mean (±SEM) of four (**a** and **c**), three (**b**) or two (**e** and **f**) independent experiments. Reproduced from [42] with permission

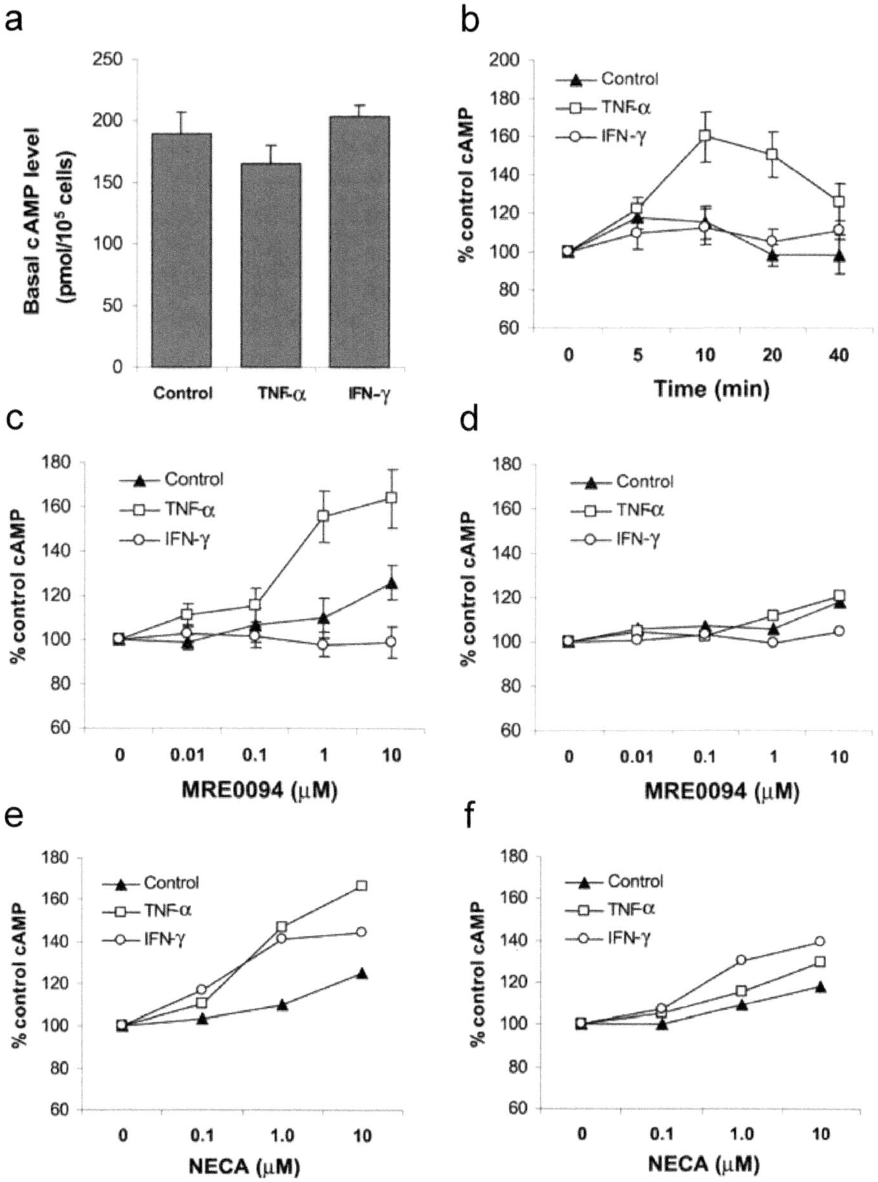

Fig. 6.1 cAMP accumulation in cytokine-treated dermal human microvascular endothelial cells (HMVEC). HMVECs were treated overnight with TNF-α or IFN-γ and thereafter stimulated with the A_{2A} adenosine receptor agonist MRE0094. The cellular cAMP content of treated HMVEC was measured by ELISA. (**a**) Basal cAMP levels in cytokine-treated HMVEC (without agonist stimulation). (**b**) Cytokine-pretreated HMVECs were stimulated with 1 μM MRE0094 for various times as indicated. (**c**) Cytokine-treated HMVECs were stimulated for 10 min with the indicated concentrations of MRE0094. (**d**) Enhanced cAMP accumulation in TNF-α-treated HMVEC by the agonist MRE0094 was attenuated in the presence of antagonist ZM-241385 (10 μM).

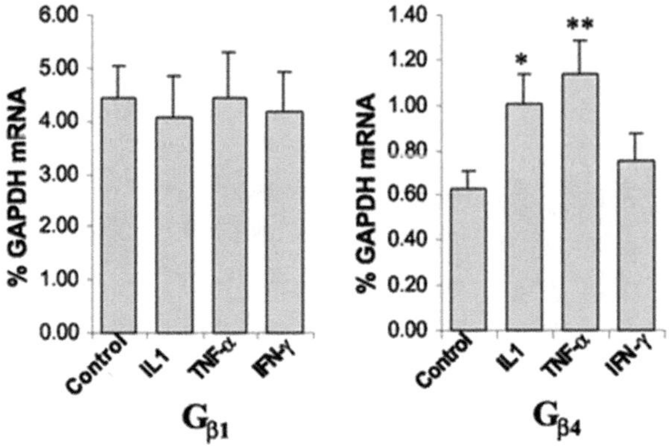

Fig. 6.2 Effect of inflammatory cytokines on the expression of $G_{\beta1}$ and $G_{\beta4}$ proteins in HMVEC. Cells were treated overnight with IL-1, TNF-α, or IFN-γ, followed by mRNA and protein analyses. Quantitative real-time PCR was performed to estimate the copy number of cDNA for $G_{\beta1}$ and $G_{\beta4}$ proteins, and data are expressed as the percentage of GAPDH cDNA. Data shown are the mean (±SEM) of four independent experiments. $^{*}p < 0.05$; $^{**}p < 0.01$. Reproduced from [42] with permission.

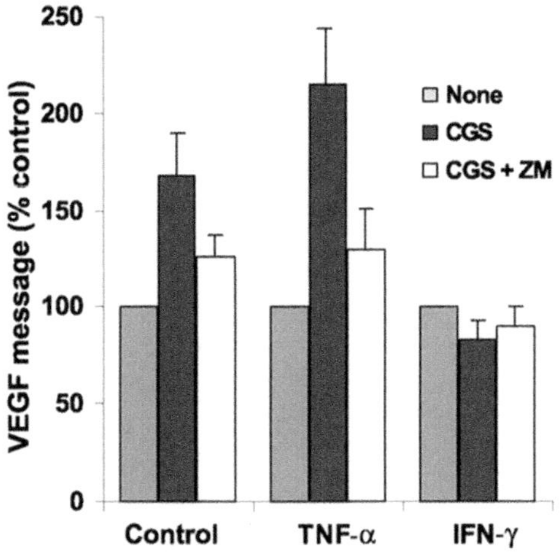

Fig. 6.3 TNF-α and IFN-γ differential modulation of the stimulatory effect of the adenosine receptor agonist CGS-2168 on VEGF mRNA expression. HMVECs were pretreated overnight with TNF-α or IFN-γ, then stimulated with CGS-21680 (1 μM) for 4 h in the absence (CGS) or the presence of 1 μM ZM-241385 (CGS + ZM). The expression of VEGF message was quantitated by real-time PCR and is presented as a percentage of the control (none). The basal levels of VEGF in unstimulated cells (without agonist or antagonist) were 1.55 ± 0.30, 2.02 ± 0.59, and 1.62 ± 0.12% of GAPDH mRNA for control, TNF-α-treated, and IFN-γ-treated cells, respectively ($n = 3$). Reproduced from [42] with permission

recruited inflammatory cells, the subsequent up-regulation of A_{2A} and A_{2B} receptors on endothelial cells and other inflammatory cells along with endogenous adenosine release constitutes a feedback loop to suppress further inflammation and to promote angiogenesis.

6.3 Adenosine Receptor Activation Promotes Angiogenesis In Vitro

Since the early studies demonstrating the angiogenic effect of adenosine in heart and skeletal muscle in rabbits [64] and the chick chorioallantoic membrane model [10], much effort has been addressed to elucidate the mechanisms and the receptors implicated in the adenosine-induced increase in vascularity. For instance, adenosine is known to stimulate endothelial cell migration and tube formation in vitro regardless the vascular origin [18, 33, 59]. On the other hand, the effect of adenosine on endothelial cell proliferation is more controversial and seems to depend on the vascular bed of origin. Thus, adenosine is mitogenic for microvascular and venular coronary and umbilical endothelial cells [35, 54], but not for aortic endothelial cells [59]. Similarly, the adenosine receptor subtypes and the intracellular mediators involved in promoting angiogenesis are still being explored.

Former studies, using venular coronary microvascular endothelial cells and adenosine analogues, suggested that the proliferative response to adenosine involves a pertussis toxin-sensitive substrate as well as an increase in cAMP, indicating that both Gi or Gs protein were involved [35]. However, as previously mentioned, most endothelial cells express mainly adenosine A_{2A} and/or A_{2B} receptors, and both receptors seem to participate of the effect of adenosine on endothelial cell proliferation, migration and release of regulatory agents such as VEGF. Studies carried out in retinal endothelial cells clearly demonstrated the participation of A_{2B} adenosine receptors on cell proliferation and migration, ERK activation, and, moreover, capillary tube formation by a mechanism involving increased angiogenic growth factor expression [18, 19]. In addition, analysis of the cloned human A_{2B} receptor identified a functional binding site for hypoxia-inducible factor (HIF) within the promoter. Additional studies in dermal endothelial cells overexpressing full-length A_{2B} receptors revealed functional phenotypes of increased barrier function and enhanced tube formation amplified during hypoxia [28].

Activation of A_{2A} receptor in human umbilical vein endothelial cells promotes endothelial cell migration and proliferation, and synthesis of the potent angiogenic factor VEGF. The mitogenic action of the A_{2A}-adenosine receptor activation is independent of protein Gs coupling, since direct activation of downstream effectors in the cyclic AMP-signalling cascade not only failed to mimic the action of receptor-activation, but even reduced cell proliferation. In contrast, activation of the A_{2A}-adenosine receptor is associated with increased levels of GTP-bound p21(ras) and MAP kinase activation, which is neither impaired following pretreatment of the cells with pertussis toxin nor affected by inhibitors of typical protein kinase C isoforms [53–55].

We have further studied the contribution of the A_{2A} receptor activation to the angiogenic effect of adenosine using a matrix of Matrigel®. We found that treatment with a selective A_{2A} receptor agonist, CGS21680, promotes an intricate tridimensional cellular network of human dermal microvasculature endothelial cells, and sprouting of aortic ring explants, both murine and from rat (Fig. 6.4). This assay was performed using fully supplemented endothelial growth media, containing different growth factors, including VEGF; therefore we thought that the promoting effect of CGS-21680 might be due to other mechanisms besides the release of angiogenic factors. When culturing the microvascular endothelial cells in presence of either a selective A_{2A} antagonist or a selective A_{2B} antagonist alone, we observed a decrease in tube formation, suggesting that endogenous released adenosine could play a role through activation of both receptors [9].

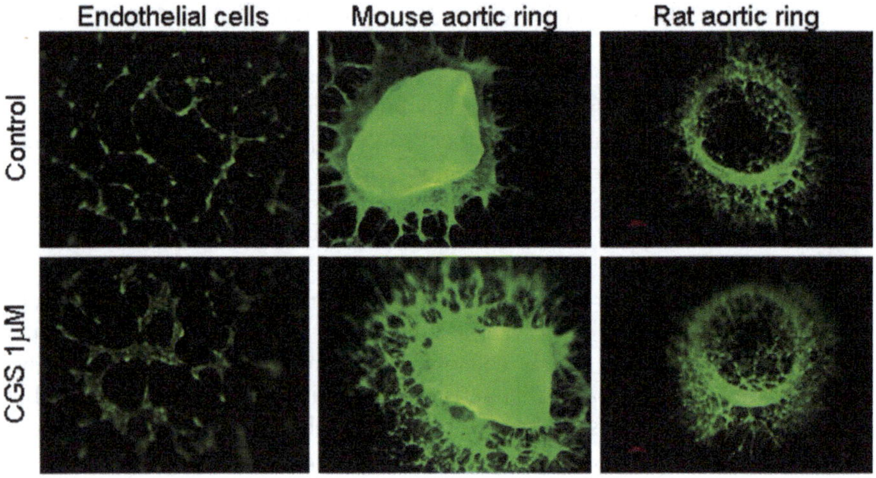

Fig. 6.4 CGS-21680 stimulated tube formation of HMVEC and from aortic ring explants: HMEC suspension (5×10^4 cells/ml) was loaded onto a Matrigel (50 μl) coated well, incubated for 16 h (37°C and 5% CO_2) in presence or absence of CGS-21680 (1 μM). Aortas were isolated from either CD31 mice or Wistar rats, cleaned, cut in 1–2 mm segments and seeded in non-polymerized Matrigel (50 μl) and incubated 30 min at 37°C and 5% CO_2 before adding fully supplemented EGM-2MV media. Aortic explants were incubated with or without CGS-21680 (1 μM) for 5 days (37°C and 5% CO_2, media was changed on day 2). Fluorophore/calcein AM (Molecular Probes) was used to stain the tubes and image was acquired with a fluorescent microscope. Representative fields of vascular tube formation examined at an ×40 magnification are shown

In order to determine the effect of adenosine A_{2A} receptor activation on endothelial cells, we generated a subtraction library by representational Display Analysis in HUVEC. We cloned several genes associated with angiogenesis, one of which was Thrombospondin-1, a glycoprotein that is a naturally occurring inhibitor of angiogenesis [23]. Thrombospondin-1 mRNA expression and protein secretion was down-regulated after treatment of human microvascular endothelial cells with the A_{2A} agonists in a dose-dependent manner, and this effect was completely abrogated by addition of a selective A_{2A} receptor antagonist, but not A_1 and A_{2B} receptor antagonists. Moreover, in the presence of antibodies to thrombospondin-1 and its receptor CD36, adenosine A_{2A} receptor agonists failed to promote tube formation by endothelial cell in Matrigel®, indicating that the angiogenic effects of adenosine A_{2A} receptor activation are, at least in part, caused by the suppression of thrombospondin-1 secretion [9]. Thus, although elevated expression of VEGF might account for the angiogenic effects of adenosine, it is more likely that adenosine also stimulates angiogenesis via other secondary angiogenic and antiangiogenic mediators or by way of an intracellular action [2].

6.4 Adenosine A_{2A} Receptor Activation Stimulates both Angiogenesis and Vasculogenesis In Vivo

Different experimental in vivo models confirmed the role of adenosine in blood vessel formation, where other cells, besides endothelial cells, are implicated in tissue neovascularization by releasing a variety of soluble regulatory cytokines, chemokines, growth factors, proteolytic enzymes and extracellulr matrix proteins, which activate local endothelial cells and facilitate endothelial progenitor cell homing [41, 44]. Thus, adenosine A_{2A} receptors activation stimulates macrophage production of VEGF synergistically with the Toll-like receptor 4 (TLR4) pathway [29] and in human mast cells, A_{2B} receptors mediate IL-8 and bFGF expression and A_3 receptors may mediate the expression of angiopoietin-2 [15].

Adenosine levels are elevated during normal retinal vascular development in neonatal dog by the activity of the ecto-5′-nucleotidase. In the canine model of oxygen-induced retinopathy, exposure of the developing retina to high oxygen haults vascular development, process known as vaso-obliteration. During this stage, 5′-nucleotidase is dramatically reduced, resulting in a sharp decline in adenosine. When animals are returned to room air, systemic levels of oxygen return to normal but the retina is hypoxic because of the lack of blood vessels. At this time, 5′-nucleotidase activity and adenosine levels are elevated well beyond normal levels. This stage is the vasoproliferative stage and A_{2A} receptor immunoreactivity was significantly elevated at the edge of forming vasculature [58]. In a similar model using mice, daily intraperitoneal injections for 5 days of pharmacologically relevant doses of a non selective adenosine receptor antagonist or the A_{2B} selective antagonists 3-N-propylxanthine (enprofylline) and 3-isobutyl-8-pyrrolidinoxanthine (IPDX) significantly reduced retinal neovascularization. By contrast, neither A_1 nor A_{2A} selective antagonists had an effect on neovascularization [36]. Further proof of the relevance of adenosine A_{2B} receptors in angiogenesis has been demonstrated by the intra-ocular injection of ribozymes designed to specifically cleave A_{2B} receptor mRNA, which significantly inhibited pre-retinal neovascularisation [3].

We have also demonstrated that adenosine A_{2A} receptor activation stimulates neovascularization in vivo. Topical application of agonists for adenosine A_{2A} receptors increased the rate of wound closure and the number of microvessels in the granulation tissue of wounds in wild-type but not A_{2A} deficient mice. Moreover, A_{2A} receptor deficient mice manifest fewer vessels in their wound granulation tissue (Fig. 6.5) [37]. Similarly to A_{2A} deficient mice, mice lacking the adapter protein in the pathway of Toll-like receptors MyD88 heal at a markedly slower rate than wounds in wild-type mice, showing delayed contraction, decreased and delayed granulation tissue formation, and reduced new blood vessel density. The A_{2A} receptor agonist CGS21680 promoted wound closure and angiogenesis in wild

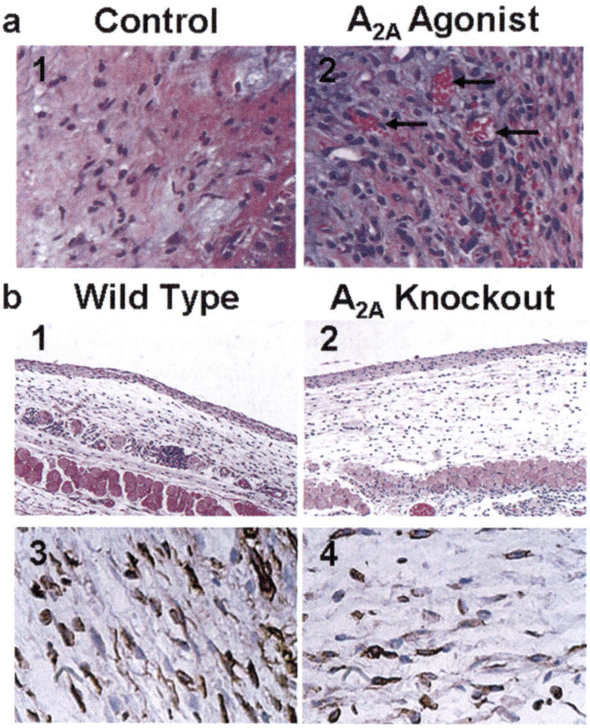

Fig. 6.5 Histological examination of representative sections of untreated and CGS-21680-treated excisional wounds in wild-type mice and air pouch wall in wild-type and A_{2A} receptor deficient mice. (**a**) On day 3 after wounding, mice were killed and the wounds dissected out, embedded in paraffin, and stained with H&E using standard techniques. Original magnifications ×400 1: Wild-type mouse treated with vehicle 2: Wild-type mouse treated with CGS-21680, arrows indicate blood vessels containing erythrocytes. (**b**) Subcutaneous vesicles were formed on the backs of wild-type and knockout mice by injecting 3 ml of steril air and were reinflated later with 1.5 ml of steril air two days. After 6 days, the mice were killed and the pouches dissected out. The dissected pouches were fixed, embedded in paraffin, and stained with H&E using standard techniques (**1** and **2**) or stained immunohistologically with a goat anti-Factor VIII antibody, visualized by immunoperoxidase-mediated conversion of diaminobenzidine, and counterstained with hematoxylin (**3** and **4**). **1:** Shown is a representative section from a pouch wall from a female wild-type mouse (original magnification, ×100). **2:** Shown is a representative section from a pouch wall from a female A_{2A} receptor knockout mouse (original magnification, ×100). **3:** Shown is a representative section from a pouch wall from a female wild-type mouse (same block as in **1**; original magnification, ×400). **4:** Shown is a representative section from a pouch wall from a female knockout mouse (same block as in **2**; original magnification ×400). Reprinted from [37] with permission

type mice, but had no significant effect on healing of MyD88(–/–) mice [34]. These results are in agreement with the synergistic interaction between Toll-like receptor and adenosine A_{2A} receptor signaling found in macrophages towards an angiogenic phenotype [29, 46].

We confirmed the role of A_{2A} receptor activation in neovascularization in vivo using a model of internal trauma and repair (air pouch model). There were fewer microvessels in the walls of healing air pouches of A_{2A} receptor deficient mice compared to wild-type controls and consequently, inflammatory vascular leakage and leukocyte accumulation in the air pouch were similarly reduced in A_{2A} knock-out mice. Since there was no change in the number of circulating leukocytes and the inflammatory peritoneal exudates in the A_{2A} deficient mice did not differ from those in the wild-type mice, it is unlikely that the induced deficiency in A_{2A} receptors is associated with a defect in the acute inflammatory response or with a difference in the number or functional capacity of the A_{2A} receptor-deficient leukocytes. Thus, the reduction in the number of microvessels in the air pouch walls of the A_{2A} knock-out mice is the most likely factor leading to reduced leukocyte accumulation in the inflamed air pouch, a finding that correlates with the diminished extravasation of Evans blue dye [37]. Pharmacological blockade of A_{2A} receptors confirmed the reduced angiogenesis in air pouch walls but, surprisingly, blockade of A_1 receptors increased the number of Factor VIII-positive cells in the healing tissue, and an A_{2B} receptor antagonist had no effect (Fig. 6.6), consistent with the role of A_{2A}

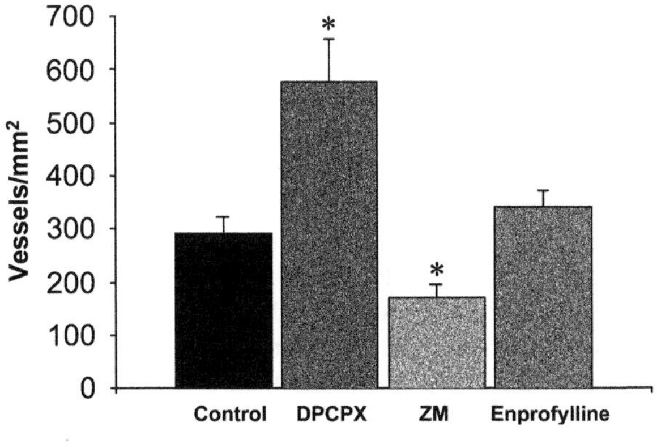

Fig. 6.6 Number of Factor VIII positive endothelial cells in the walls of air pouches of mice treated with adenosine A_1, A_{2A} and A_{2B} receptor antagonists. Osmotic pumps containing either DMSO (control), 8-Cyclopentyl-1,3-dipropylxanthine (DPCPX, an adenosine A_1 receptor selective antagonist, 15 mg/kg/d), ZM 241385 (an adenosine A_{2A} receptor selective antagonist, 50 mg/kg/d) or enprofylline (an adenosine A_{2B} receptor selective antagonist, 50 mg/kg/d) were implanted in the peritoneum of BALB/C mice. Air pouches were induced and harvested on BALB/C mice, as described for Fig. 6.5, and the number of Factor VIII positive endothelial cells were enumerated in the immunohistologically-labeled sections, as described. Shown are the means (±SEM) of 5 different fields counted on 5 mice in each group. *p<0.005 vs. Wild-type mice, 1-Way ANOVA (Tukey Test)

receptors in promoting angiogenesis. The reduction in vascularity by paharmaclogical blockade of A_{2A} receptors corresponded as well with a reduction in Evans blue extravasation and in leukocyte accumulation after injection of carrageenan into the airpouch (unpublished results).

In the mouse ischemic hind limb model, antagonism of adenosine receptors with the non selective antagonist caffeine abrogated VEGF up-regulation induced by local injection of the non selective agonist NECA into the hind limb and produced a 46% reduction of neovascularization. Stimulation of adenosine receptors not only contributed to the overall effect of hypoxia but also had additional actions in the regulation of angiogenic factors. Thus, adenosine actions appear to be complementary to the direct effects of hypoxia rather than redundant [47].

6.4.1 Adenosine Receptor Activation Promotes Wound Vasculogenesis

In addition to the classic mechanisms of angiogenesis, bone marrow derived endothelial precursor cells, also referred to as angioblasts, have the ability to incorporate into sites of active neovascularization in vivo, contributing to the development of new vessels in the adult [8]. Adenosine A_{2A} receptors have been detected in forming blood vessels and endothelial progenitor cells in the nerve fiber layer of peripheral retina, both in neonatal dogs and during the vasoproliferative stage of canine oxygen induced retinopathy [58]. We have shown that topical application of an adenosine A_{2A} receptor agonist promotes endothelial precursor cells homing to newly forming vessels in wound granulation tissue. We inflicted two full thickness wounds in mice that had undergone bone marrow transplantation with whole non-fractionated bone marrow cells from transgenic mice that express the marker gene green fluorescent protein (GFP) under the endothelial specific Tie2 promoter. Tie2 is a specific receptor only found in vascular endothelial cells therefore the recipient mice express GFP only in those vascular endothelial cells that were originated in the bone marrow. Topical treatment with the selective adenosine A_{2A} agonist CGS-21680 induced a three-fold increase in CD-31 positive, GFP negative endothelial cells and a higher than 10-fold increase in GFP cells in the granulation tissue of 3-day wounds [39]. Adenosine A_{2A} receptor activation also increases dermal recruitment of other bone marrow originated and peripheral blood circulating cells, fibrocytes, which migrate into injured tissue where they display fibroblast-like properties. In a bleomycin-induced model of dermal fibrosis, increased fibrocytes recruitment to the dermis of bleomycin-treated mice was abrogated by genetic deletion or pharmacological blockade of adenosine A_{2A} receptors, suggesting that this receptor subtype is implicated in bone marrow derived fibrocyte recruitment [24].

Bone marrow-derived mesenchymal stem cells are characterized by their ability to form colonies comprising spindle-shaped cells deriving from a single cell, and they express all four adenosine receptors and the enzyme ecto-5′ nucleotidase (CD73) that catalyzes the extracellular conversion of AMP to adenosine [25]. Mesenchymal stem cells isolated from bone marrow of either A_{2A} deficient or ecto-5′ nucleotidase (CD73)-deficient mice formed significantly fewer colonies, and this effect was mimic by culture with the adenosine A_{2A} receptor antagonist ZM241385, although culture of wild type cells with the A_{2A} agonist CGS21680 did not enhance colony formation. Interestingly, there was a significant decrease in CD73 expression in cells from A_{2A} receptor deficient mice and a marked reduction in A_{2A} receptor expression in CD73 deficient mice, confirming a reciprocal regulation of CD73 by A_{2A} receptors and vice versa. Moreover, adenosine A_{2A} receptors regulate the expression of differentiation antigens on the mesenchymal stem cells, such as CD90, CD105, and procollagen alpha2 type I.

These findings indicate that adenosine via A_{2A} receptor plays a critical role in promoting the proliferation and differentiation of mouse bone marrow mesenchymal stem cell [25].

A recent study also demonstrates a role of adenosine in vasculogenesis of the myocardium, although in this territory the adenosine receptors involved appear to differ. Stimulation of adenosine receptors increased the adhesion of murine embryonic endothelial progenitor cells to cardiac endothelial cells rapidly, within minutes. Furthermore, adenosine increased cell retention in isolated mouse hearts perfused with embryonic endothelial precursor cells. Similarly, adhesion of human adult culture-expanded endothelial progenitor cells to human cardiac endothelial cells was increased by stimulation of adenosine receptors. These investigators showed that murine embryonic endothelial precursor cells preferentially express functional A_1 adenosine receptors, while the A_{2B} receptor subtype is preferentially expressed by cardiac endothelial cells, and both subtypes are involved in the regulation of progenitor cell adhesion to the endothelium by the interaction between P-selectin and its ligand [48].

6.5 Contribution of the Fibrinolytic System to the Angiogenic Effect of Adenosine

Besides of degrading the fibrin clot, the fibrinolytic system is an essential constituent of the proteolytic machinery involved in the breakdown of extracellular matrix, which enables migration of fibroblasts and endothelial cells at sites of injury. The serine proteases urokinase (uPA) and tissue (tPA) plasminogen activators degrade plasminogen to plasmin, which in turn degrades components of the wound matrix, fibrinogen and fibrin, and activate several matrix metalloproteinases. Plasminogen activation is tightly regulated by the balance between plasminogen activators tPA and uPA and their principal inhibitors, plasminogen activator inhibitor-1 (PAI-1) and PAI-2 [30, 50]. Since adenosine receptor agonists increase plasminogen activator release from rabbit alveolar macrophages [21], we are currently studying the effect of a selective adenosine A_{2A} receptor agonist on wound closure in wild type, tPA deficient and uPA deficient mice, and the effect of adenosine A_{2A} receptor activation on the production and secretion of t-PA, u-PA, and PAI-1 by dermal microvascular endothelial cells and the monocytic cell line THP-1 [60, 61]. Annexin II is a proposed endothelial cell co-receptor for plasminogen and t-PA [20] and in vitro studies have suggested that this endothelial cell surface protein can stimulate t-PA–mediated plasminogen activation in the complete absence of fibrin [31]. We have found that promotion of wound healing by topical application of a selective adenosine A_{2A} receptor agonist requires the expression of tissue (tPA) but not urokinase plasminogen activator (uPA). Our preliminary results in human dermal microvascular endothelial cells point to a new hypothesis, that tPA participates of the proangiogenic effect of A_{2A} receptor activation through a complex mechanism that involves its main inhibitor PAI-1 and its cell receptor annexin II (Fig. 6.7).

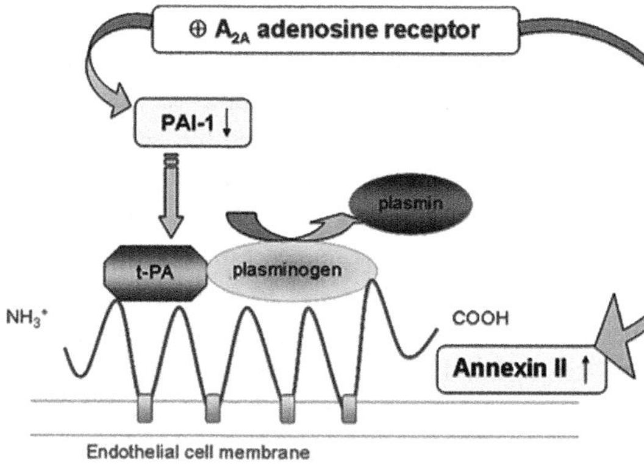

Fig. 6.7 Proposed mechanism for adenosine A_{2A} receptor stimulation of plasmin generation on the surface of endothelial cells and its implication in angiogenesis. Adenosine A2A receptor activation in endothelial cells diminishes the release of the plasminogen activator inhibitor-1 (PAI-1) and promotes the production of annexin II, which acts as a co-receptor for plasminogen and its activator tPA

6.6 Conclusion

Adenosine, through interaction with specific membrane receptors, promotes the formation of new vessels by increasing both local vessel sprouting and recruitment of endothelial progenitor cells from the bone marrow. The contribution of the different receptor subtypes to the angiogenic effect of adenosine depends mainly on the pathophysiological conditions of the organ affected. While endothelial cell A_{2B} receptor activation appears to prevail in revascularization of hypoxic tissue and in retinopathy, A_{2A} receptors play a relevant role in physiological revascularization of damaged tissues. In addition, all the adenosine receptors subtypes expressed by resident and migrating cells modulate the release of regulatory molecules that provide a proper environment for neovascularization.

References

1. Abeles D, Kwei S, Stavrakis G, Zhang Y, Wang ET, Garcia-Cardena G. (2006) Gene expression changes evoked in a venous segment exposed to arterial flow. J Vasc Surg 44:863–70.
2. Adair TH. (2005) Growth regulation of the vascular system: an emerging role for adenosine. Am J Physiol Regul Integr Comp Physiol 289:R283–96.
3. Afzal A, Shaw LC, Caballero S, Spoerri PE, Lewin AS, Zeng D, Belardinelli L, Grant MB. (2003) Reduction in preretinal neovascularization by ribozymes that cleave the A2B adenosine receptor mRNA. Circ Res 93:500–6.
4. Asahara T, Murohara T, Sullivan A, Silver M, van der Zee R, Li T, Witzenbichler B, Schatteman G, Isner JM. (1997) Isolation of putative progenitor endothelial cells for angiogenesis. Science 275:964–7.
5. Auchampach JA. (2007) Adenosine receptors and angiogenesis. Circ Res 101:1075–7.
6. Braun M, Lelieur K, Kietzmann M. (2006) Purinergic substances promote murine keratinocyte proliferation and enhance impaired wound healing in mice. Wound Repair Regen 14:152–61.
7. Chan ES, Fernandez P, Merchant AA, Montesinos MC, Trzaska S, Desai A, Tung CF, Khoa DN, Pillinger MH, Reiss AB, Tomic-Canic M, Chen JF, Schwarzschild MA, Cronstein BN. (2006) Adenosine A2A receptors in diffuse dermal fibrosis: pathogenic role in human dermal fibroblasts and in a murine model of scleroderma. Arthritis Rheum 54:2632–42.
8. Conway EM, Collen D, Carmeliet P. (2001) Molecular mechanisms of blood vessel growth. Cardiovasc Res 49:507–21.
9. Desai A, Victor-Vega C, Gadangi S, Montesinos MC, Chu CC, Cronstein BN. (2005) Adenosine A2A receptor stimulation increases angiogenesis by down-regulating production of the antiangiogenic matrix protein thrombospondin 1. Mol Pharmacol 67:1406–13.
10. Dusseau JW, Hutchins PM, Malbasa DS. (1986) Stimulation of angiogenesis by adenosine on the chick chorioallantoic membrane. Circ Res 59:163–70.
11. Eltzschig HK, Faigle M, Knapp S, Karhausen J, Ibla J, Rosenberger P, Odegard KC, Laussen PC, Thompson LF, Colgan SP. (2006) Endothelial catabolism of extracellular adenosine during hypoxia: the role of surface adenosine deaminase and CD26. Blood 108:1602–10.
12. Eltzschig HK, Ibla JC, Furuta GT, Leonard MO, Jacobson KA, Enjyoji K, Robson SC, Colgan SP. (2003) Coordinated adenine nucleotide phosphohydrolysis and nucleoside signaling in posthypoxic endothelium: role of ectonucleotidases and adenosine A2B receptors. J Exp Med 198:783–96.
13. Falanga V. (2005) Wound healing and its impairment in the diabetic foot. Lancet 366: 1736–43.
14. Feoktistov I, Goldstein AE, Ryzhov S, Zeng D, Belardinelli L, Voyno-Yasenetskaya T, Biaggioni I. (2002) Differential expression of adenosine receptors in human endothelial cells: role of A2B receptors in angiogenic factor regulation. Circ Res 90:531–8.
15. Feoktistov I, Ryzhov S, Goldstein AE, Biaggioni I. (2003) Mast cell-mediated stimulation of angiogenesis: cooperative interaction between A2B and A3 adenosine receptors. Circ Res 92:485–92.
16. Feoktistov I, Ryzhov S, Zhong H, Goldstein AE, Matafonov A, Zeng D, Biaggioni I. (2004) Hypoxia modulates adenosine receptors in human endothelial and smooth muscle cells toward an A2B angiogenic phenotype. Hypertension 44:649–54.
17. Fredholm BB, AP IJ, Jacobson KA, Klotz KN, Linden J. (2001) International Union of Pharmacology. XXV. Nomenclature and classification of adenosine receptors. Pharmacol Rev 53:527–52.
18. Grant MB, Davis MI, Caballero S, Feoktistov I, Biaggioni I, Belardinelli L. (2001) Proliferation, migration, and ERK activation in human retinal endothelial cells through A(2B) adenosine receptor stimulation. Invest Ophthalmol Vis Sci 42:2068–73..

19. Grant MB, Tarnuzzer RW, Caballero S, Ozeck MJ, Davis MI, Spoerri PE, Feoktistov I, Biaggioni I, Shryock JC, Belardinelli L. (1999) Adenosine receptor activation induces vascular endothelial growth factor in human retinal endothelial cells. Circ Res 85:699–706..
20. Hajjar KA, Jacovina AT, Chacko J. (1994) An endothelial cell receptor for plasminogen/tissue plasminogen activator. I. Identity with annexin II. J Biol Chem 269:21191–7.
21. Hasday JD, Sitrin RG. (1987) Adenosine receptors on rabbit alveolar macrophages: binding characteristics and effects on cellular function. J Lab Clin Med 110:264–73.
22. Hasko G, Cronstein BN. (2004) Adenosine: an endogenous regulator of innate immunity. Trends Immunol 25:33–9.
23. John M. Rhodes MS. (2007) The extracellular matrix and blood vessel formation: not just a scaffold. Journal of Cellular and Molecular Medicine 11:176–205.
24. Katebi M, Fernandez P, Chan ES, Cronstein BN. (2008) Adenosine A(2A) receptor blockade or deletion diminishes fibrocyte accumulation in the skin in a murine model of scleroderma, bleomycin-induced fibrosis. Inflammation 31:299–303.
25. Katebi M, Soleimani M, Cronstein BN. (2009) Adenosine A2A receptors play an active role in mouse bone marrow-derived mesenchymal stem cell development. J Leukoc Biol 85:438–44.
26. Khakh BS, Burnstock G, Kennedy C, King BF, North RA, Seguela P, Voigt M, Humphrey PPA. (2001) International Union of Pharmacology. XXIV. Current Status of the Nomenclature and Properties of P2X Receptors and Their Subunits. Pharmacol Rev 53:107–118.
27. Khoa ND, Montesinos MC, Reiss AB, Delano D, Awadallah N, Cronstein BN. (2001) Inflammatory cytokines regulate function and expression of adenosine A(2A) receptors in human monocytic THP-1 cells. J Immunol 167:4026–32.
28. Kong T, Westerman KA, Faigle M, Eltzschig HK, Colgan SP. (2006) HIF-dependent induction of adenosine A2B receptor in hypoxia. Faseb J 20:2242–50.
29. Leibovich SJ, Chen JF, Pinhal-Enfield G, Belem PC, Elson G, Rosania A, Ramanathan M, Montesinos C, Jacobson M, Schwarzschild MA, Fink JS, Cronstein B. (2002) Synergistic up-regulation of vascular endothelial growth factor expression in murine macrophages by adenosine A(2A) receptor agonists and endotoxin. Am J Pathol 160:2231–44.
30. Li WY, Chong SS, Huang EY, Tuan TL. (2003) Plasminogen activator/plasmin system: a major player in wound healing? Wound Repair Regen 11:239–47.
31. Ling Q, Jacovina AT, Deora A, Febbraio M, Simantov R, Silverstein RL, Hempstead B, Mark WH, Hajjar KA. (2004) Annexin II regulates fibrin homeostasis and neoangiogenesis in vivo. J Clin Invest 113:38–48.
32. Liu M, Kluger MS, D′Alessio A, Garcia-Cardena G, Pober JS. (2008) Regulation of arterial-venous differences in tumor necrosis factor responsiveness of endothelial cells by anatomic context. Am J Pathol 172:1088–99.
33. Lutty GA, Mathews MK, Merges C, McLeod DS. (1998) Adenosine stimulates canine retinal microvascular endothelial cell migration and tube formation. Curr Eye Res 17:594–607.
34. Macedo L, Pinhal-Enfield G, Alshits V, Elson G, Cronstein BN, Leibovich SJ. (2007) Wound Healing Is Impaired in MyD88-Deficient Mice: A Role for MyD88 in the Regulation of Wound Healing by Adenosine A2A Receptors. Am J Pathol 171:1774–88.
35. Meininger CJ, Granger HJ. (1990) Mechanisms leading to adenosine-stimulated proliferation of microvascular endothelial cells. Am J Physiol 258:H198–H206.
36. Mino RP, Spoerri PE, Caballero S, Player D, Belardinelli L, Biaggioni I, Grant MB. (2001) Adenosine receptor antagonists and retinal neovascularization in vivo. Invest Ophthalmol Vis Sci 42:3320–24.
37. Montesinos MC, Desai A, Chen JF, Yee H, Schwarzschild MA, Fink JS, Cronstein BN. (2002) Adenosine promotes wound healing and mediates angiogenesis in response to tissue injury via occupancy of A(2A) receptors. Am J Pathol 160:2009–18.
38. Montesinos MC, Gadangi P, Longaker M, Sung J, Levine J, Nilsen D, Reibman J, Li M, Jiang CK, Hirschhorn R, Recht PA, Ostad E, Levin RI, Cronstein BN. (1997) Wound healing is accelerated by agonists of adenosine A2 (G alpha s- linked) receptors. J Exp Med 186:1615–20.

39. Montesinos MC, Shaw JP, Yee H, Shamamian P, Cronstein BN. (2004) Adenosine A2A Receptor Activation Promotes Wound Neovascularization by Stimulating Angiogenesis and Vasculogenesis. Am J Pathol 164:1887–92.
40. Murphree LJ, Marshall MA, Rieger JM, MacDonald TL, Linden J. (2002) Human A(2A) adenosine receptors: high-affinity agonist binding to receptor-G protein complexes containing Gbeta(4). Mol Pharmacol 61:455–62.
41. Naik MU, Mousa SA, Parkos CA, Naik UP. (2003) Signaling through JAM-1 and {alpha}v{beta}3 is required for the angiogenic action of bFGF: dissociation of the JAM-1 and {alpha}v{beta}3 complex. Blood 102:2108–14.
42. Nguyen DK, Montesinos MC, Williams AJ, Kelly M, Cronstein BN. (2003) Th1 cytokines regulate adenosine receptors and their downstream signaling elements in human microvascular endothelial cells. J Immunol 171:3991–8.
43. Peirce SM, Skalak TC, Rieger JM, Macdonald TL, Linden J. (2001) Selective A(2A) adenosine receptor activation reduces skin pressure ulcer formation and inflammation. Am J Physiol Heart Circ Physiol 281:H67–H74.
44. Pepper MS. (2001) Role of the matrix metalloproteinase and plasminogen activator-plasmin systems in angiogenesis. Arterioscler Thromb Vasc Biol 21:1104–17.
45. Petzelbauer P, Bender JR, Wilson J, Pober JS. (1993) Heterogeneity of dermal microvascular endothelial cell antigen expression and cytokine responsiveness in situ and in cell culture. J Immunol 151:5062–72.
46. Ramakers BP, Riksen NP, Rongen GA, van der Hoeven JG, Smits P, Pickkers P. (2006) The effect of adenosine receptor agonists on cytokine release by human mononuclear cells depends on the specific Toll-like receptor subtype used for stimulation. Cytokine 35: 95–9.
47. Ryzhov S, McCaleb JL, Goldstein AE, Biaggioni I, Feoktistov I. (2007) Role of adenosine receptors in the regulation of angiogenic factors and neovascularization in hypoxia. J Pharmacol Exp Ther 320:565–72.
48. Ryzhov S, Solenkova NV, Goldstein AE, Lamparter M, Fleenor T, Young PP, Greelish JP, Byrne JG, Vaughan DE, Biaggioni I, Hatzopoulos AK, Feoktistov I. (2008) Adenosine receptor-mediated adhesion of endothelial progenitors to cardiac microvascular endothelial cells. Circ Res 102:356–63.
49. Sands WA, Palmer TM. (2005) Adenosine receptors and the control of endothelial cell function in inflammatory disease. Immunol Lett 101:1–11.
50. Schafer BM, Maier K, Eickhoff U, Todd RF, Kramer MD. (1994) Plasminogen activation in healing human wounds. Am J Pathol 144:1269–80.
51. Schaper NC, Huijberts M, Pickwell K. (2008) Neurovascular control and neurogenic inflammation in diabetes. Diabetes Metab Res Rev 24(Suppl 1):S40–S44.
52. Schulte G, Fredholm BB. (2003) Signalling from adenosine receptors to mitogen-activated protein kinases. Cell Signal 15:813–s27.
53. Seidel MG, Klinger M, Freissmuth M, Holler C. (1999) Activation of Mitogen-activated Protein Kinase by the A2A-adenosine Receptor via a rap1-dependent and via a p21ras-dependent Pathway. J Biol Chem 274:25833–41.
54. Sexl V, Mancusi G, Baumgartner-Parzer S, Schutz W, Freissmuth M. (1995) Stimulation of human umbilical vein endothelial cell proliferation by A2-adenosine and beta 2-adrenoceptors. Br J Pharm 114:1577–86.
55. Sexl V, Mancusi G, Holler C, Gloria-Maercker E, Schutz W, Freissmuth M. (1997) Stimulation of the mitogen-activated protein kinase via the A2A-adenosine receptor in primary human endothelial cells. J.Biol.Chem. 272:5792–9.
56. Shen J, DiCorleto PE. (2008) Adenosine prompts the heart to recruit endothelial progenitors. Circ Res 102:280–82.
57. Swerlick RA, Lawley TJ. (1993) Role of microvascular endothelial cells in inflammation. J Invest Dermatol 100:111S-5S.

58. Taomoto M, McLeod DS, Merges C, Lutty GA. (2000) Localization of adenosine A2a receptor in retinal development and oxygen-induced retinopathy. Invest Ophthalmol Vis Sci 41:230–43.
59. Teuscher E, Weidlich V. (1985) Adenosine nucleotides, adenosine and adenine as angiogenesis factors. Biomed Biochim Acta 44:493–95.
60. Valls M, Montesinos M. (2007) Adenosine A2a receptor activation modulates the fibrinolytic system in Human Microvascular Endothelial Cells. (HMVEC). Meth Find 29:39.
61. Valls MD, Cronstein BN, Montesinos MC. (2006) Adenosine A2A receptor activation promotes production of tissue Plasminogen activator. (tPA) by endothelial cells. Meth Find 28:138.
62. Valls MD, Cronstein BN, Montesinos MC. (2009) Adenosine receptor agonists for promotion of dermal wound healing. Biochem Pharmacol 77:1117–24.
63. Victor-Vega C, Desai A, Montesinos MC, Cronstein BN. (2002) Adenosine A2A receptor agonists promote more rapid wound healing than recombinant human platelet-derived growth factor (Becaplermin gel). Inflammation 26:19–24.
64. Ziada AMAR, Hudlicka O, Tyler KR, Wright AJA. (1984) The effect of long-term vasodilatation on capillary growth and performance in rabbit heart and skeletal muscle. Cardiovasc Res 18:724–32.

Chapter 7
Hypoxia-Inducible Factors and Adenosine Signaling in Vascular Growth

Aftab Ahmad, Carl W. White, and Shama Ahmad

Abstract Low oxygen environment or hypoxia is conducive towards vascular growth and endothelial proliferation. Therefore it is an essential element in both disease and development. Hypoxia stabilizes the hypoxia-inducible transcription factors -1α and -2α and also increases adenosine levels thereby activating the adenosine receptor signaling pathways. While these pathways have been described independently to a greater extent, increasing evidence suggests that there is significant crosstalk. Here we will summarize the contributions and interdependence of hypoxia-inducible transcription factors and the adenosine receptor pathways in vascular growth.

Keywords Angiogenesis · adenosine · A_{2A} receptor · A_{2B} receptor · Adenosine deaminase · ATP · CD73 · CD39 · Hypoxia-inducible transcription factors · HIF-1α · HIF-2α · Prolyl 4-hydroxylases · Endothelial cells · Vascular growth · VEGF · VEGFR-1 · VEGFR-2 · DMOG (dimethyloxalylglycine) · DFO (desferrioxamine) · VHL (von Hippel-Lindau tumor suppressor gene product)

7.1 Introduction

Low oxygen environment or hypoxia plays a central role in complex processes such as fetal development and diseases like cancer, sepsis and ischemic heart disease. Physiological and pathological responses to acute and chronic hypoxia are mediated primarily through the hypoxia-inducible transcription factors HIF-1α and HIF-2α. Such conditions also often cause an increase in both intracellular and extracellular levels of adenosine, and its receptors. Accumulation of extracellular adenosine following hypoxia and/or ischemia causes activation of signaling cascades that are geared towards maintaining homeostasis and limiting cellular damage [29]. In recent

A. Ahmad (✉)
Department of Pediatrics, National Jewish Health, 1400 Jackson Street, Denver, CO 80206, USA
e-mail: ahmada@njhealth.org

E. Gerasimovskaya, E. Kaczmarek (eds.), *Extracellular ATP and Adenosine as Regulators of Endothelial Cell Function*, DOI 10.1007/978-90-481-3435-9_7,

years, the role of adenosine as a potential stimulator of tissue oxygenation and vessel growth has emerged [1]. In this chapter we will discuss the angiogenic potential of adenosine receptors as driven by hypoxia in endothelial cells.

7.2 Endothelial Cells in Vascular Growth and Development

Endothelium constitutes a highly specialized organ that lines the vascular system and lymphatic channels. Endothelial cells exhibit a great degree of heterogeneity that arises from their location as well as their specialization in response to environmental cues (autocrine and paracrine factors, shear stress and pressure) and pathophysiological stresses [5, 41]. Characterization of this heterogeneity and study of associated functional changes is key to the development of therapeutic approaches, such as those targeting angiogenesis. This point is well illustrated by one such study that describes the gene expression profile of 53 cultured human endothelial cell lines from arteries, veins and microvessels from 14 different tissues [12]. In this study the patterns of gene expression that could distinguish between endothelial cells of large and small vessels, and between those of veins and arteries, were evaluated. Expression patterns of gene products from specialized endothelia were related to functional roles such as transport of molecules, migration of immune cells, neurogenesis, tracheal branching, and the establishment of the left/right asymmetry of the body. Thus the transcriptional programs of endothelial cells from different tissues and organs are specifically adapted during development to assume distinct roles at each site. The pulmonary circulation represents a unique vascular bed that receives 100% of the cardiac output. Although endothelial cells of pulmonary artery and those of capillaries express classical endothelial cell markers, they differ in critical roles such as barrier function and angiogenic potential [68]. Endothelial cell populations also contain significant numbers of proliferating progenitor cells. Although each endothelial population may have a distinct number of progenitor cells, they are phenotypically related to their vascular origin. Thus it is also important to consider the contents and molecular programming of the endothelial progenitor cells as they could play an essential role in vascular development, homeostasis and repair after injury. With an improved understanding of signaling in normal diverse endothelia one could develop strategies to target altered responses in diseased or abnormal vessels.

Vascular growth and proliferation requires an early phase of vasculogenesis followed by angiogenesis. Vasculogenesis involves the de novo formation of vessels from progenitor cells. On the other hand, angiogenesis is a complex multistep process involving endothelial cell invasion, migration and cell proliferation resulting in formation of new blood vessels from pre-existing ones. Under physiological conditions, angiogenesis is required for normal growth and development during fetal life and also in tissue repair during adult life. Fetal lung develops in a hypoxic environment (3–5% O_2). Such hypoxic environments activate signaling pathways that increase vascular cell proliferation.

7.3 Hypoxia and Hypoxia-Inducible Transcription Factors

Hypoxia stabilizes hypoxia-inducible transcription factors (HIFs) HIF-1α and HIF-2α. Active HIFs are heterodimeric proteins that dimerize with HIF-1β (ARNT), translocate to the nucleus and bind to specific, hypoxia response element or HRE DNA sequences of target genes, promoting their transcription. Under room air conditions, HIFs are hydroxylated by prolyl 4-hydoxylases (PHDs) using molecular oxygen, iron (Fe^{2+}), α-ketoglutarate and ascorbate as substrates or cofactors (Fig. 7.1). Thus absence of oxygen (hypoxia), iron chelation, use of analogs of α-ketoglutarate or siRNA mediated knockdown of prolyl 4-hydroxylases can each inhibit hydroxylation of HIFs (Fig. 7.1) [6, 39, 65]. Hydroxylation of proline residues P402 and P562 of human HIF-1α (P405 and P531 of HIF-2α) tags it for pVHL (von Hippel-Lindau tumor suppressor gene product) binding, ubiq-

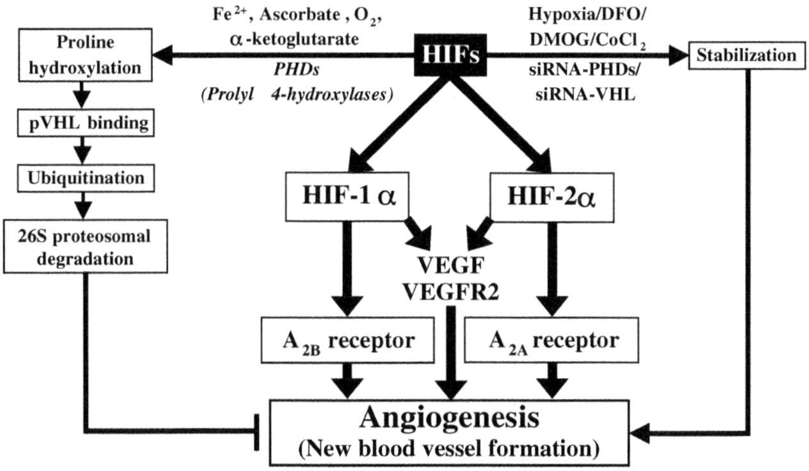

Fig. 7.1 Schematic model of hypoxia-induced pathways leading to angiogenic responses in endothelial cells. Under normal physiological conditions, hypoxia-inducible transcription factors, HIF-1α and HIF-2α are degraded through a series of steps, starting with hydroxylation of the transcription factor by prolyl 4-hydroxylases (PHDs). This degradation pathway leading to destabilization of HIFs has a negative effect on angiogenesis. On the other hand, inhibiting PHD activity using DFO, DMOG, hypoxia or siRNA-mediated knockdown of the enzyme can stabilize HIFs. Inhibiting the binding of pVHL to the hydroxylated HIFs using $CoCl_2$ or siRNA-mediated knockdown of VHL can also stabilize HIFs. Both HIF-1α and HIF-2α regulate vascular growth by common as well as unique pathways. VEGF-VEGFR2 pathway is a known inducer of angiogenesis and is regulated both by HIF-1α and HIF-2α in endothelial cells. Adenosine A_{2A} and A_{2B} receptors are also known to promote angiogenesis (or new blood vessel formation). While HIF-1α regulates expression of A_{2B} receptor in human dermal microvascular endothelial cells, HIF-2α regulates expression of A_{2A} receptor in human lung microvascular endothelial cells. DFO: desferrioxamine (Iron chelator); DMOG: dimethyloxalylglycine (analog of α-ketoglutarate); pVHL (von Hippel-Lindau tumor suppressor gene product)

uitination and subsequent degradation through the proteasomal pathway. Thus siRNA-mediated knockdown of VHL or absence of functional VHL as in naturally occurring mutations, can also prevent degradation of HIFs [13, 58]. In addition, $CoCl_2$ mediated inhibition of pVHL binding to the hydroxylated HIFs can also prevent degradation [74]. HIFs also can be hydroxylated by FIHs (Factor Inhibiting HIFs) at specific asparagine residues (N803 of HIF-1α and N847 of HIF-2α) located in the C-terminal transactivation domain. Hydroxylation of asparagine residues under non-hypoxic conditions inhibits transactivation function of HIFs by preventing interaction of HIFs with CBP/p300. Thus both stabilization and functional activity of HIFs are maintained under hypoxic conditions.

Hypoxia-inducible transcription factors HIF-1α and HIF-2α bind the same consensus DNA binding element and can upregulate a common set of genes involved in cell growth, proliferation and angiogenesis. These include among others the vascular endothelial growth factor (VEGF), its receptors including VEGFR-1 (Flt-1) and VEGFR-2 (KDR or flk-1) (Fig. 7.1). While both HIF-1α and HIF-2α are important in vascular growth and proliferation, HIF-2α is perceived to have a greater role because of its abundance in endothelial cells.

7.4 HIFs in Vascular Growth and Development

Endothelial cells are essential in embryonic vascular growth and development. A number of receptors on endothelial cells regulate endothelial cell function. Mitogens that act on these receptors are either secreted by endothelial cells themselves or contributed by other cell types. Further, most of these receptors and their ligands are regulated by the HIFs. Thus, altering the HIF pathway can profoundly affect vascular growth in a number of models. For example, a HIF-1α knockout mouse shows severely reduced vascular growth that results in embryonic lethality by E11 [37]. A HIF-2α null mouse, in 129 SV/ICR background, exhibited severe vascular defects that died by E12.5 [60]. Similar observations were obtained in other HIF-2α knockout or knockdown models [14, 17, 50]. Interestingly, HIF-2α null mouse in some backgrounds did not show vascular defects [66, 72].

Several genes downstream of the HIFs can influence vascular growth. The most common among these targets is the vascular endothelial growth factor (VEGF), a known inducer of angiogenesis. Loss of even a single allele of VEGF, caused abnormal vascular growth and embryonic lethality by E12 [9, 27]. Receptors for VEGF are also regulated by HIFs in endothelial cells. VEGF has a higher affinity for VEGFR-1 than for VEGFR-2. Disruption of VEGFR-1 (Flt-1) and VEGFR-2 (flk-1 or KDR) in mice leads to embryonic lethality due to vascular overgrowth in the former [28] and reduced vascular growth in the latter [67]. Therefore multiple parallel pathways operate to fine tune vascular growth and development during embryonic life. Importantly, HIFs appear to be central regulators of angiogenesis.

7.5 Hypoxia and Adenosine

Hypoxia, in addition to stabilization of HIF-1α and HIF-2α, also increases intracellular and extracellular levels of adenosine (Fig. 7.2). Extracellularly, adenosine is produced by degradation of ATP by ectonucleotidases like CD39 (ATPDase) and CD73 (ecto-5′-nucleotidase). Hypoxia, through an SP1 dependent mechanism, can increase expression of CD39, which catalyzes hydrolysis of ATP and ADP [20]. HIF-1α increases the expression of CD73 that catalyzes the conversion of AMP to adenosine [70]. Intracellular adenosine, under normoxic resting conditions, is either converted to AMP by adenosine kinase or degraded to inosine by adenosine deaminase. Hypoxia can inhibits both adenosine kinase and adenosine deaminase [51] resulting in intracellular adenosine accumulation. Additionally, hypoxia also regulates expression of adenosine transporters ENT1 and ENT2, through a HIF-1α dependent mechanism [21]. Therefore multiple parallel pathways contribute to increased adenosine concentrations in hypoxia.

Biological activity of adenosine is mediated through its binding to four different adenosine receptors, A_1, A_{2A}, A_{2B} and A_3 [4]. While A_1 and A_3 receptors inhibit

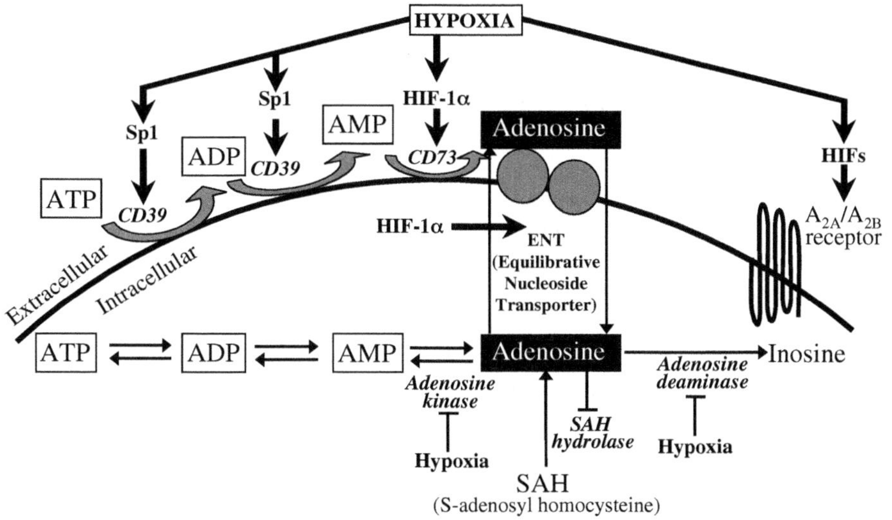

Fig. 7.2 Hypoxic regulation of adenosinergic signaling in a typical cell. Hypoxia causes both intracellular and extracellular levels of adenosine to increase. Extracellularly, adenosine is formed by the hydrolysis of ATP to adenosine. Two enzymes in this pathway, CD39 and CD73 are transcriptionally upregulated by hypoxia. While CD39 is regulated by hypoxia through an Sp1 dependent mechanism, CD73 is regulated by HIF-1α. Besides hydrolysis of ATP, intracellular levels of adenosine are also affected by adenosine kinase and adenosine deaminase that convert adenosine to AMP and inosine respectively. Hypoxia inhibits both of these enzymes leading to increased accumulation of adenosine. Intracellular and extracellular levels of adenosine are also maintained through the hypoxia and HIF-1α regulatable nucleoside transporters. Cell surface receptors for adenosine, the A_{2A} and A_{2B} receptors, are also regulated by hypoxia and HIFs. Solid arrows indicate transcriptional upregulation by hypoxia and/or HIFs

Table 7.1 Characterization of adenosine receptor subtypes. Adenosine receptors are characterized into four subtypes based on its affinity towards adenosine and coupling to G-proteins, which activate or inhibit adenylyl cyclase

Adenosine receptors	Effect on adenyly cyclase activity	Effect on cAMP production	Affinity for adenosine	References
A1	Inhibitory	Decreases	High ($\sim 10^{-8}$M)	[15, 52]
A2A	Stimulatory	Increases	High ($\sim 10^{-7}$M)	[15, 52]
A2B	Stimulatory	Increases	Low ($\sim 10^{-5}$M)	[8]
A3	Inhibitory	Decreases	Low ($\sim 10^{-5}$M)	[15]

adenylyl cyclase and are Gi (inhibitory G-protein)-coupled, the A_{2A} and A_{2B} receptors are Gs (stimulatory G protein)-coupled (Table 7.1). Both the high affinity A_{2A} and the low affinity A_{2B} receptors activate adenylyl cyclase, resulting in increases in intracellular cAMP. Among the adenosine receptors, both adenosine A_{2A} and the related A_{2B} receptors have unique properties, and are distinctly regulated in endothelial as well as other cell types. The activated adenosine A_{2A} receptor exhibits anti-inflammatory properties [54] and can protect against tissue injury in a number of organ systems [10, 42, 46, 55, 59, 62]. On the other hand adenosine A_{2B} receptor can exhibit both pro- and anti-inflammatory properties depending on the cell type [11, 19, 34, 53].

HIFs also regulate expression of adenosine receptors A_{2A} and A_{2B} in endothelial cells (Fig. 7.1). Endothelial cells can express either A_{2A} receptor or A_{2B} receptor or both [25, 56]. Adenosine A_{2B} receptor expression is regulated by HIF-1α in microvascular endothelial cells of the dermis [40]. By contrast, A_{2B} receptor expression was shown to be unresponsive to hypoxia and HIFs in microvascular endothelial cells derived from the pulmonary circulation [3]. In addition, adenosine A_{2A} receptor expression can be regulated by HIF-2α but not HIF-1α in these lung microvascular cells. It is interesting that adenosine receptors are regulated so differently even in microvascular endothelial cells of different tissue origins. This indicates fundamental differences in growth regulation of the microvasculature in the pulmonary versus the systemic circulation, similar to differences in micro- versus macrovessels [38]. Adenosine A_{2A} receptor also promotes proliferation of lung microvascular endothelial cells upon its activation [3, 45]. This is in contrast to studies in other cell types, including PC12 (pheochromocytoma) and smooth muscle cells, where activation of adenosine A_{2A} receptor caused decreased proliferation [45, 67, 69]. Similarly other studies with PC12 cells showed that there was a decrease in cell viability upon activation of the receptor in one investigation [73], whereas inhibition of apoptosis was observed in another [36]. Taken together, these studies point to a differential regulation and function of the receptor in cells and tissues. These studies show that either adenosine A_{2A} receptor or the related adenosine

A_{2B} receptor can be regulated by hypoxia in specific cell types. Cells expressing either of these receptors may have differential sensitivities based, in part, on local adenosine concentrations. Specifically, (a) adenosine A_{2B} receptor has a much lower affinity for adenosine relative to the adenosine A_{2A} receptor and would likely be activated in areas where adenosine concentrations are high, and (b) higher adenosine concentrations, while activating A_{2B} receptor, can simultaneously desensitize the high affinity A_{2A} receptor [57]. Thus, adenosine A_{2A} receptors would likely be more important and active at lower adenosine concentrations. Consistent with this hypothesis, constitutively high levels of adenosine were found to have an adverse effect on lung growth and maturation. In a study using $ADA^{-/-}$ mice (adenosine deaminase; ADA), global increases in adenosine levels were linked to abnormal alveolar development [7]. Interestingly there were decreased A_{2A} receptor mRNA levels in the $ADA^{-/-}$ mice compared to the control mice suggesting a likely role for the A_{2A} receptor. Thus, optimal levels of adenosine, together with activation of specific receptors like A_{2A}, are likely to be critical in supporting lung growth and development.

7.6 Adenosine and Adenosine Receptors in Vascular Growth and Development

Several studies have shown that adenosine and its analogs can promote endothelial cell proliferation and migration [18, 22, 23, 30, 44, 47]. Adenosine and its analogues can also promote angiogenesis in a number of models through upregulation of pro-angiogenic cytokines. Adenosine can induce expression of VEGF in a number of cell types including aortic smooth muscle cells [61], coronary artery smooth muscle cells [32] and myocardial myoblasts [33]. Adenosine has also been shown to increase expression of VEGF in chick embryo heart and skeletal muscle and also in humans (reviewed in [2]). NECA, an adenosine analog and activator of adenosine A_{2A} and A_{2B} receptor increased levels of VEGF in human retinal endothelial cells [31], umbilical vein endothelial cells [26], skin microvascular endothelial cells [25] mast cells [24], macrophages [35, 43] and retinal pericytes [71]. These effects are predominantly mediated through the A_{2B} receptor and in some cases through the A_{2A} receptor as well. Additionally, activation of A_3 receptor also increased expression of VEGF through a HIF-1α–dependent mechanism [48, 49]. Other than VEGF, adenosine and its analogues have also been shown to increase expression of insulin-like growth factor-1 (IGF-1), basic fibroblast growth factor (bFGF), interleukin-8 (IL-8) and angiopoietin-2 (Ang-2) [31, 63, 64]. Thus multiple angiogenic pathways are regulated by adenosine and its analogues through activation of its receptors.

In human dermal microvascular endothelial cells, activation of A_{2B}, but not A_{2A}, receptor promotes angiogenesis [25]. By contrast, activation of A_{2A}, but not A_{2B}, receptor promotes angiogenesis in human umbilical vein endothelial cells and human lung microvascular endothelial cells (HLMVEC) [3, 16]. Therefore differential expression and function of adenosine receptor subtypes could contribute to functional heterogeneity in different types of endothelia.

Our understanding on the role of adenosine and its receptors during embryonic development is still not clear. Adenosine A_1, A_{2A}, A_{2B} and A_3 receptor knockouts are not embryonically lethal and have normal survival. Among the adenosine receptors the adenosine A_{2B} receptor is present largely in the vasculature. While both A_{2A} and A_{2B} knockout mice exhibit increased proinflammatory responses, the A_{2B} knockout mouse also exhibits increased leukocyte adhesion to the vasculature. Interestingly, these receptors are regulated differently in mouse and humans. Unlike in mice where adenosine A_{2A} receptor is not regulated by hypoxia and HIFs, this receptor is regulated by hypoxia in rat derived PC12 cells and human pulmonary endothelial cells [3]. It appears that these receptors may play a more important role in postnatal vascular growth and vascular diseases during adult life. Additionally, species-specific differences can also be important and relevant in understanding vascular growth processes in disease and development.

7.7 Conclusions

In summary, hypoxia promotes vascular growth and proliferation, in part through hypoxia-inducible transcription factors, adenosine and adenosine receptors. Furthermore, part of the HIF-mediated vascular growth is mediated through the adenosine signaling pathway where HIFs can modulate adenosine related receptors and enzymes at multiple locations. These can have important implications in tumor growth and other diseases involving vascular growth. Understanding of these pathways will help us target them for therapeutic benefits in diseases.

References

1. Adair TH. (2004) An emerging role for adenosine in angiogenesis. Hypertension 44:618–20.
2. Adair TH. (2005) Growth regulation of the vascular system: an emerging role for adenosine. Am J Physiol Regul Integr Comp Physiol 289:R283–96.
3. Ahmad A, Ahmad S, Glover L, Miller S, Shannon JM, Guo X, Franklin WA, Bridges JP, Schaack JB, Colgan SP, White CW. (2009) Adenosine A(2A) receptor is a unique angiogenic target of HIF-2α in pulmonary endothelial cells. Proc Natl Acad Sci USA 106: 10684–9.
4. Ahmad S, Ahmad A, White CW. (2006) Purinergic signaling and kinase activation for survival in pulmonary oxidative stress and disease. Free Radic Biol Med 41:29–40.
5. Aird WC. (2008) Endothelium in health and disease. Pharmacol Rep 60:139–43.
6. Asikainen TM, Ahmad A, Schneider BK, Ho WB, Arend M, Brenner M, Gunzler V, White CW. (2005) Stimulation of HIF-1alpha, HIF-2alpha, and VEGF by prolyl 4-hydroxylase inhibition in human lung endothelial and epithelial cells. Free Radic Biol Med 38:1002–13.
7. Banerjee SK, Young HW, Barczak A, Erle DJ, Blackburn MR. (2004) Abnormal alveolar development associated with elevated adenine nucleosides. Am J Respir Cell Mol Biol 30:38–50.

8. Bruns RF. (1980) Adenosine receptor activation in human fibroblasts: nucleoside agonists and antagonists. Can J Physiol Pharmacol 58:673–91.
9. Carmeliet P, Ferreira V, Breier G, Pollefeyt S, Kieckens L, Gertsenstein M, Fahrig M, Vandenhoeck A, Harpal K, Eberhardt C, Declercq C, Pawling J, Moons L, Collen D, Risau W, Nagy A. (1996) Abnormal blood vessel development and lethality in embryos lacking a single VEGF allele. Nature 380:435–9.
10. Cassada DC, Tribble CG, Laubach VE, Nguyen BN, Rieger JM, Linden J, Kaza AK, Long SM, Kron IL, Kern JA. (2001) An adenosine A2A agonist, ATL-146e, reduces paralysis and apoptosis during rabbit spinal cord reperfusion. J Vasc Surg 34:482–8.
11. Chen H, Yang D, Carroll SH, Eltzschig HK, Ravid K. (2009) Activation of the macrophage A2b adenosine receptor regulates tumor necrosis factor-alpha levels following vascular injury. Exp Hematol 37:533–8.
12. Chi JT, Chang HY, Haraldsen G, Jahnsen FL, Troyanskaya OG, Chang DS, Wang Z, Rockson SG, van de Rijn M, Botstein D, Brown PO. (2003) Endothelial cell diversity revealed by global expression profiling. Proc Natl Acad Sci USA 100:10623–8.
13. Clifford SC, Cockman ME, Smallwood AC, Mole DR, Woodward ER, Maxwell PH, Ratcliffe PJ, Maher ER. (2001) Contrasting effects on HIF-1alpha regulation by disease-causing pVHL mutations correlate with patterns of tumourigenesis in von Hippel-Lindau disease. Hum Mol Genet 10:1029–38.
14. Compernolle V, Brusselmans K, Acker T, Hoet P, Tjwa M, Beck H, Plaisance S, Dor Y, Keshet E, Lupu F, Nemery B, Dewerchin M, Van Veldhoven P, Plate K, Moons L, Collen D, Carmeliet P. (2002) Loss of HIF-2alpha and inhibition of VEGF impair fetal lung maturation, whereas treatment with VEGF prevents fatal respiratory distress in premature mice. Nat Med 8:702–10.
15. Daly JW, Padgett WL. (1992) Agonist activity of 2- and 5′-substituted adenosine analogs and their N6-cycloalkyl derivatives at A1- and A2-adenosine receptors coupled to adenylate cyclase. Biochem Pharmacol 43:1089–93.
16. Desai A, Victor-Vega C, Gadangi S, Montesinos MC, Chu CC, Cronstein BN. (2005) Adenosine A2A receptor stimulation increases angiogenesis by down-regulating production of the antiangiogenic matrix protein thrombospondin 1. Mol Pharmacol 67:1406–13.
17. Duan LJ, Zhang-Benoit Y, Fong GH. (2005) Endothelium-intrinsic requirement for Hif-2alpha during vascular development. Circulation 111:2227–32.
18. Dubey RK, Gillespie DG, Jackson EK. (2002) A(2B) adenosine receptors stimulate growth of porcine and rat arterial endothelial cells. Hypertension 39:530–5.
19. Eckle T, Grenz A, Laucher S, Eltzschig HK. (2008) A2B adenosine receptor signaling attenuates acute lung injury by enhancing alveolar fluid clearance in mice. J Clin Invest 118:3301–15.
20. Eltzschig HK, Kohler D, Eckle T, Kong T, Robson SC, Colgan SP. (2009) Central role of Sp1-regulated CD39 in hypoxia/ischemia protection. Blood 113:224–32.
21. Eltzschig HK, Abdulla P, Hoffman E, Hamilton KE, Daniels D, Schonfeld C, Loffler M, Reyes G, Duszenko M, Karhausen J, Robinson A, Westerman KA, Coe IR, Colgan SP. (2005) HIF-1-dependent repression of equilibrative nucleoside transporter (ENT) in hypoxia. J Exp Med 202:1493–505.
22. Ethier MF, Dobson JG, Jr. (1997) Adenosine stimulation of DNA synthesis in human endothelial cells. Am J Physiol 272:H1470–9.
23. Ethier MF, Chander V, Dobson JG, Jr. (1993) Adenosine stimulates proliferation of human endothelial cells in culture. Am J Physiol 265:H131–8.
24. Feoktistov I, Ryzhov S, Goldstein AE, Biaggioni I. (2003) Mast cell-mediated stimulation of angiogenesis: cooperative interaction between A2B and A3 adenosine receptors. Circ Res 92:485–92.
25. Feoktistov I, Goldstein AE, Ryzhov S, Zeng D, Belardinelli L, Voyno-Yasenetskaya T, Biaggioni I. (2002) Differential expression of adenosine receptors in human endothelial cells: role of A2B receptors in angiogenic factor regulation. Circ Res 90:531–8.

26. Feoktistov I, Ryzhov S, Zhong H, Goldstein AE, Matafonov A, Zeng D, Biaggioni I. (2004) Hypoxia modulates adenosine receptors in human endothelial and smooth muscle cells toward an A2B angiogenic phenotype. Hypertension 44:649–54.
27. Ferrara N, Carver-Moore K, Chen H, Dowd M, Lu L, O'Shea KS, Powell-Braxton L, Hillan KJ, Moore MW. (1996) Heterozygous embryonic lethality induced by targeted inactivation of the VEGF gene. Nature 380:439–42.
28. Fong GH, Rossant J, Gertsenstein M, Breitman ML. (1995) Role of the Flt-1 receptor tyrosine kinase in regulating the assembly of vascular endothelium. Nature 376:66–70.
29. Gorlach A. (2005) Control of adenosine transport by hypoxia. Circ Res 97:1–3.
30. Grant MB, Davis MI, Caballero S, Feoktistov I, Biaggioni I, Belardinelli L. (2001) Proliferation, migration, and ERK activation in human retinal endothelial cells through A(2B) adenosine receptor stimulation. Invest Ophthalmol Vis Sci 42:2068–73.
31. Grant MB, Tarnuzzer RW, Caballero S, Ozeck MJ, Davis MI, Spoerri PE, Feoktistov I, Biaggioni I, Shryock JC, Belardinelli L. (1999) Adenosine receptor activation induces vascular endothelial growth factor in human retinal endothelial cells. Circ Res 85:699–706.
32. Gu JW, Brady AL, Anand V, Moore MC, Kelly WC, Adair TH. (1999) Adenosine upregulates VEGF expression in cultured myocardial vascular smooth muscle cells. Am J Physiol 277:H595–602.
33. Gu JW, Ito BR, Sartin A, Frascogna N, Moore M, Adair TH. (2000) Inhibition of adenosine kinase induces expression of VEGF mRNA and protein in myocardial myoblasts. Am J Physiol Heart Circ Physiol 279:H2116–23.
34. Ham J, Rees DA. (2008) The adenosine a2b receptor: its role in inflammation. Endocr Metab Immune Disord Drug Targets 8:244–54.
35. Hashimoto E, Kage K, Ogita T, Nakaoka T, Matsuoka R, Kira Y. (1994) Adenosine as an endogenous mediator of hypoxia for induction of vascular endothelial growth factor mRNA in U-937 cells. Biochem Biophys Res Commun 204:318–24.
36. Huang NK. (2003) Adenosine A2A receptors regulate oxidative stress formation in rat pheochromocytoma PC12 cells during serum deprivation. Neurosci Lett 350:127–31.
37. Iyer NV, Kotch LE, Agani F, Leung SW, Laughner E, Wenger RH, Gassmann M, Gearhart JD, Lawler AM, Yu AY, Semenza GL. (1998) Cellular and developmental control of O_2 homeostasis by hypoxia inducible factor 1 alpha. Genes Dev 12:149–62.
38. King J, Hamil T, Creighton J, Wu S, Bhat P, McDonald F, Stevens T. (2004) Structural and functional characteristics of lung macro- and microvascular endothelial cell phenotypes. Microvasc Res 67:139–51.
39. Koditz J, Nesper J, Wottawa M, Stiehl DP, Camenisch G, Franke C, Myllyharju J, Wenger RH, Katschinski DM. (2007) Oxygen-dependent ATF-4 stability is mediated by the PHD3 oxygen sensor. Blood 110:3610–7.
40. Kong T, Westerman KA, Faigle M, Eltzschig HK, Colgan SP. (2006) HIF-dependent induction of adenosine A2B receptor in hypoxia. Faseb J 20:2242–50.
41. Langenkamp E, Molema G. (2009) Microvascular endothelial cell heterogeneity: general concepts and pharmacological consequences for anti-angiogenic therapy of cancer. Cell Tissue Res 335:205–22.
42. Lasley RD, Jahania MS, Mentzer RM, Jr.. (2001) Beneficial effects of adenosine A(2a) agonist CGS-21680 in infarcted and stunned porcine myocardium. Am J Physiol Heart Circ Physiol 280:H1660–6.
43. Leibovich SJ, Chen JF, Pinhal-Enfield G, Belem PC, Elson G, Rosania A, Ramanathan M, Montesinos C, Jacobson M, Schwarzschild MA, Fink JS, Cronstein B. (2002) Synergistic up-regulation of vascular endothelial growth factor expression in murine macrophages by adenosine A(2A) receptor agonists and endotoxin. Am J Pathol 160:2231–44.
44. Lutty GA, Mathews MK, Merges C, McLeod DS. (1998) Adenosine stimulates canine retinal microvascular endothelial cell migration and tube formation. Curr Eye Res 17: 594–607.
45. Martin PL. (1999) Adenosine agonists for the prevention of restenosis? IDrugs 2:44–51.

46. McPherson JA, Barringhaus KG, Bishop GG, Sanders JM, Rieger JM, Hesselbacher SE, Gimple LW, Powers ER, Macdonald T, Sullivan G, Linden J, Sarembock IJ. (2001) Adenosine A(2A) receptor stimulation reduces inflammation and neointimal growth in a murine carotid ligation model. Arterioscler Thromb Vasc Biol 21:791–6.
47. Meininger CJ, Schelling ME, Granger HJ. (1988) Adenosine and hypoxia stimulate proliferation and migration of endothelial cells. Am J Physiol 255:H554–62.
48. Merighi S, Benini A, Mirandola P, Gessi S, Varani K, Leung E, Maclennan S, Borea PA. (2006) Adenosine modulates vascular endothelial growth factor expression via hypoxia-inducible factor-1 in human glioblastoma cells. Biochem Pharmacol 72:19–31.
49. Merighi S, Benini A, Mirandola P, Gessi S, Varani K, Leung E, MacLennan S, Baraldi PG, Borea PA. (2005) A3 adenosine receptors modulate hypoxia-inducible factor-1alpha expression in human A375 melanoma cells. Neoplasia 7:894–903.
50. Morita M, Ohneda O, Yamashita T, Takahashi S, Suzuki N, Nakajima O, Kawauchi S, Ema M, Shibahara S, Udono T, Tomita K, Tamai M, Sogawa K, Yamamoto M, Fujii-Kuriyama Y. (2003) HLF/HIF-2alpha is a key factor in retinopathy of prematurity in association with erythropoietin. EMBO J 22:1134–46.
51. Morote-Garcia JC, Rosenberger P, Kuhlicke J, Eltzschig HK. (2008) HIF-1-dependent repression of adenosine kinase attenuates hypoxia-induced vascular leak. Blood 111:5571–80.
52. Muller CE, Scior T. (1993) Adenosine receptors and their modulators. Pharm Acta Helv 68:77–111.
53. Nemeth ZH, Lutz CS, Csoka B, Deitch EA, Leibovich SJ, Gause WC, Tone M, Pacher P, Vizi ES, Hasko G. (2005) Adenosine augments IL-10 production by macrophages through an A2B receptor-mediated posttranscriptional mechanism. J Immunol 175:8260–70.
54. Ohta A, Sitkovsky M. (2001) Role of G-protein-coupled adenosine receptors in downregulation of inflammation and protection from tissue damage. Nature 414:916–20.
55. Okusa MD, Linden J, Huang L, Rieger JM, Macdonald TL, Huynh LP. (2000) A(2A) adenosine receptor-mediated inhibition of renal injury and neutrophil adhesion. Am J Physiol Renal Physiol 279:F809–18.
56. Olanrewaju HA, Qin W, Feoktistov I, Scemama JL, Mustafa SJ. (2000) Adenosine A(2A) and A(2B) receptors in cultured human and porcine coronary artery endothelial cells. Am J Physiol Heart Circ Physiol 279:H650–56.
57. Palmer TM, Stiles GL. (1997) Identification of an A2a adenosine receptor domain specifically responsible for mediating short-term desensitization. Biochemistry 36:832–8.
58. Pan J, Mestas J, Burdick MD, Phillips RJ, Thomas GV, Reckamp K, Belperio JA, Strieter RM. (2006) Stromal derived factor-1. (SDF-1/CXCL12) and CXCR4 in renal cell carcinoma metastasis. Mol Cancer 5:56.
59. Peirce SM, Skalak TC, Rieger JM, Macdonald TL, Linden J. (2001) Selective A(2A) adenosine receptor activation reduces skin pressure ulcer formation and inflammation. Am J Physiol Heart Circ Physiol 281:H67–74.
60. Peng J, Zhang L, Drysdale L, Fong GH. (2000) The transcription factor EPAS-1/hypoxia-inducible factor 2alpha plays an important role in vascular remodeling. Proc Natl Acad Sci USA 97:8386–91.
61. Pueyo ME, Chen Y, D'Angelo G, Michel JB. (1998) Regulation of vascular endothelial growth factor expression by cAMP in rat aortic smooth muscle cells. Exp Cell Res 238:354–8.
62. Ross SD, Tribble CG, Linden J, Gangemi JJ, Lanpher BC, Wang AY, Kron IL. (1999) Selective adenosine-A2A activation reduces lung reperfusion injury following transplantation. J Heart Lung Transplant 18:994–1002.
63. Ryzhov S, McCaleb JL, Goldstein AE, Biaggioni I, Feoktistov I. (2007) Role of adenosine receptors in the regulation of angiogenic factors and neovascularization in hypoxia. J Pharmacol Exp Ther 320:565–72.
64. Ryzhov S, Zaynagetdinov R, Goldstein AE, Matafonov A, Biaggioni I, Feoktistov I. (2009) Differential role of the carboxy-terminus of the A(2B) adenosine receptor in stimulation of adenylate cyclase, phospholipase Cbeta, and interleukin-8. Purinergic Signal 5:289–98.

65. Schultz K, Murthy V, Tatro JB, Beasley D. (2009) Prolyl hydroxylase 2 deficiency limits proliferation of vascular smooth muscle cells by hypoxia-inducible factor-1{alpha}-dependent mechanisms. Am J Physiol Lung Cell Mol Physiol 296:L921–7.
66. Scortegagna M, Morris MA, Oktay Y, Bennett M, Garcia JA. (2003) The HIF family member EPAS1/HIF-2alpha is required for normal hematopoiesis in mice. Blood 102:1634–40.
67. Shalaby F, Rossant J, Yamaguchi TP, Gertsenstein M, Wu XF, Breitman ML, Schuh AC. (1995) Failure of blood-island formation and vasculogenesis in Flk-1-deficient mice. Nature 376:62–66.
68. Stevens T, Phan S, Frid MG, Alvarez D, Herzog E, Stenmark KR. (2008) Lung vascular cell heterogeneity:endothelium, smooth muscle, and fibroblasts. Proc Am Thorac Soc 5:783–91.
69. Sun CN, Cheng HC, Chou JL, Lee SY, Lin YW, Lai HL, Chen HM, Chern Y. (2006) Rescue of p53 blockage by the A(2A) adenosine receptor via a novel interacting protein, translin-associated protein X. Mol Pharmacol 70:454–66.
70. Synnestvedt K, Furuta GT, Comerford KM, Louis N, Karhausen J, Eltzschig HK, Hansen KR, Thompson LF, Colgan SP. (2002) Ecto-5′-nucleotidase. (CD73) regulation by hypoxia-inducible factor-1 mediates permeability changes in intestinal epithelia. J Clin Invest 110:993–1002.
71. Takagi H, King GL, Robinson GS, Ferrara N, Aiello LP. (1996) Adenosine mediates hypoxic induction of vascular endothelial growth factor in retinal pericytes and endothelial cells. Invest Ophthalmol Vis Sci 37:2165–76.
72. Tian H, Hammer RE, Matsumoto AM, Russell DW, McKnight SL. (1998) The hypoxia-responsive transcription factor EPAS1 is essential for catecholamine homeostasis and protection against heart failure during embryonic development. Genes Dev 12:3320–4.
73. Trincavelli ML, Falleni A, Chelli B, Tuscano D, Costa B, Gremigni V, Lucacchini A, Martini C. (2003) A(2A) adenosine receptor ligands and proinflammatory cytokines induce PC 12 cell death through apoptosis. Biochem Pharmacol 66:1953–62.
74. Yuan Y, Hilliard G, Ferguson T, Millhorn DE. (2003) Cobalt inhibits the interaction between hypoxia-inducible factor-alpha and von Hippel-Lindau protein by direct binding to hypoxia-inducible factor-alpha. J Biol Chem 278:15911–6.

Chapter 8
Regulated Extracellular Nucleotide Metabolism and Function at the Mucosa

Christopher F. MacManus, Holger K. Eltzschig, and Sean P. Colgan

Abstract Within mucosal tissues, regulation of nucleotide metabolism may be directed at various physiological/pathophysiological levels by myriad cell types. Adenosine (Ado) can regulate a number of processes at the mucosa, e.g. dampening of the inflammatory response and resultant tissue protection. Ado receptors are widely expressed on a variety of organs and cell types and have been demonstrated to be central to the modulation of inflammation. In many instances, local Ado levels are dictated by extracellular sources of adenine nucleotides. Activated neutrophils and platelet degranulation are two such important sources of ATP and ADP. Sequential phosphohydrolysis of adenine nucleotides by CD39 and CD73 results in the conversion of ATP and ADP to AMP, and the subsequent conversion of AMP to Ado, whereby signaling through Ado receptors can elicit effects such as increase in endothelial barrier function, increase in epithelial electrogenic Cl^- secretion and water transport. Further regulation of Ado signaling has been observed under hypoxic conditions, due to hypoxia-inducible factor-1 (HIF-1)-regulated expression of A2BR and CD73, whereby additional SP-1 transcription factor activity has been identified as being required for hypoxic induction of CD39 expression in endothelial cells. Here we describe the different levels of regulation of adenine nucleotide metabolism and the resultant functional cell-cell signaling at the mucosa.

8.1 Nucleotide Metabolism at the Mucosa

The mucosal surfaces (tissues or organs lined by an epithelium) represent a complex environment wherein a diverse array of cell types may interact, including endothelia, epithelia, hematological cells and stromal cells. It is therefore important to understand the complexity of signaling networks that exist to orchestrate physiological and pathophysiological events within these tissues. Nucleotides and

---

C.F. MacManus (✉)
Mucosal Inflammation Program, 12700 E. 19th Ave, Aurora, CO 80045, USA
e-mail: christopher.macmanus@ucdenver.edu

The authors declare no financial interests in any of the work submitted here.

E. Gerasimovskaya, E. Kaczmarek (eds.), *Extracellular ATP and Adenosine as Regulators of Endothelial Cell Function*, DOI 10.1007/978-90-481-3435-9_8,

their metabolites have been long identified as playing a crucial role in cell-cell signaling in a number of different tissues, ranging from early observations in the context of neurotransmission [53] to more recent observations in mucosal tissues [20, 40, 49]. Circulating nucleotides and nucleosides may be metabolized under the control of a variety of cell types and the end products of these metabolic processes may signal in an autocrine or paracrine manner. Nucleotide regulation takes place via a cascade of hydrolyzing enzymes that serve to sequentially cleave phosphate groups from adenine nucleotides (adenosine triphosphate [ATP], adenosine diphosphate [ADP] and adenosine monophosphate [AMP]), ultimately producing nucleoside molecules such as adenosine. The nucleotide intermediates themselves may signal through a subset of purinergic receptors of the P2 family (further subdivided into P2Y and P2X receptors). The end product of these hydrolysis reactions, adenosine may then elicit its functions by binding one of its cognate G-protein coupled receptor (GPCR), namely the A1, A2A, A2B and A3 receptors. It is at this level that adenine nucleotides and adenosine may act to modulate cellular processes such as enhancing barrier function, inducing chloride secretion and dampening of the immune response, to name but a few.

8.2 Role of NTPDases (CD39 and CD39-Like)/ 5′-Ectonucleotidase (CD73) Enzymes

One of the levels at which extracellular nucleotides may be regulated is under the control of a number of cell-surface molecules called ectonucleotidases, whose enzymatic activity is to cleave phosphate groups from circulating nucleotides with varying degrees of specificity for their substrates. CD39 is one such ectonucleotidase initially observed as an activation marker in paracortical lymphocytes, macrophages, and dendritic cells resident within lymphoid tissue. Cloning of CD39 revealed the presence of apyrase conserved regions (ACR) and striking sequence homology with yeast guanosine diphosphatase, an enzyme involved in catalyzing the removal of a phosphate from GDP after sugar transfer within the Golgi apparatus [43]. Further evidence for the ATP-hydrolyzing capabilities of CD39 came with the cloning and sequencing of a potato tuber ATP apyrase, which shared sequence homology with both mouse and human CD39 [24]. The ecto-pyrase activity of CD39 was demonstrated in EBV transformed B-cells and CD39 overexpressing COS-7 cells, assessed by the ability of CD39 to liberate free $[^{32}P]P_i$ from $[\gamma^{32}P]$ATP [70].

Further identification of a number of CD39-like nucleoside triphosphate diphosphohydrolases (NTPDases) have since revealed a family of 8 related proteins, denoted NTPdase1-8 (NTPDase1 representing CD39 under this nomenclature). NTPDase1,2,3 and 8 are transmembrane proteins with 5 ACRs situated on the extracellular region, conferring nucleotidase activity to the enzyme and allowing for hydrolysis of extracellular nucleotides [60]. Differential regulation of hydrolysis is evident between each NTPDase subtype, with NTPDase1 having similar hydrolyzing activity on both ATP and ADP, and NTPDases3 & 8 preferentially hydrolyzing

ATP over ADP. NTPDase2 has been identified for acting almost exclusively to hydrolyze ATP [51, 62]. Together, these NTPDases act in a concerted manner to regulate the production of extracellular adenosine monophosphate (AMP).

The final step in generation of extracellular adenosine is conferred by the glycosyl phosphatidylinositol (GPI)-anchored membrane protein, ecto-5′ nucleotidase (CD73). CD73 is the predominant source for accumulation of extracellular adenosine from released adenine nucleotides [75]. CD73 metabolizes AMP to adenosine, which is then either free to act as a ligand to one of four adenosine receptors (A1, A2A, A2B, A3), or transported into the cell by dipyridamole-sensetive channels and degraded by the purine salvage pathway. As with the NTPDases, CD73 is a widely distributed cell surface enzyme, whose expression has been shown across a variety of tissues. It has been demonstrated that the AMP hydrolyzing activity of CD73 varies widely from tissue to tissue, with the rank order of tissue activity as follows: colon > kidney = brain > liver > lung > heart >> muscle [69]. As high levels of activity are reported at a number of mucosal sites, it is important to understand the physiological role played at this terminal step of the nucleotide hydrolysis pathway.

8.3 Physiological and Pathophysiologic Influences of Nucleotide Metabolites

Mucosal tissues such as the kidney, lung and intestine are lined with an epithelial layer that serves as both a barrier to the external environment and a conduit for nutrients, ions and gas exchange. During mucosal inflammation, the major sources of extracellular nucleotides are considered to be platelet- and leukocyte-derived [44]. Of particular interest is the regulated release of adenine nucleotides by polymorphonuclear cells (PMN, neutrophils) as they cross the endothelium and epithelium to a site of infection and/or inflammation. A number of years ago, it was appreciated that when interacting with endothelia and epithelia, PMN actively release adenine nucleotides. Original studies by Madara et al. examining biological properties of soluble mediators derived from activated inflammatory cells (e.g. neutrophils and eosinophils) identified a small, protease-resistant fraction termed neutrophil-derived secretagogue (NDS), which when incubated on epithelia, activated electrogenic chloride secretion and fluid transport (responsible for the hydration of mucosal surfaces). Subsequent biophysical analysis of NDS identified this molecule to be AMP [42]. With no known AMP receptor, studies turned toward defining potential metabolic pathways for adenosine generation. Biochemical and pharmacologic studies demonstrated the polarized expression of CD73 on the apical surface of cultured and primary intestinal epithelial cells [63]. Further biochemical and morphological studies revealed that CD73 exists in both a GPI-linked surface fraction as well as in a sub-apical caveolin-rich domain within the epithelium. Such expression patterns have subsequently been shown in a variety of mucosal epithelial cell types.

Subsequent studies have addressed the role of adenine nucleotides in the regulation of tissue barrier function. Indeed, as PMN cross both epithelial and endothelial barriers, they must effectively deform a tightly sealed cellular barrier. Since the active paracellular movement of PMN can cause disruptions to epithelial/endothelial cell layers, intrinsic pathways must exist to "reseal" the disrupted cells in order to prevent further tissue damage. Original studies revealed that PMN migrating across an endothelial barrier secreted 5′ AMP which resulted in increased barrier function as demonstrated by a decrease in paracellular permeability [39]. Furthermore, blockade of CD73 activity using a neutralizing antibody (1E9), or pharmacological blockade of CD73 using α,β-methylene ADP (APCP) resulted in an impaired ability for the migrating neutrophils to reseal both epithelial and endothelial barriers [39]. This observation strongly suggested the necessity for extracellular nucleotide metabolism in the regulation of cell barrier function. In addition to the regulation of barrier function from neutrophil-derived AMP, autocrine ATP signaling has been shown to induce leukocyte sequestration in ischemic cerebral tissue through $P2X_7$-mediated upregulation of $\alpha_M\beta_2$ integrin. Interestingly, this phenomenon of leukosequestration is inhibited by CD39 directed catalysis of ATP [28]. Once neutrophils have tethered to the endothelium, they respond to inflammatory damage by flowing a chemical gradient set up by a number of chemoattractants secreted at sites of tissue injury or inflammation. One important mediator of neutrophil chemotaxis has been identified as interleukin-8 (IL-8). ATP has been shown to stimulate production of IL-8 by astrocytes and eosinophils [29, 32]. Furthermore, ATP, through P2Y activation has been demonstrated to synergistically induce IL-8 mediated chemotaxis in human neutrophils in vitro [37]. Recent studies have identified the importance of ATP metabolism in regulating the rate of IL-8, C5a and formyl-methionyl-leucyl-phenylalanine (fMLP)-mediated PMN chemotaxis. Corriden et al demonstrated the ability for NTPDase1 at the leading edge of migrating neutrophils to hydrolyze PMN-derived ATP to AMP, where further metabolism to adenosine, potentially by alkaline phosphatase controls the rate of PMN migration [9]. Additionally, PMN from NTPDase knockout mice exhibited ablation of migration towards IP injected murine fMLP receptor ligand (W-peptide) and a reduced migration rate in vitro towards an fMLP gradient [9].

More recent studies have revealed that the active fraction originally defined as PMN-derived 5′ AMP may actually be ATP. Indeed, studies directed at understanding vascular barrier function during conditions of inflammation or hypoxia identified the existence of a soluble fraction derived from activated PMN which amplified the resealing of vascular barrier function following subjection of cultured endothelial cells to periods of hypoxia. A screen of HPLC purified fractions identified this activity as PMN-derived ATP [15]. Like 5′-AMP, this activity was functionally linked to CD73 and to Ado receptors. This was puzzling, since CD73 was not known to utilize ATP as a substrate. Subsequent studies directed at understanding mechanisms of ATP release by activated PMN considered several potential mechanisms, including exocytosis of ATP containing vesicles, transport via connexin hemichannels, transport through nucleoside transporters, or direct transport through ATP-binding cassette (ABC) proteins [46]. Initially, it was determined that ATP does not localize with known granule markers in PMN. While isolated granules from resting PMNs contained greater than 95% of proteolytic enzyme activity markers, ATP levels within the isolated granules were nearly undetectable. Moreover, PMN cytosolic fractions contained ATP concentrations that were higher than 5 mM, thereby suggesting that activation-dependent ATP release likely occurs independent of classical PMN degranulation. Based on these findings, a pharmacologic approach was employed to examine potential mechanisms. Brefeldin A (BFA), a general vesicular secretion inhibitor did not influence activated PMN ATP secretion. Likewise, neither the nucleoside transport inhibitor dipyridamole nor the general ABC transport inhibitor verapamil significantly influenced PMN ATP secretion [14].

Based on previous reports suggesting that connexin hemichannels may serve as ATP release channels [23] and the observation that PMN express surface connexins [74], the non-specific gap junction inhibitor 18-α-glycyrrhetinic acid (18αGA) was examined. These studies revealed that 18αGA inhibited ATP release in a concentration-dependent manner. Likewise, connexin-mimetic peptides specifically directed against Cx43 [23], but not Cx40, significantly blocked ATP liberation from activated PMN. Cx43 molecules can assemble as hexadimers (so called "connexons") that form junctional connections between different cell types. In addition to their role as gap-junction proteins, recent studies indicate that Cx43 connexons are also active in single plasma membranes and can function in intercellular signaling as ATP release channels [23]. The conductance and permeability of such Cx43 hemichannels is regulated by modification of their cytoplasm domain, with phosphorylation of Ser-368 causing a conformational change resulting in decreased connexon permeability [5]. Studies addressing Cx43 Ser-368 phosphorylation in intact PMN showed prominent phosphorylation in resting PMN and protein phosphatase 2A-dependent dephosphorylation within minutes of PMN activation. As a proof of principle, PMN were isolated from tamoxifen-inducible Cx43 conditionally deleted mice [12] and examined for ATP release. These studies revealed that ATP release correlated with the degree of Cx43 expression. In total, these studies provide strong evidence that ATP release occurs through a conformational opening of membrane Cx43 hemichannels in response to PMN activation [14].

In addition to neutrophil-derived extracellular nucleotides, there is increasing evidence to suggest that epithelial cells hydrolyze platelet-derived ATP as platelets migrate alongside neutrophils across an epithelial barrier. In a recent study by Weissmuller et al., it was revealed that platelets "piggyback" on the surface of PMN during transmigration in vitro and in biopsies from patients with inflammatory bowel disease [71]. Together, these cells represent a rich source of ATP at the luminal aspect of the epithelium. Indeed, translocated platelets were observed to release large quantities of ATP, which is metabolized to adenosine via a 2-step enzymatic reaction mediated by ectonucleotideases, including CD73 and NTPDases expressed on the luminal membrane of the epithelium. Measurement of electrogenic Cl secretion in human epithelial cells demonstrated the ability for platelet/PMN derived ATP to induce copious amounts of fluid transport, which was blocked by the pan-NTPDase inhibitor POM-1. Futhermore, blockade of CD73 using APCP also abrogated ATP-mediated Cl^- secretion, as did inhibition of the adenosine A2B receptor using a specific antagonist, MRS1754. Further studies using a $Cd73^{-/-}$ knockout mouse model, it was demonstrated that ATP induced water transport (along an osmotic gradient driven by the active secretion of Cl^- ions) was dependent upon the 5′ ectonucleotidase [71]. Thus, it was concluded that luminal ATP-derived adenosine represents a tissue-adaptive response during inflammation which may represent a mechanism for flushing of potentially harmful pathogens at from the surface of the mucosa (Fig. 8.1).

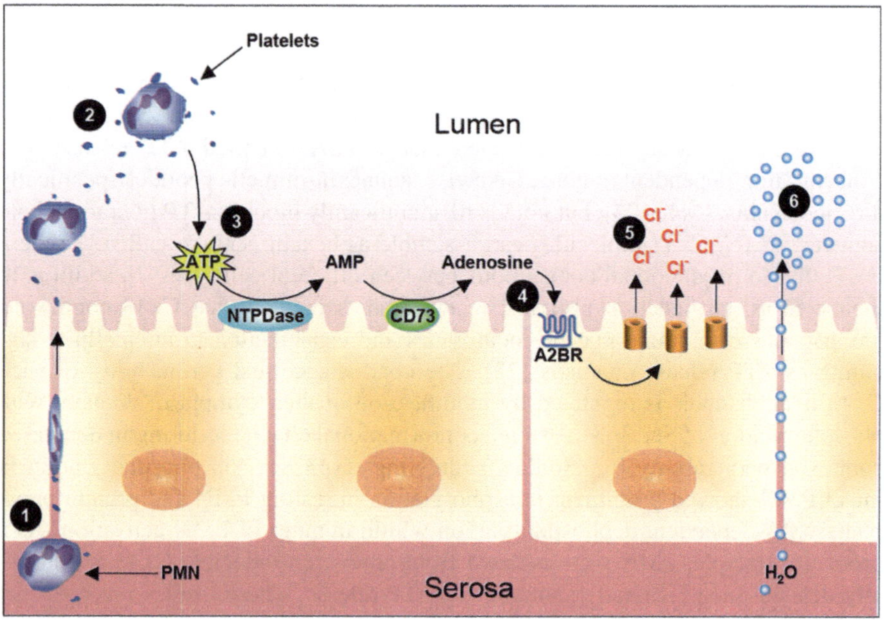

Fig. 8.1 Model of facilitated platelet translocation and activation of epithelial electrogenic Cl secretion during PMN transmigration. During active inflammation, platelets are caught in the flow of PMN transmigration, resulting in platelet translocation across the apical side of mucosal epithelial cells (1). PMN and platelet-derived ATP (2) is selectively metabolized to adenosine by a two-step enzymatic reaction involving ecto-apyrase and ecto-nucleotidase (CD73) (3). Adenosine binding to apical adenosine A2B receptors (4) results in activation of electrogenic Cl secretion and the paracellular movement of water (6). Such platelet/PMN – epithelial crosstalk pathway may serve as a defensive response by which mucosal surfaces are flushed from bacteria and bacterial products under inflammatory conditions. Figure adapted from Weissmuller et al. [71]

8.4 Transcriptional Regulation of CD39, CD73 and Adenosine Receptors

It is well appreciated that under levels of low oxygen tension (pO_2), extracellular adenosine levels are increased [17, 59]. It is in this hypoxic microenvironment that a wide range of genes is upregulated under the control of the transcription factor hypoxia-inducible factor (HIF) [55, 56]. HIF-mediated signaling represents a relatively rapid adaptive mechanism to low oxygen conditions within the ischemic or inflamed tissue [58, 68]. Classical HIF-regulated pathways have been identified in developmental pathways where gene activation may be required as cell mass and number increase in the developing organism [57]. This may be coupled with an increased requirement for regulation of glycolytic genes to meet the energy requirements of the cells, and an increase in oxygen delivery in tissues through erythropoietin (EPO) induction and an increase in vascular networks through the induction of vascular endothelial growth factor (VEGF). HIF activation may contribute to inappropriate adaptation to hypoxic areas such as in the core of solid tumors whereby HIF signaling permits evasion from apoptosis, increased cell proliferation and increased angiogenesis [57]. Conversely, HIF-1 activation may serve to resolve potentially injurious levels of either physiological tissue hypoxia, or the high oxygen demand that is associated with pathologies such as inflammation [33, 67, 68]. Accordingly, it has been demonstrated that adenosine signaling is protective in inflammatory diseases such as glomerulonephritis [21] colitis [20, 41] and during ischaemic preconditioning models of intestinal, hepatic and lung injury [13, 25, 26]. This protective phenotype afforded by increased levels of adenosine is achieved through an upregulation of the ecto-nucleotideases and their subsequent activity on extracellular ATP, ADP and AMP. Both CD39 and CD73 have been shown to be upregulated in intestinal epithelial cells incubated under hypoxic conditions (pO_2 20 torr) and the functional activity of both ecto-enzymes was shown to increase under hypoxic conditions as demonstrated by increased conversion of etheno-ATP (E-ATP) to etheno-AMP (E-AMP), and increased E-AMP conversion to etheno-adenosine (E-Ado) for CD39 and CD73 activity, respectively. Furthermore, mice subjected to ambient hypoxia (8% O_2, 92% N_2) demonstrated an increase in CD73 expression in intestinal mucosal scrapings [65]. Further analysis demonstrated the ability for HIF-1 to associate with a HIF consensus site within the CD73 gene promoter region, whose deletion resulted in a loss of luciferase reporter activity [65].

Although CD39 appears to be functionally upregulated under hypoxic conditions, there is no current evidence that it is directly regulated by interactions between the HIF-1 transcription factor and the promoter region of CD39. There is evidence, however, that other transcription factors may be involved in the regulation of genes under hypoxic conditions, such as Sp1, early growth response-1 (Egr-1), activating transcription factor-4 (ATF-4) and erythroblastosis virus E26 oncogene homolog 1 (Ets-1) [4, 48, 66, 73]. Truncation analysis of the promoter region of CD39 has revealed the presence of a Sp1 binding site, which when deleted or mutated, results in loss of promoter reporter activity using a firefly luciferase based assay [16]. Studies have identified HIF-1 dependent induction of Sp1, leading to indirect activation of hypoxic target genes. Mutation of either Egr1 or Sp1 binding sites, but not the HIF-1 response element (HRE) in the promoter region of the gene encoding the epinephrine-synthesizing enzyme, phenylethanolamine N-methyltransferase (PNMT) results in an abrogation of HIF induction of PNMT [66]. More direct evidence of HIF-1/Sp1 coactivation has been demonstrated with hypoxic regulation of the stress response protein Redd1. In this model, both Sp1 and HIF-1 are required for maximal activity of the Redd1 promoter, suggesting a coordinated response of

HIF-1 and Sp1 under hypoxic conditions [31]. It remains unclear whether hypoxia-mediated increases in CD39 are due to HIF-induced Sp1 or through a more complex transcriptional coordination.

While metabolic control of adenosine generation at sites of tissue hypoxia / inflammation is now well established, it is only recently appreciated that subsequent signaling through adenosine receptors can also be influenced by low pO_2 environments. A number of studies have strongly implicated enhanced adenosine A2B receptor (A2BR) signaling by hypoxia and inflammation within the mucosa [15, 25, 26, 34–36, 64]. As an interesting caveat, it was recently shown that the neuronal guidance molecule netrin-1 binds to and activates A2BR as a mechanism of attenuating inflammation within the mucosa [52]. With regard to adenosine receptor regulation, analysis of the cloned human A2BR promoter identified a functional hypoxia-responsive region, including a functional binding site for hypoxia-inducible factor (HIF) within the A2BR promoter [36]. Further studies examining HIF-1alpha DNA binding and HIF-1alpha gain and loss of function confirmed strong dependence of A2BR induction by HIF-1alpha in vitro and in vivo mouse models. Additional studies in endothelia over-expressing full-length A2BR revealed functional phenotypes of increased endothelial barrier function and enhanced angiogenesis [36]. More recently, studies in pulmonary endothelial cells revealed that the A2AR is induced selectively through HIF-2-dependent mechanisms. Adenoviral vector over-expression and siRNA-mediated repression of HIF-2alpha demonstrated prominent upregulation of A2AR that correlated with increased angiogenesis and increased A2AR in lung tumor samples [1]. Together, these studies indicate the central role of hypoxia in mediating a heightened capacity for extracellular nucleotide metabolism, coupled with an adaptive physiological response through the regulation of adenosine receptors.

8.5 Therapeutic Considerations for Targeting Nucleotide Metabolism

The design and implementation of adenosine receptor agonists and antagonists is currently an area of intense investigation [8, 11]. While drugs targeting adenosine receptors hold great promise in a variety of diseases, specificity and pharmacodynamics have been significant challenges. An alternative to modulating adenosine receptor signaling may be pharmacologically regulate extracellular nucleotide metabolism.

One strategy may be to enhance the rate of extracellular adenosine formation using soluble nucleotidases. Some work has been done in animal models using soluble 5′ nucleotidase and apyrase. Administration of 5′-nucleotidase has been shown to be beneficial in a number of experimental scenarios. For example, enzyme administration promotes vascular barrier function and decreases neutrophil accumulation in inflammatory models [18, 69]. A significant limitation for this line of work has been the identification of a reliable source of purified protein. Indeed, one of the major sources of 5′-nucleotidase may be snake venoms, particularly those from rattlesnakes (genera *Crotalus* and others) [3]. It is thought that snakes utilize purines as an effective, multifunctional means of envenomation [2].

CD73 directed therapies have not been well-developed. In our own experience, we have documented use of the CD73 inhibitor APCP in various murine models [18, 65, 69]. APCP is well tolerated, biologically available through the oral route, and non-toxic in mice up to 60 mg/kg/day. Interestingly, we have noticed that APCP treatment of mice appears to promote subtle increases in aggressive behavior and anxiety (unpublished observation). It is possible that this behavior is related to previous findings of increased aggression in adenosine A1R- and A2AR-deficient mice [22, 38], and may suggest that APCP is biologically available across the central nervous system. While many studies have suggested that adenosine is beneficial for most host responses, there may be examples where inhibition of adenosine generation (e.g. by inhibition of CD39 or CD73) is warranted. As alluded to above, during enteric pathogen infection, adenosine may promote water transport across intestinal epithelia and the symptoms of secretory diarrhea. For these purposes, CD39 and CD73 inhibitors such as POM-1 and APCP, respectively, could prove effective as anti-diarrheals. In this same context, it is possible that pulmonary edema related to infectious pneumonia or inflammation may benefit from the use of these inhibitors. Studies directed at defining these principles are currently underway.

Given the established association between angiogenesis and adenosine A2 receptor activation, the controlled regulation of CD39 / CD73 activity could influence angiogenesis. It is possible, for example, that systemic administration and/or targeting of inhibitors could prove beneficial for inhibition of tumor angiogenesis, currently an area of intense interest in cancer research. This is particularly compelling given the known association between hypoxia and the tumor microenvironment [57], wherein hypoxia and HIF-1 activation are potent transcriptional stimuli for CD73 expression[65] (also see above). Under such circumstances, it is reasonable that many tumors might over-express CD73. Thus, inhibition of adenosine production by CD39 / CD73 could be a therapeutic target for the prevention of tumor angiogenesis and metastasis [6, 61, 72].

Finally, given the strong association between hypoxia/HIF and extracellular adenosine generation, it may be possible to mimic hypoxia through pharmacologic means. Of great interest in this regard are agents which active HIF, primarily through the inhibition of HIF prolyl hydroxylases. These enzymes were identified on the principle that other mammalian prolyl-hydroxylases such as those which target extracellular collagen were 2-oxoglutarate-dependent [7], and it was predicted that the HIF prolyl hydroxylases would also belong to this family of enzymes. Based on conserved structural features [7], a candidate molecular approach was used to define HIF-modifying enzymes. This approach identified the HIF prolyl hydroxylases as the products of genes related to *C. elegans* egl-9, a gene that was first described in the context of an EGg-Laying abnormal (EGL) phenotype [19]. In mammalian cells, three PHD isoforms were identified (PHD 1-3), and shown to hydroxylate HIF-α in vitro [27, 30]. These enzymes have an absolute requirement for oxygen as substrate. The overall reaction results in insertion of one oxygen atom into the HIF-α

peptide substrate at the prolyl residue, with the other oxygen molecule generating succinate from 2-OG with the release of CO_2. A number of PHD inhibitors have been described, including direct inhibitors of the prolyl-hydroxylases [47], analogs of naturally occurring cyclic hydroxamates [54], as well as antagonists of alpha-keto-glutarate [45]. We and others have hypothesized that pharmacologic activation of HIF would provide a protective adaptation to murine mucosal disease [68]. For these purposes, we have used prolyl hydroxylase inhibitors which stabilize HIF and subsequently drive the expression of downstream HIF target genes (such as CD73 and adenosine A2BR). Our results show that the PHD inhibition-mediated induction of HIF-1α provides an overall beneficial influence on clinical symptoms (weight loss, colon length, tissue TNF α/IFNγ) in murine TNBS or DSS colitis models, most likely due to their barrier protective function and wound healing during severe tissue hypoxia at the site of inflammation [10, 50]. These findings emphasize the role of HIF during mucosal inflammatory diseases may provide the basis for a therapeutic use of PHD inhibitors in the regulation of nucleotide metabolism.

8.6 Conclusion

The dynamic interplay of leukocytes and endothelial/epithelial cells defines a complex and elegant lesson in biology. Studies of model systems incorporating cultured cells and purified PMN, for example, have allowed for the identification of functional determinants now well accepted in the scientific literature. The identification of soluble metabolites, such as adenine nucleotides, which in turn are metabolized to active nucleosides, will continue to contribute important information regarding the regulation of cell-cell interactions. Such information will provide previously unappreciated insight into the pathogenesis of inflammatory diseases. Targeting the mechanisms by which PMN liberate adenine nucleotides may permit insights into new approaches to influence inflammatory cascades.

Acknowledgments This work was supported by National Institutes of Health grants DK50189, DE016191, HL60569, HL092188 and by grants from the Crohn's and Colitis Foundation of America and the Foundation for Anesthesia Education and Research.

References

1. Ahmad A, Ahmad S, Glover LE, Miller S, Shannon JM, Guo X, Schaack JB, Colgan SP, White CW. (2009) Adenosine A2A receptor is a unique angiogenic target of HIF-2α in pulmonary endothelial cells. Proc Nat Acad Sci 106: 10684–89.
2. Aird SD. (2002) Ophidian envenomation strategies and the role of purines. Toxicon 40:335–93.
3. Aird SD. (2005) Taxonomic distribution and quantitative analysis of free purine and pyrimidine nucleosides in snake venoms. Comp Biochem Physiol B Biochem Mol Biol 140:109–26.
4. Ameri K, Lewis CE, Raida M, Sowter H, Hai T, Harris AL. (2004) Anoxic induction of ATF-4 through HIF-1-independent pathways of protein stabilization in human cancer cells. Blood 103:1876–82.

5. Bao X, Reuss L, Altenberg GA. (2004) Regulation of purified and reconstituted connexin 43 hemichannels by protein kinase C-mediated phosphorylation of Serine 368. J Biol Chem 279:20058–66.
6. Blay J, White TD, Hoskin DW. (1997) The extracellular fluid of solid carcinomas contains immunosuppressive concentrations of adenosine. Cancer Res 57:2602–5.
7. Bruick RK. (2003) Oxygen sensing in the hypoxic response pathway: regulation of the hypoxia-inducible transcription factor. Genes Dev 17:2614–23.
8. Burnstock G. (2002) Purinergic signaling and vascular cell proliferation and death. Arterioscler Thromb Vasc Biol 22:364–73.
9. Corriden R, Chen Y, Inoue Y, Beldi G, Robson SC, Insel PA, Junger WG. (2008) Ecto-nucleoside triphosphate diphosphohydrolase 1 (E-NTPDase1/CD39) regulates neutrophil chemotaxis by hydrolyzing released ATP to adenosine. J Biol Chem 283:28480–6.
10. Cummins EP, Seeballuck F, Keely SJ, Mangan NE, Callanan JJ, Fallon PG, Taylor CT. (2008) The hydroxylase inhibitor dimethyloxalylglycine is protective in a murine model of colitis. Gastroenterology 134:156–65.
11. Di Virgilio F, Chiozzi P, Ferrari D, Falzoni S, Sanz JM, Morelli A, Torboli M, Bolognesi G, Baricordi OR. (2001) Nucleotide receptors: an emerging family of regulatory molecules in blood cells. Blood 97:587–600.
12. Eckardt D, Theis M, Degen J, Ott T, van Rijen HV, Kirchhoff S, Kim JS, de Bakker JM, Willecke K. (2004) Functional role of connexin43 gap junction channels in adult mouse heart assessed by inducible gene deletion. J Mol Cell Cardiol 36:101–10.
13. Eckle T, Faigle M, Grenz A, Laucher S, Thompson LF, Eltzschig HK. (2008) A2B adenosine receptor dampens hypoxia-induced vascular leak. Blood 111:2024–35.
14. Eltzschig HK, Eckle T, Mager A, Kuper N, Karcher C, Weissmuller T, Boengler K, Schulz R, Robson SC, Colgan SP. (2006) ATP release from activated neutrophils occurs via connexin 43 and modulates adenosine-dependent endothelial cell function. Circ Res 99:1100–8.
15. Eltzschig HK, Ibla JC, Furuta GT, Leonard MO, Jacobson KA, Enjyoji K, Robson SC, Colgan SP. (2003) Coordinated adenine nucleotide phosphohydrolysis and nucleoside signaling in posthypoxic endothelium: role of ectonucleotidases and adenosine A_{2B} receptors. J Ex Med 198:783–96.
16. Eltzschig HK, Kohler D, Eckle T, Kong T, Robson SC, Colgan SP. (2009) Central role of Sp1-regulated CD39 in hypoxia/ischemia protection. Blood 113:224–32.
17. Eltzschig HK, Macmanus CF, Colgan SP. (2008) Neutrophils as sources of extracellular nucleotides: functional consequences at the vascular interface. Trends Cardiovasc Med 18:103–7.
18. Eltzschig HK, Thompson LF, Karhausen J, Cotta RJ, Ibla JC, Robson SC, Colgan SP. (2004) Endogenous adenosine produced during hypoxia attenuates neutrophil accumulation: Coordination by extracellular nucleotide metabolism. Blood 104:3986–92.
19. Epstein AC, Gleadle JM, McNeill LA, Hewitson KS, O'Rourke J, Mole DR, Mukherji M, Metzen E, Wilson MI, Dhanda A, Tian YM, Masson N, Hamilton DL, Jaakkola P, Barstead R, Hodgkin J, Maxwell PH, Pugh CW, Schofield CJ, Ratcliffe PJ. (2001) *C. elegans* EGL-9 and mammalian homologs define a family of dioxygenases that regulate HIF by prolyl hydroxylation. Cell 107:43–54.
20. Frick JS, MacManus CF, Scully M, Glover LE, Eltzschig HK, Colgan SP. (2009) Contribution of adenosine A2B receptors to inflammatory parameters of experimental colitis. J Immunol 182:4957–64.
21. Garcia GE, Truong LD, Li P, Zhang P, Du J, Chen JF, Feng L. (2008) Adenosine A2A receptor activation and macrophage-mediated experimental glomerulonephritis. Faseb J 22:445–54.
22. Gimenez-Llort L, Fernandez-Teruel A, Escorihuela RM, Fredholm BB, Tobena A, Pekny M, Johansson B. (2002) Mice lacking the adenosine A1 receptor are anxious and aggressive, but are normal learners with reduced muscle strength and survival rate. Eur J Neurosci 16:547–50.
23. Goodenough DA, Paul DL. (2003) Beyond the gap: functions of unpaired connexon channels. Nat Rev Mol Cell Biol 4:285–94.

24. Handa M, Guidotti G. (1996) Purification and cloning of a soluble ATP-diphosphohydrolase (apyrase) from potato tubers (Solanum tuberosum). Biochem Biophys Res Commun 218:916–23.
25. Hart ML, Jacobi B, Schittenhelm J, Henn M, Eltzschig HK. (2009) Cutting Edge: A2B Adenosine receptor signaling provides potent protection during intestinal ischemia/reperfusion injury. J Immunol 182:3965–68.
26. Hart ML, Much C, Gorzolla IC, Schittenhelm J, Kloor D, Stahl GL, Eltzschig HK. (2008) Extracellular adenosine production by ecto-5′-nucleotidase protects during murine hepatic ischemic preconditioning. Gastroenterology 135:1739–50 e3.
27. Hon WC, Wilson MI, Harlos K, Claridge TD, Schofield CJ, Pugh CW, Maxwell PH, Ratcliffe PJ, Stuart DI, Jones EY. (2002) Structural basis for the recognition of hydroxyproline in HIF-1 alpha by pVHL. Nature 417:975–8.
28. Hyman MC, Petrovic-Djergovic D, Visovatti SH, Liao H, Yanamadala S, Bouis D, Su EJ, Lawrence DA, Broekman MJ, Marcus AJ, Pinsky DJ. (2009) Self-regulation of inflammatory cell trafficking in mice by the leukocyte surface apyrase CD39. J Clin Invest 119:1136–49.
29. Idzko M, Panther E, Bremer HC, Sorichter S, Luttmann W, Virchow CJ, Jr., Di Virgilio F, Herouy Y, Norgauer J, Ferrari D. (2003) Stimulation of P2 purinergic receptors induces the release of eosinophil cationic protein and interleukin-8 from human eosinophils. Br J Pharmacol 138:1244–50.
30. Jaakkola P, Mole DR, Tian YM, Wilson MI, Gielbert J, Gaskell SJ, Kriegsheim A, Hebestreit HF, Mukherji M, Schofield CJ, Maxwell PH, Pugh CW, Ratcliffe PJ. (2001) Targeting of HIF-alpha to the von Hippel-Lindau ubiquitylation complex by O_2-regulated prolyl hydroxylation. Science 292:468–72.
31. Jin HO, An S, Lee HC, Woo SH, Seo SK, Choe TB, Yoo DH, Lee SB, Um HD, Lee SJ, Park MJ, Kim JI, Hong SI, Rhee CH, Park IC. (2007) Hypoxic condition- and high cell density-induced expression of Redd1 is regulated by activation of hypoxia-inducible factor-1alpha and Sp1 through the phosphatidylinositol 3-kinase/Akt signaling pathway. Cell Signal 19:1393–403.
32. John GR, Simpson JE, Woodroofe MN, Lee SC, Brosnan CF. (2001) Extracellular nucleotides differentially regulate interleukin-1beta signaling in primary human astrocytes: implications for inflammatory gene expression. J Neurosci 21:4134–42.
33. Karhausen J, Haase VH, Colgan SP. (2005) Inflammatory hypoxia: Role of hypoxia-inducible factor. Cell Cycle 4:256–8.
34. Kolachala V, Asamoah V, Wang L, Obertone TS, Ziegler TR, Merlin D, Sitaraman SV. (2005) TNF-alpha upregulates adenosine 2b (A2b) receptor expression and signaling in intestinal epithelial cells: a basis for A2bR overexpression in colitis. Cell Mol Life Sci 62:2647–57.
35. Kolachala VL, Vijay-Kumar M, Dalmasso G, Yang D, Linden J, Wang L, Gewirtz A, Ravid K, Merlin D, Sitaraman SV. (2008) A2B adenosine receptor gene deletion attenuates murine colitis. Gastroenterology 135:861–70.
36. Kong T, Westerman KA, Faigle M, Eltzschig HK, Colgan SP. (2006) HIF-dependent induction of adenosine A2B receptor in hypoxia. FASEB J 20:2242–50.
37. Kukulski F, Ben Yebdri F, Lecka J, Kauffenstein G, Levesque SA, Martin-Satue M, Sevigny J. (2009) Extracellular ATP and P2 receptors are required for IL-8 to induce neutrophil migration. Cytokine 46:166–70.
38. Ledent C, Vaugeois JM, Schiffmann SN, Pedrazzini T, El Yacoubi M, Vanderhaeghen JJ, Costentin J, Heath JK, Vassart G, Parmentier M. (1997) Aggressiveness, hypoalgesia and high blood pressure in mice lacking the adenosine A2a receptor. Nature 388:674–8.
39. Lennon PF, Taylor CT, Stahl GL, Colgan SP. (1998) Neutrophil-derived 5′-adenosine monophosphate promotes endothelial barrier function via CD73-mediated conversion to adenosine and endothelial A_{2B} receptor activation. J Exp Med 188:1433–43.
40. Louis NA, Hamilton KE, Canny G, Shekels LL, Ho SB, Colgan SP. (2006) Selective induction of mucin-3 by hypoxia in intestinal epithelia. J Cell Biochem 99:1616–27.

41. Louis NA, Robinson AM, Macmanus CF, Karhausen J, Scully M, Colgan SP. (2008) Control of IFN-{alpha}A by CD73: Implications for mucosal inflammation. J Immunol 180:4246–55.
42. Madara JL, Patapoff TW, Gillece-Castro B, Colgan SP, Parkos CA, Delp C, Mrsny RJ. (1993) 5′-adenosine monophosphate is the neutrophil-derived paracrine factor that elicits chloride secretion from T84 intestinal epithelial cell monolayers. J Clin Invest 91:2320–5.
43. Maliszewski CR, Delespesse GJ, Schoenborn MA, Armitage RJ, Fanslow WC, Nakajima T, Baker E, Sutherland GR, Poindexter K, Birks C, Alpert A, Friend D, Gimpel SD, Gayle RBI. (1994) The CD39 lymphoid cell activation antigen. Molecular cloning and structural characterization. J Immunol 153:3574–83.
44. Marcus AJ, Broekman MJ, Drosopoulos JH, Islam N, Pinsky DJ, Sesti C, Levi R. (2003) Metabolic control of excessive extracellular nucleotide accumulation by CD39/ecto-nucleotidase-1: implications for ischemic vascular diseases. J Pharmacol Exp Ther 305:9–16.
45. Mole DR, Schlemminger I, McNeill LA, Hewitson KS, Pugh CW, Ratcliffe PJ, Schofield CJ. (2003) 2-oxoglutarate analogue inhibitors of HIF prolyl hydroxylase. Bioorg Med Chem Lett 13:2677–80.
46. Novak I. (2003) ATP as a signaling molecule: the exocrine focus. News Physiol Sci 18:12–17.
47. Nwogu JI, Geenen D, Bean M, Brenner MC, Huang X, Buttrick PM. (2001) Inhibition of collagen synthesis with prolyl 4-hydroxylase inhibitor improves left ventricular function and alters the pattern of left ventricular dilatation after myocardial infarction. Circulation 104:2216–21.
48. Oikawa M, Abe M, Kurosawa H, Hida W, Shirato K, Sato Y. (2001) Hypoxia induces transcription factor ETS-1 via the activity of hypoxia-inducible factor-1. Biochem Biophys Res Commun 289:39–43.
49. Reutershan J, Vollmer I, Stark S, Wagner R, Ngamsri KC, Eltzschig HK. (2009) Adenosine and inflammation: CD39 and CD73 are critical mediators in LPS-induced PMN trafficking into the lungs. Faseb J 23:473–82.
50. Robinson A, Keely S, Karhausen J, Gerich ME, Furuta GT, Colgan SP. (2008) Mucosal protection by hypoxia-inducible factor prolyl hydroxylase inhibition. Gastroenterology 134:145–55.
51. Robson SC, Wu Y, Sun X, Knosalla C, Dwyer K, Enjyoji K. (2005) Ectonucleotidases of CD39 family modulate vascular inflammation and thrombosis in transplantation. Semin Thromb Hemost 31:217–33.
52. Rosenberger P, Schwab JM, Mirakaj V, Masekowsky E, Mager A, Morote-Garcia JC, Unertl K, Eltzschig HK. (2009) Hypoxia-inducible factor-dependent induction of netrin-1 dampens inflammation caused by hypoxia. Nat Immunol 10:195–202.
53. Satchell DG, Lynch A, Bourke PM, Burnstock G. (1972) Potentiation of the effects of exogenously applied ATP and purinergic nerve stimulation on the guinea-pig taenia coli by dipyridamole and hexobendine. Eur J Pharmacol 19:343–50.
54. Schlemminger I, Mole DR, McNeill LA, Dhanda A, Hewitson KS, Tian YM, Ratcliffe PJ, Pugh CW, Schofield CJ. (2003) Analogues of dealanylalahopcin are inhibitors of human HIF prolyl hydroxylases. Bioorg Med Chem Lett 13:1451–4.
55. Semenza GL. (2000) HIF-1 and human disease: one highly involved factor. Genes Dev 14:1983–91.
56. Semenza GL. (2001) HIF-1, O(2), and the 3 PHDs: how animal cells signal hypoxia to the nucleus. Cell 107:1–3.
57. Semenza GL. (2009) Regulation of cancer cell metabolism by hypoxia-inducible factor 1. Semin Cancer Biol 19:12–6.
58. Semenza GL, Agani F, Feldser D, Iyer N, Kotch L, Laughner E, Yu A. (2000) Hypoxia, HIF-1, and the pathophysiology of common human diseases. Adv Exp Med Biol 475:123–30.
59. Sitkovsky M, Lukashev D. (2005) Regulation of immune cells by local-tissue oxygen tension: HIF1 alpha and adenosine receptors. Nat Rev Immunol 5:712–21.
60. Smith TM, Kirley TL. (1999) Site-directed mutagenesis of a human brain ecto-apyrase: evidence that the E-type ATPases are related to the actin/heat shock 70/sugar kinase superfamily. Biochemistry 38:321–8.

61. Spychala J, Lazarowski E, Ostapkowicz A, Ayscue LH, Jin A, Mitchell BS. (2004) Role of estrogen receptor in the regulation of ecto-5′-nucleotidase and adenosine in breast cancer. Clin Cancer Res 10:708–17.
62. Stella J, Bavaresco L, Braganhol E, Rockenbach L, Farias PF, Wink MR, Azambuja AA, Barrios CH, Morrone FB, Oliveira Battastini AM. (2009) Differential ectonucleotidase expression in human bladder cancer cell lines. Urol Oncol. 2009 Apr 15. [Epub ahead of print] PMID: 19372055.
63. Strohmeier GR, Lencer WI, Patapoff TW, Thompson LF, Carlson SL, Moe SJ, Mrsny RJ, Madara JL. (1997) Surface expression, polarization, and functional significance of CD73 in human intestinal epithelia. J Clin Invest 99:2588–601.
64. Sun CX, Zhong H, Mohsenin A, Morschl E, Chunn JL, Molina JG, Belardinelli L, Zeng D, Blackburn MR. (2006) Role of A2B adenosine receptor signaling in adenosine-dependent pulmonary inflammation and injury. J Clin Invest 116:2173–82.
65. Synnestvedt K, Furuta GT, Comerford KM, Louis N, Karhausen J, Eltzschig HK, Hansen KR, Thompson LF, Colgan SP. (2002) Ecto-5′-nucleotidase (CD73) regulation by hypoxia-inducible factor-1 (HIF-1) mediates permeability changes in intestinal epithelia. J Clin Invest 110:993–1002.
66. Tai TC, Wong-Faull DC, Claycomb R, Wong DL. (2009) Hypoxic stress-induced changes in adrenergic function: role of HIF1 alpha. J Neurochem 109:513–24.
67. Taylor CT, Colgan SP. (1999) Therapeutic targets for hypoxia-elicited pathways. Pharm Res 16:1498–505.
68. Taylor CT, Colgan SP. (2008) Hypoxia and gastrointestinal disease. J Mol Med 85:1295–300.
69. Thompson LF, Eltzschig HK, Ibla JC, Van De Wiele CJ, Resta R, Morote-Garcia JC, Colgan SP. (2004) Crucial role for ecto-5′-nucleotidase (CD73) in vascular leak during hypoxia. J Exp Med 200:1395–405.
70. Wang TF, Guidotti G. (1996) CD39 is an ecto-(Ca2+,Mg2+)-apyrase. J Biol Chem 271:9898–901.
71. Weissmuller T, Campbell EL, Rosenberger P, Scully M, Beck PL, Furuta GT, Colgan SP. (2008) PMNs facilitate translocation of platelets across human and mouse epithelium and together alter fluid homeostasis via epithelial cell-expressed ecto-NTPDases. J Clin Invest 118:3682–92.
72. Xiang HJ, Liu ZC, Wang DS, Chen Y, Yang YL, Dou KF. (2006) Adenosine A(2b) receptor is highly expressed in human hepatocellular carcinoma. Hepatol Res 36:56–60.
73. Yan SF, Lu J, Zou YS, Soh-Won J, Cohen DM, Buttrick PM, Cooper DR, Steinberg SF, Mackman N, Pinsky DJ, Stern DM. (1999) Hypoxia-associated induction of early growth response-1 gene expression. J Biol Chem 274:15030–40.
74. Zahler S, Hoffmann A, Gloe T, Pohl U. (2003) Gap-junctional coupling between neutrophils and endothelial cells: a novel modulator of transendothelial migration. J Leukoc Biol 73:118–26.
75. Zimmermann H. (2000) Extracellular metabolism of ATP and other nucleotides. Naunyn Schmiedebergs Arch Pharmacol 362:299–309.

Chapter 9
Cell Surface ATP Synthase: A Potential Target for Anti-Angiogenic Therapy

Yvonne M. Mowery and Salvatore V. Pizzo

Abstract Since the presence of F_1F_0 adenosine triphosphate (ATP) synthase was discovered on the surface of human cells, numerous studies have elucidated new functions for this molecule that classically functions as part of the electron transport chain in mitochondria. Like mitochondrial ATP synthase, the cell surface form of this molecule functions as an energy-producing proton pump, coupling the production of extracellular ATP with the transport of protons outside the cell. ATP synthase also acts as an endothelial cell surface receptor for angiostatin, which has led several groups to examine this molecule as a target for inhibiting angiogenesis. Given its involvement in proton transport and the pH-dependent activity of angiostatin and antibodies directed against it, ATP synthase likely functions in part to regulate pH for endothelial cells existing in an acidic microenvironment. In addition, it catalyzes the synthesis and hydrolysis of ATP in the extracellular milieu, potentially affecting purinergic signaling. ATP synthase also plays a role in endothelial cell signaling in response to flow-induced shear stress. Recently, ATP synthase on the surface of endothelial cells has been identified as the receptor for two additional proteins: coupling factor 6 and endothelial monocyte-activating polypeptide II. This chapter will focus on the role of endothelial cell surface ATP synthase in angiogenesis, shear stress response, and as a receptor.

Keywords Cell surface ATP synthase · Alpha subunit of ATP synthase · Beta subunit of ATP synthase · Endothelial cell · Angiostatin · Angiogenesis · Anti-angiogenic treatment · ATP synthesis · ATP hydrolysis · Purinergic signaling · pH regulation · Coupling factor 6 · Shear stress · Endothelial monocyte-activating polypeptide II · Heparan sulfate · Caveolae · Sangivamycin · Monoclonal antibody therapy

---

Y.M. Mowery (✉)
Department of Pathology, Duke University, DUMC Box 3712, Durham, NC 27710, USA
e-mail: ymm3@duke.edu

E. Gerasimovskaya, E. Kaczmarek (eds.), *Extracellular ATP and Adenosine as Regulators of Endothelial Cell Function*, DOI 10.1007/978-90-481-3435-9_9,

9.1 Introduction

9.1.1 F_1F_0 ATP Synthase Structure and Mechanism

F_1F_0 ATP synthase is a multi-subunit enzyme complex that is best known for its ATP-synthesizing role as Complex V in the electron transport chain in mitochondria. The F_0 portion of the protein acts as a proton channel situated within the inner mitochondrial membrane. The soluble F_1 portion is composed of five subunits ($\alpha_3\beta_3\gamma\delta\varepsilon$) and oriented toward the mitochondrial matrix. The α and β subunits are arranged in an alternating hexameric pattern that is attached to F_0 by a central and peripheral stalk. The γ and ε subunits form the rotating central stalk that extends into the middle of the $\alpha_3\beta_3$ ring, while the δ subunit combines with the b subunit of F_0 to form the static peripheral stalk (Fig. 9.1a) [8, 10].

ATP synthase couples the energy generated by protons translocating down their gradient through F_0 with the formation of a high-energy phosphoanhydride bond. The movement of protons through the channel induces an internal rotation of a ring of c subunits in F_0, leading to rotation of the attached γ subunit in the central stalk of F_1. This in turn triggers a series of conformational changes of the catalytic β subunits, which synthesize ATP from adenosine diphosphate (ADP) and inorganic phosphate (P_i). This "binding change mechanism" proposed by Paul Boyer drives the generation of ATP by each of the three β subunits in succession. The β subunits can also catalyze ATP hydrolysis, which engenders conformational changes that drive rotation of the central stalk of F_1 in the opposite direction. Rotation in this direction is associated with proton translocation against the proton gradient, utilizing the energy provided by hydrolyzing ATP to ADP and P_i [9].

9.1.2 Cell Surface F_1F_0 ATP Synthase

In 1994 Das et al. discovered the presence of the β subunit of ATP synthase on the cell surface of three human tumor cell lines [16]. ATP synthase had previously been known to exist in the plasma membrane of bacteria, inner mitochondrial membrane, and the thylakoid membrane of chloroplasts; however, this report marked the first identification of a component of ATP synthase in the cell membrane of a eukaryotic cell (Fig. 9.1b). Five years later, our laboratory identified cell surface F_1F_0 ATP synthase as a receptor for angiostatin on human endothelial cells [26]. Confocal microscopy of nonpermeabilized endothelial cells demonstrated colocalization of the α and β subunits, distributed in a punctate pattern on the cell surface, consistent with arrangement in caveolae (Fig. 9.2) [25]. These findings were initially controversial due to the long-held belief that this protein existed only as an intracellular enzyme in eukaryotes; however, ATP synthase has since then been detected on the surface of several other normal cell types, including hepatocytes, adipocytes, and keratinocytes, as well as additional cancer cell lines. In hepatocytes, it functions as a receptor for apolipoprotein A-I, which upon binding stimulates ATP hydrolysis

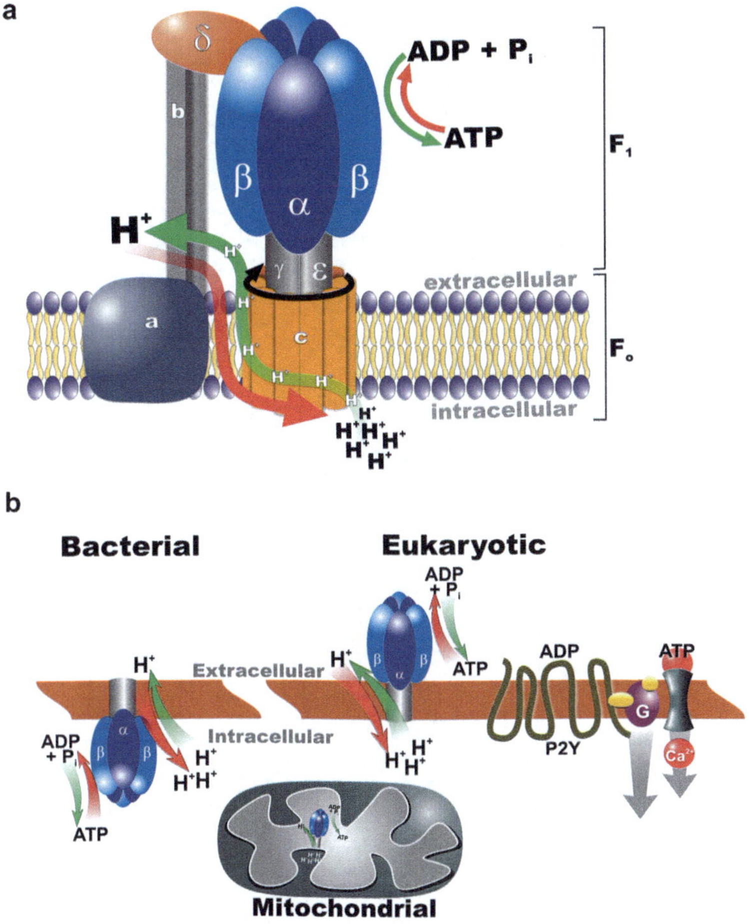

Fig. 9.1 Basic structure and cellular locations of F_1F_0 ATP synthase. (**a**) F_1F_0 ATP synthase consists of a membrane-embedded component (F_0) and a soluble portion (F_1). Flow of protons down their concentration gradient through F_0 causes rotation of the central stalk, leading to conformational changes of the β subunits and catalysis of ATP synthesis. The complex can also function in reverse, coupling ATP hydrolysis with proton flow against the electrochemical gradient. (**b**) ATP synthase is oriented in opposing directions in the plasma membrane of bacteria versus eukaryotic cells. Cell surface ATP synthase in eukaryotes catalyzes ATP synthesis and hydrolysis in the extracellular milieu. The resulting ATP can trigger cation influx into the cell through ATP-gated ion channels (P2X purinoreceptors). ADP or ATP can bind to G-protein coupled receptors (P2Y purinoreceptors) that can activate a variety of downstream signaling pathways. In mitochondria, the electron transport chain generates a proton gradient across the inner mitochondrial membrane. ATP synthase located within this membrane couples proton translocation with ATP synthesis in the mitochondrial matrix. (Modified from Chi and Pizzo [14])

and high-density lipoprotein (HDL) endocytosis [24]. Other researchers have also demonstrated colocalization of ATP synthase with caveolin-1 on the plasma membrane of endothelial cells, supporting the theory proposed by Moser et al. that the α

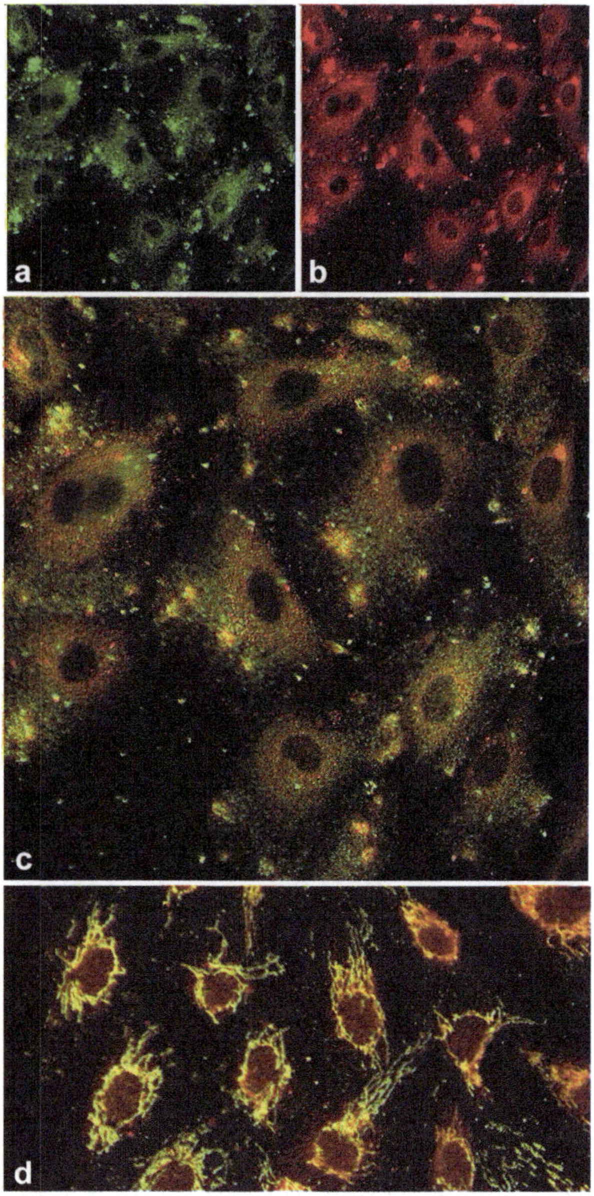

Fig. 9.2 Confocal images of α and β subunits of ATP synthase on HUVEC surface. (**a**) Nonpermeabilized HUVEC were immunostained with a murine monoclonal antibody directed against the α subunit of ATP synthase (*green*). (**b**) The same field of HUVEC were immunostained with a rabbit polyclonal antibody directed against the β subunit of ATP synthase (*red*). (**c**) A digital overlay of panels **a** and **b** demonstrates colocalization (*yellow*) of α and βsubunits in punctate pattern on surface of HUVEC. (**d**) HUVEC were permeabilized with 100% ethanol prior to immunostaining with antibodies against the α and β subunits of ATP synthase. The staining pattern is characteristic of mitochondrial ATP synthase staining. (Reprinted by permission of Proceedings of the National Academy of Sciences, Copyright 2001 National Academy of Sciences, USA [25])

and β subunits are localized to caveolae [46, 47]. Furthermore, several other research groups have confirmed the presence of enzymatically active F_1F_0 ATP synthase on the surface of endothelial cells [2, 47, 54]. It is worth noting that the orientation of ATP synthase in the plasma membrane of eukaryotic cells is opposite to the orientation in bacteria. Thus, ATP hydrolysis and synthesis occurs in the extracellular milieu of endothelial cells, potentially influencing extracellular ADP/ATP levels and affecting signaling via purinergic receptors (Fig. 9.1b).

9.2 ATP Synthase and Angiogenesis

9.2.1 Receptor for Angiostatin

Angiostatin is an endogenous anti-angiogenic molecule formed by the proteolytic cleavage of plasminogen. Its structure consists of the first three or four disulfide-linked loops (kringles 1–3 or 1–4) of plasminogen. In 1994, Judah Folkman and his group reported the isolation of this circulating 38 kDa proteolytic fragment from the serum and urine of mice bearing Lewis lung carcinoma tumors [28]. They demonstrated that angiostatin specifically inhibits endothelial cell proliferation in vitro and prevents angiogenesis inthe chicken chorioallantoic membrane (CAM) model. Additionally, they found that angiostatin inhibits growth of tumor metastases. Later studies also showed that systemic treatment of immunodeficient mice with angiostatin induces a dramatic size reduction in primary human tumors, followed by the onset of dormancy [27]. These findings generated interest in the use of angiostatin as a cancer therapeutic; however, the encouraging effects of angiostatin in animal tumor models did not translate into success in human clinical trials. The lack of efficacy in these trials was largely due to the short serum half-life of angiostatin (~20 minutes), which necessitates the use of high doses and frequent administration to achieve a therapeutic effect [4, 43].

The discovery of ATP synthase as one of the receptors of angiostatin shed light on at least one component of the mechanism of action of angiostatin. Our laboratory utilized affinity chromatography to identify proteins located in the plasma membrane of human umbilical vein endothelial cells (HUVEC) that bound to angiostatin. A single ~55-kDa protein was eluted from the angiostatin-Sepharose column. This protein was identified as either the α or β subunit of ATP synthase based on amino-terminal sequencing, mass spectrometry, and peptide mass fingerprinting. Incubation of HUVEC with a polyclonal antibody against the α subunit inhibited binding of angiostatin to the cells by ~56% [26]. The lack of complete binding inhibition could in part be due to the presence of other receptors for angiostatin on endothelial cells. Additionally, angiostatin also binds to the β subunit of ATP synthase, which is 23% homologous and 57% similar to the α subunit [26]. The activity of angiostatin actually depends upon this β subunit interaction, as the use of a polyclonal antibody against the β subunit inhibits endothelial proliferation like angiostatin, whereas an antibody against the α subunit does not [25, 26]. Similarly, a monoclonal antibody against the β subunit inhibits endothelial cell tube formation, while six monoclonal antibodies against the α subunit tested by our laboratory do not induce inhibition in this in vitro assay predictive of anti-angiogenic activity [15]. Thus, use of the anti-α subunit polyclonal antibody may have only partially inhibited the interaction of angiostatin with the adjacent β subunit via steric hindrance and/or binding of the polyclonal antibody to epitope(s) on the β subunit that are shared with the α subunit.

Use of an antibody against ATP synthase to inhibit interaction of angiostatin with the receptor largely inhibits the anti-proliferative effect of angiostatin on endothelial cells (~80%) [26]. Another group demonstrated that kringles 1–5 from plasminogen also bind to cell surface ATP synthase, activating caspases-8, -9, and -3 and leading to endothelial cell apoptosis [41]. These results indicate that cell surface ATP synthase is involved in the anti-angiogenic effect of angiostatin. A therapeutic mimetic able to bind this receptor could potentially overcome the pharmacokinetic disadvantages associated with recombinant angiostatin.

Further studies of angiostatin revealed that it blocks the enzymatic activity of ATP synthase [25]. In the absence of angiostatin, HUVEC synthesize approximately 40 pmol of ATP per million cells. Treatment with 1 μM angiostatin reduces this extracellular ATP production by 81%, while treatment with polyclonal antibodies against the α and β subunits of ATP synthase inhibits ATP generation by 65 and 57%, respectively. Oligomycin, a known inhibitor of ATP synthase enzymatic activity, decreases ATP synthesis by 84% (Fig. 9.3). Purified bovine F_1 ATP synthase was also utilized to test the effect on ATP hydrolysis, as the F_1 subunit in the absence of the F_0 proton channel can only catalyze ATP hydrolysis. Angiostatin and polyclonal

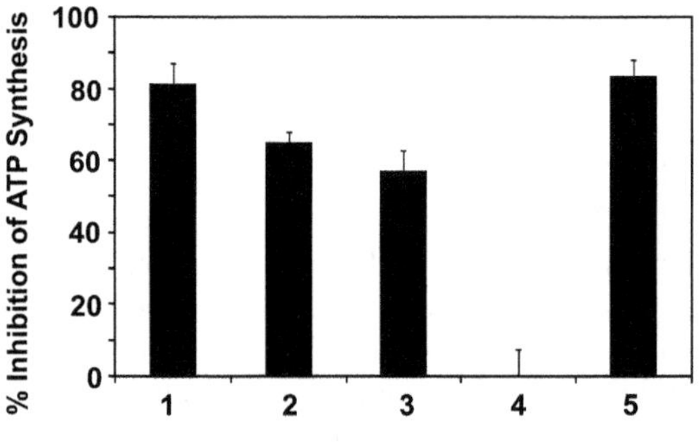

Fig. 9.3 Inhibition of HUVEC extracellular ATP synthesis by polyclonal antibodies against ATP synthase. Extracellular ATP generation by HUVEC in the presence of 50 μM ADP was measured by bioluminescent luciferase assay. Cells were treated with 1 μM angiostatin (**1**), 1 mg/ml polyclonal antibody against the α subunit of ATP synthase (**2**), 0.5 mg/ml polyclonal antibody against the β subunit of ATP synthase (**3**), 1 mg/ml pre-immune serum (**4**), or 50 μg/ml oligomycin (**5**). Inhibition was calculated relative to cells incubated in media only. Antibodies directed against ATP synthase inhibited extracellular ATP generation, although to a lesser extent than angiostatin and oligomycin, a known F_0 inhibitor. (Adapted from [25], reprinted by permission of Proceedings of the National Academy of Sciences, Copyright 2001 National Academy of Sciences, USA [25])

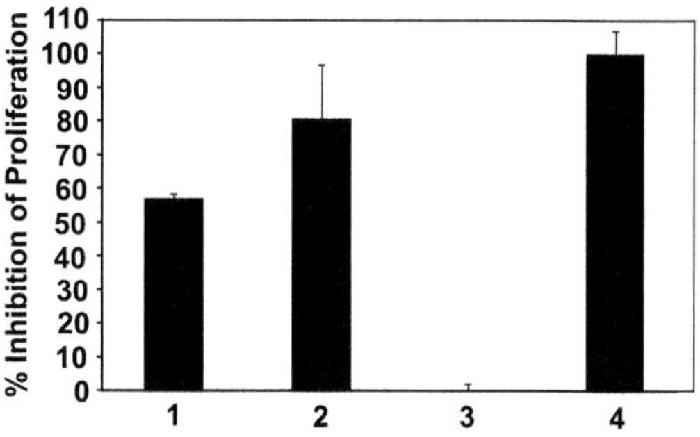

Fig. 9.4 Inhibition of HUVEC proliferation by polyclonal antibodies against ATP synthase. HUVECproliferation was induced by basic fibroblast growth factor (bFGF, 10 ng/ml) and vascular endothelial growth factor (VEGF, 3 ng/ml). Cells were treated with 1 μM angiostatin (**1**), 100 μg/ml polyclonal antibody against the β-subunit of ATP synthase (**2**), 100 μg/ml pre-immune serum (**3**), or 10 μg/ml cycloheximide as a positive control (**4**). The antibody against the β subunit inhibited endothelial cell proliferation to a greater degree than angiostatin. (Adapted from [25], reprinted by permission of *Proceedings of the National Academy of Sciences*, Copyright 2001 National Academy of Sciences, USA [25])

antibodies against the α and β subunits also inhibit this reverse reaction, while an antibody against the γ subunit does not affect ATP hydrolysis. In addition, treatment with angiostatin or anti-β-ATP synthase polyclonal antibody inhibits HUVEC proliferation (Fig. 9.4) [25]. A later publication by the Higuti laboratory supported these results, demonstrating decreased extracellular ATP synthesis and inhibited proliferation of endothelial cells in the presence of other ATP synthase inhibitors [2]. These results indicated that molecules targeting ATP synthase could potentially mimic the anti-angiogenic activity of angiostatin. Data from the Moser et al. study suggest that an inhibitory antibody against the β catalytic subunit of ATP synthase could mimic the in vivo anti-angiogenic activity of angiostatin without the potential risk posed by other small molecule inhibitors of ATP synthase, e.g. piceatannol, that could enter the cell and inhibit oxidative phosphorylation [25].

9.2.2 Target for Anti-Angiogenic Treatment

In 2007 Chi et al. further demonstrated that cell surface ATP synthase is a promising therapeutic target by demonstrating the anti-angiogenic activity of a monoclonal antibody (MAb3D5AB1) against its β subunit. The antibody utilized in this study binds to the β subunit with 25-fold greater affinity than angiostatin and inhibits both ATP synthesis and hydrolysis by the enzyme. As with angiostatin, the inhibition of ATP synthase activity occurs specifically under conditions of low external pH (pH_e), which will be discussed in greater detail in the following section. Furthermore, MAb3D5AB1 disrupts endothelial cell tube formation under acidic conditions and inhibits bFGF-stimulated angiogenesis in vivo in the CAM model (Fig. 9.5) [15].

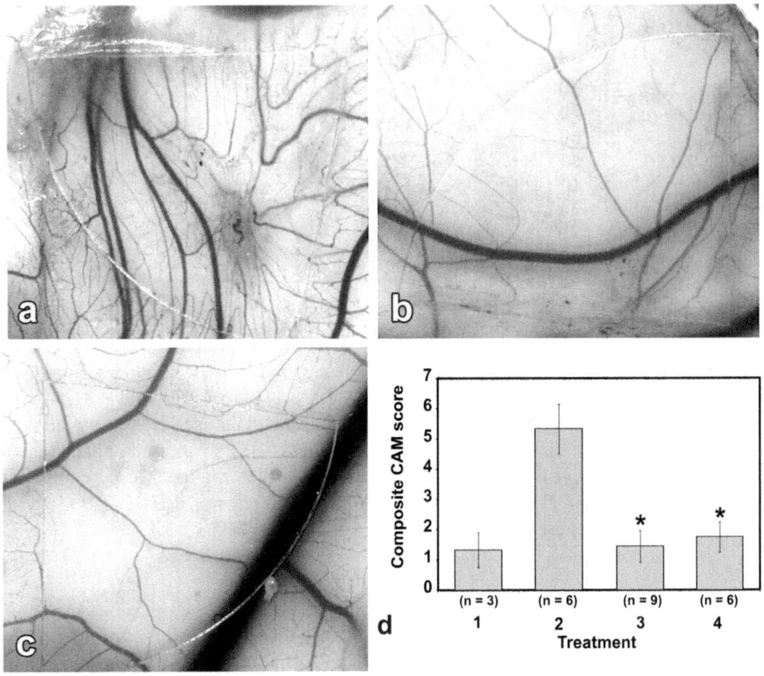

Fig. 9.5 Inhibition of angiogenesis in the CAM assay by MAb3D5AB1, a monoclonal antibody against the α-subunit of ATP synthase. Chicken chorioallantoic membranes (CAM) were treated with 100 ng bFGF with or without ATP synthase inhibitors dissolved on a Thermanox disc. Vascular growth, branching, and density were assessed after three days, and each CAM was assigned a composite score between one and six by an investigator blinded to the treatment groups. A score of one indicates no increased vascularization. Representative images of membranes treated with 100 ng bFGF alone (**a**), bFGF + 8 μg MAb3D5AB1 (**b**), or bFGF + 30 μg angiostatin (**c**) are shown. **d**) Composite CAM scores for membranes treated with PBS (**1**), bFGF (**2**), bFGF + MAb3D5AB1 (**3**), and bFGF + angiostatin (**4**) indicate inhibition of bFGF-induced angiogenesis by MAb3D5AB1 and angiostatin. (Figure 6 reprinted with permission from Chi et al. [15])

The antibiotic sangivamycin (4-amino-5-carboxamide-7-(D-ribofuranosyl) pyrrolo[2,3-*d*]pyrimidine), which has previously been shown to specifically inhibit endothelial cell proliferation, also inhibits angiogenesis, purportedly through binding to cell surface ATP synthase [22, 29]. Sangivamycin inhibits cell surface ATP production and blood vessel formation in the CAM model. In addition, the drug decreases angiogenesis stimulated by Lewis lung carcinoma cells in the murine dorsal air sac (DAS) assay. Komi et al. hypothesize that sangivamycin, which is structurally similar to adenosine, acts as a competitive antagonist of F_1 ATP synthase. This group does not address whether sangivamycin also enters mitochondria and inhibits mitochondrial ATP synthase, given that the antibiotic can cross the plasma membrane. Sangivamycin also has known inhibitory effects on protein kinase C, rhodopsin kinase, protein kinase A, β-adrenergic receptor kinase, and nuclear protein kinases-I and -II [23, 30, 34, 36]. Given this lack of specificity and its potential effects on oxidative phosphorylation, sangivamycin is not an ideal candidate for use as an anti-angiogenic therapeutic.

Two additional research groups reported developing monoclonal antibodies against the β subunit of ATP synthase in 2008. Zhang et al. reported inhibition of cell surface ATP synthesis and endothelial cell tube formation with the monoclonal antibody McAb178-5G10. In addition to blocking endothelial cell proliferation and migration in vitro, this antibody decreased proliferation of MDA-MB-231 breast cancer cells, which also express cell surface ATP synthase [54]. Most recently, the Fan laboratory produced the murine monoclonal antibody mAb6F2C4 and demonstrated its potential as a cancer therapeutic in vivo. Like McAb178-5G10, this antibody decreases endothelial cell growth and tube formation, as well as proliferation of an ATP synthase-expressing hepatoma cell line (SMMC-7721). Furthermore, intra-tumoral injection of this antibody every three days inhibits in vivo growth of SMMC-7721 flank tumors in nude mice. Although the authors show in vitro binding of the antibody to endothelial cells, immunohistochemical staining of the tumors demonstrates mAb6F2C4 binding to the tumor cells only, with no apparent vascular staining. Despite this puzzling lack of staining, the tumor growth delay is likely attributable to a combination of the direct anti-tumorigenic effects of the antibody and its anti-angiogenic effects [45]. The direct inhibitory effect of these two monoclonal antibodies on tumor growth also mimics angiostatin, since angiostatin is cytotoxic under acidic conditions to tumor cells expressing cell surface ATP synthase [13]. Monoclonal antibodies targeting ATP synthase provide several advantages over the use of angiostatin as a cancer therapeutic. They have increased specificity and binding affinity for the target, limited side effect profiles, and a longer serum half-life.

The potential for stimulation of angiogenesis by extracellular production of ATP points to one potential mechanism by which targeting the enzymatic activity of ATP synthase can inhibit angiogenesis. A recent report demonstrates that extracellular ATP induces a pro-angiogenic effect on vasa vasorum endothelial cells (VVEC) isolated from pulmonary artery adventitia [18]. These cells demonstrated a dose-dependent increase in proliferation in response to exogenous addition of ATP. Additionally, they were stimulated to migrate and to form endothelial cell tube networks in the presence of 100 μM ATP. Thus, inhibition of extracellular ATP production by ATP synthase may reduce pro-angiogenic purinergic signaling.

9.2.3 pH Regulation: Potential Mechanism for the Anti-Angiogenic Effect of ATP Synthase Inhibition

Inhibition of ATP synthase disrupts translocation of protons out of the cell and decreases extracellular ATP production. Therefore, the anti-angiogenic effect of blocking this enzyme is likely due to changes in intracellular pH regulation and/or disruption cell signaling through ATP binding to cell surface purinoreceptors. The relative contributions of these two actions have not yet been elucidated, however most attention has focused on pH homeostasis due to the differential effects of angiostatin and anti-ATP synthase antibodies at neutral and low pH_e. It has been reported that angiostatin disrupts endothelial cell tube formation under acidic conditions (pH_e 6.7), but not at a normal physiologic pH of 7.3 [42]. Additionally, angiostatin treatment of endothelial cells under conditions of low pH_e results in increased toxicity of angiostatin to the cells [44]. A pH-dependent effect on angiostatin-induced cell death has also been shown in tumor cells expressing cell surface ATP synthase [13].

In both endothelial cells and ATP synthase-expressing tumor cells, exposure to acidic conditions causes a greater decrease in intracellular pH (pH_i) in cells treated with angiostatin than untreated cells. These results prompted our laboratory to hypothesize that angiostatin induces cell death by disrupting pH_i homeostasis. ATP synthase is more active in generating ATP at pH_e 6.7 than at 7.2, and this enzymatic activity likely tempers the pH_i decrease that occurs in acidic conditions by transporting protons out of the cell [13, 14]. This activity is puzzling, given that it involves the energetically unfavorable synthesis of ATP under conditions when protons are being moved against their concentration gradient in order to regulate pH_i. This confounding activity could be explained by the localization of cell surface ATP synthase in caveolae, where pH can potentially differ from other areas along the plasma membrane due to the distinct composition of proteins present in these invaginations [14]. Regardless of the mechanism, the experimental evidence discussed above indicates increased ATP synthesis by cell surface ATP synthase under acidic conditions.

Like angiostatin, the first inhibitory monoclonal antibody against the β subunit of ATP synthase, MAb3D5AB1, reduces extracellular ATP synthesis under low pH_e conditions (Fig. 9.6a). Exposure of HUVEC to a drop in pH_e of 7.3 to 6.7 resulted in a pH_i decrease of 0.41 when challenged in the presence of 67 nM MAb3D5AB1, relative to a decrease of only 0.25 in the absence of the antibody (Fig. 9.6b) [15]. This effect by the monoclonal antibody is more potent (~15-fold) than that of angiostatin, which must be present at a concentration of ~1 μM to achieve this degree of pH_i decrease [3]. Again mimicking the properties of angiostatin, MAb3D5AB1 inhibits endothelial cell tube formation by 60% under acidic conditions (pH_e 6.7), but not at the normal physiologic pH_e of 7.4. In vivostudies also demonstrate anti-angiogenic activity in the acidic microenvironment of the chicken CAM model, but not in the rat corneal neovascularization assay, where the pH is ~7.4 as would be expected in this normal tissue [15]. The two other anti-β-ATP synthase antibodies (McAb178-5G10 and mAb6F2C4) recently reported in the literature also inhibit extracellular ATP synthesis, HUVEC proliferation, and endothelial cell tube formation preferentially in an acidic environment [45, 54]. These results have garnered particular interest, given that this pH-dependent effect would likely target the anti-angiogenic effects of antibodies against the β subunit to acidic microenvironments such as tumors, while sparing normal tissues.

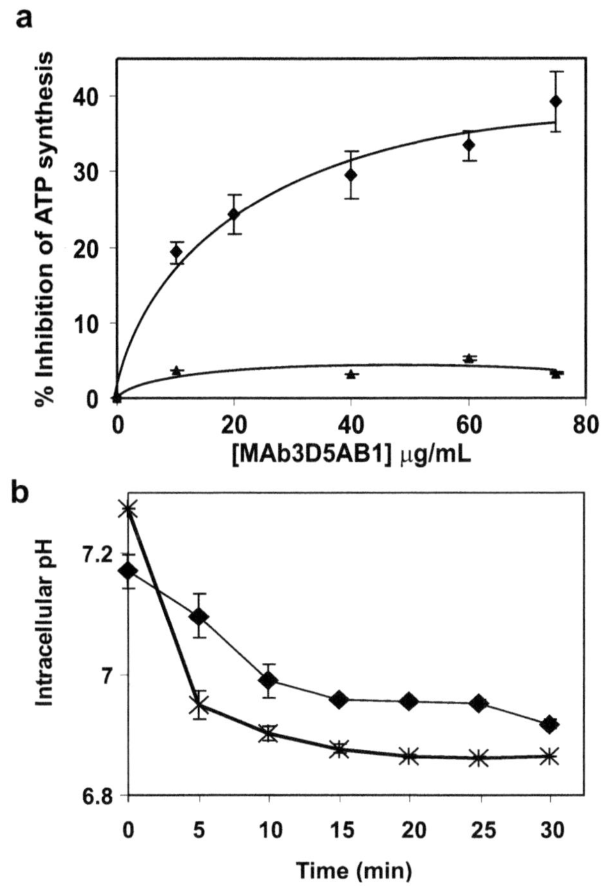

Fig. 9.6 Inhibition of ATP synthase activity and intracellular pH regulation by MAb3D5AB1. (**a**) Extracellular ATP generation by HUVEC was measured at pH_e 6.7 (♦) and pH_e 7.2 (▲). MAb3D5AB1 inhibited cell surface ATP synthesis in a dose-dependent manner at low pH_e but not at pH_e 7.2. (**b**) The intracellular pH of HUVEC was monitored by fluorescence of the pH-sensitive fluoroprobe carboxy snarf-1 acetoxymethyl ester after acute extracellular acidification from pH_e 7.4 to 6.7. Intracellular acidification occurred to a greater degree in the presence of 10 μg/ml MAb3D5AB1 (×) than in media alone (♦). (Figure 4 reprinted with permission from Chi et al. [15])

9.3 ATP Synthase and Endothelial Cell Response to Shear Stress

The luminal surface of endothelial cells is exposed to shear stress imposed by blood flow through the vasculature. Transduction of this mechanical force induces a variety of responses in endothelial cells that are involved in vasoregulation and vascular remodeling. Extracellular ATP production by ATP synthase has recently been

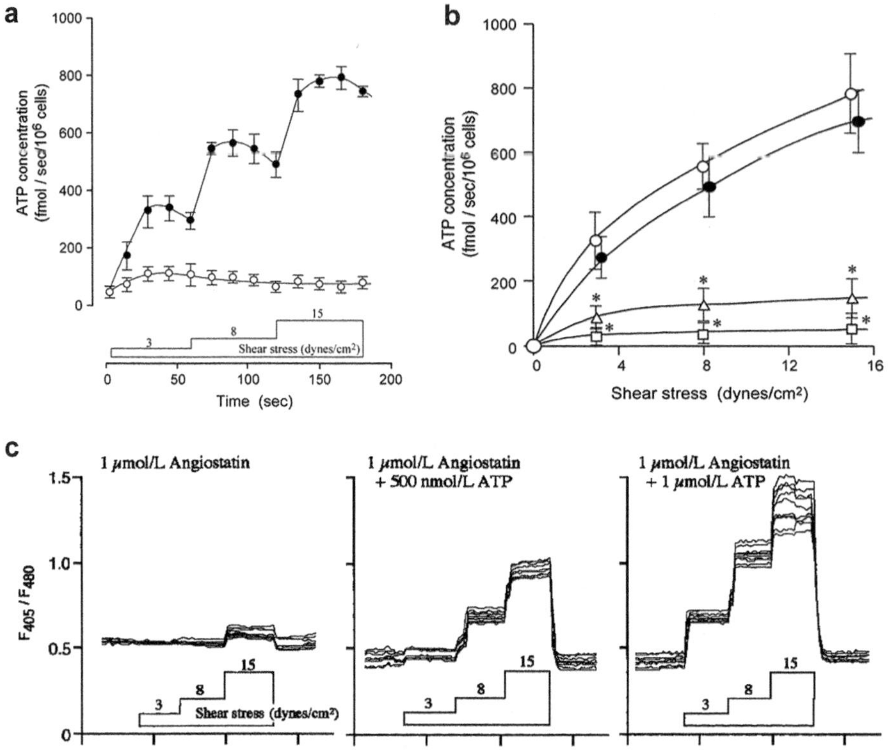

Fig. 9.7 Effect of angiostatin on shear stress-dependent ATP production and calcium influx in HPAEC. (**a**) Extracellular ATP generation by HPAEC was measured under static conditions (○) and in response to a stepwise increase in shear stress (•). ATP concentration increased in a dose-dependent manner with the rise in shear stress. (**b**) Extracellular ATP generation was measured after a one-hour incubation in the presence of media alone (○), 0.5 μmol/L angiostatin (△), or 1.0 μmol/L angiostatin (□). Treatment with angiostatin caused a decrease in shear stress-dependent ATP synthesis that was reversed by flushing away the angiostatin (•). The effect of angiostatin indicates that ATP synthase is responsible for the shear stress-dependent ATP generation. (**c**) Changes inintracellular Ca^{2+} levels of HPAEC in response to shear stress were measured using the ratiometric Ca^{2+} indicator indo 1-acetoxymethyl ester. Relative Ca^{2+} levels are indicated by the 405-to-480-nm fluorescence ratio (F_{405}/F_{480}). In the presence of angiostatin, a shear stress-dependent rise in intracellular Ca^{2+} does not occur. Addition of exogenous ATP restores the dose-dependent rise in shear stress-induced Ca^{2+} levels normally observed in the absence of angiostatin. (Reprinted with modification from Yamamoto et al. [51], used with permission of the American Physiological Society [51])

implicated in endothelial responses to shear stress [49, 51]. Yamamoto et al. demonstrated that flow-induced shear stress causes an influx of calcium into human pulmonary artery endothelial cells (HPAEC) that is largely dependent on an increase in extracellular ATP level. Cultured HPAEC release ATP in response to shear stress in a dose-dependent manner (Fig. 9.7a). Similar results have been observed in freshly-isolated endothelial cells from rabbit thoracic aorta [6]. Degradation of ATP by apyrase prevents the shear stress-dependent influx of Ca^{2+} in HPAEC, indicating that Ca^{2+} transport into these cells depends on the presence of extracellular ATP [51]. Yamamato et al. propose that cell surface F_1F_0 ATP synthase is the source for shear stress-induced ATP production, since incubation of HPAEC with the ATP synthase inhibitors angiostatin, oligomycin, piceatannol, or antibody directed against the β subunit of ATP synthase prevents the rise in extracellular ATP levels and Ca^{2+} influx without affecting intracellular ATP levels (Fig. 9.7b) [49, 51]. Addition of exogenous extracellular ATP overcomes the inhibitory effect of ATP synthase inhibitors on Ca^{2+} flow into the endothelial cells (Fig. 9.7c).

Yamamoto et al. went on to demonstrate that ATP induces Ca^{2+} influx through $P2X_4$—a purinoreceptor with greater sensitivity to ATP than ADP. siRNA knockdown of this ATP-gated cation channel inhibits shear flow-induced Ca^{2+} responses in HPAEC. The G protein-coupled P2Y purinoreceptors do not appear to be involved, as the shear stress-induced calcium rise is due to influx of external Ca^{2+} rather than Ca^{2+} release from internal stores. P2Y receptor activation would have stimulated inositol 1,4,5-trisphosphate (IP_3) production, leading to release of Ca^{2+} from intracellular stores. Thus, ATP synthase production of extracellular ATP plays a role in flow-induced pulmonary artery endothelial cell signaling by activating Ca^{2+} influx through the $P2X_4$ cation channel. Differences do exist in the level of shear stress-induced ATP production by endothelial cells from distinct sources, indicating that involvement of ATP synthase in endothelial response to shear stress may vary in different vascular beds. For example, although flow-stimulated Ca^{2+} influx occurs in HUVEC, this response requires the addition of exogenous ATP [1, 48]. The authors attribute this incongruent finding to the relatively lower shear-stress induced endogenous ATP production by HUVEC relative to HPAEC [51]. Alternatively, they propose a difference in P2X receptor ATP affinity in HUVEC versus HPAEC due to potential variations in P2X subunit multimerization.

Regardless of the potentially disparate roles of ATP in different vascular beds, a recent study in mice deficient in the $P2X_4$ receptor ($P2rx4^{-/-}$) indicates that purine signaling through this ligand-gated cation channel does have in vivo relevance to vascular regulation. These knockout mice display abnormal flow-induced calcium responses and impaired vasodilation in response to increased blood flow. The authors attribute this decreased vasodilation to the lack of nitric oxide release by $P2rx4^{-/-}$ mice in response to shear stress, in contrast to the dose-dependent increase in NO production by endothelial cells from wild-type mice with normal vasodilatory response. The $P2X_4$ receptor also appears to be involved in vascular remodeling as demonstrated by the response of $P2rx4^{-/-}$ mice to a chronic decrease in blood flow via external carotid artery ligation. While the common carotid artery diameter decreases in wild-type mice, it does not change in the knockout mice [50]. Further work is required to elucidate how shear stress stimulates ATP production by ATP synthase, however current evidence does suggest that flow-dependent increase in ATP is important in acute and chronic vascular changes. It should be noted that other groups have demonstrated flow-induced release of ATP from intracellular stores, and the relative contributions of this released ATP versus ATP generated by ATP synthase remains unclear [7, 20].

9.4 Coupling Factor 6

ATP synthase on the surface of endothelial cells has recently been identified as the receptor for another molecule—coupling factor 6 (CF6). Interestingly, CF6 itself is actually an integral subunit in the stalk of ATP synthase, required for oxidative phosphorylation [17]. In 1998 this peptide was isolated from cardiac tissue of spontaneously hypertensive rats and identified as an inhibitor of prostacyclin synthesis. CF6 is thought to act by blocking the Ca^{2+}-dependent activity of phospholipase A_2, which normally catalyzes the release of the precursor of the vasodilator prostacyclin—arachidonic acid (AA)—from the plasma membrane [31]. The concentration of this peptide is significantly greater in the plasma of spontaneously hypertensive rats than in normotensive control rats (294 ± 126 pg/ml versus 2310 ± 813 pg/ml). Furthermore, intravenous injection of recombinant CF6 results in decreased plasma prostacyclin levels and increased blood pressure, indicating a potential role in systemic vasoregulation and hypertension [31]. CF6 is present on the surface of endothelial cells, from which it is released into systemic circulation. Expression and release of this peptide are increased in response to shear stress, thus opposing the synthesis of prostacyclin that is also induced by shear stress [33].

Cell surface ATP synthase serves as at least one receptor for CF6 on endothelial cells. CF6 binds to the β subunit—at a distinct location from angiostatin—and causes increased ATP hydrolysis by ATP synthase. The intracellular pH of HUVEC also decreases upon treatment with CF6, as ATP hydrolysis is coupled with the transport of protons from the extracellular milieu into the cytoplasm (Fig. 9.1a). Pre-treatment of rats with anti-β subunit antibody reduces the CF6-induced rise in blood pressure by ~50%. Evidence supports the involvement of ATP synthase in the pathway through which CF6 inhibits prostacyclin synthesis. CF6 suppresses the release of AA in HUVEC, but this effect is reduced by pre-treating with anti-β subunit antibody. ATP also stimulates ATP hydrolysis and suppresses AA release, though to a lesser degree than CF6. This effect is not additive with the suppression by CF6, indicating that a common signaling pathway is likely. The mechanism by which CF6 binding to the β subunit of ATP synthase causes inhibition of AA release remains unclear, but Osanai et al. propose that the decrease in pH induced by CF6 might inhibit the activity of cytosolic phospholipase A_2, which has been shown to be pH-sensitive. The P2X and P2Y receptors do not appear to be involved, as treatment with pyridoxal-phosphate-6-azophenyl-2′,4′-disulphonic acid (PPAD), an inhibitor of these receptors, does not prevent the suppression of AA release induced by ATP [32]. Regardless of the precise post-receptor signaling pathway, CF6 and ATP synthase could play a role in the development or maintenance of hypertension through effects on prostacyclin levels, which are decreased in many models of hypertension [47].

9.5 Endothelial Monocyte-Activating Polypeptide II

Endothelial monocyte-activating polypeptide II (EMAP II) is a tumor-derived proinflammatory cytokine produced by cleavage of the C-terminus from p43, a protein associated with several aminoacyl-tRNA synthetases [35]. This protein has many effects on endothelial cells, including the induction of von Willebrand factor release and tissue factor activation, as well as the upregulation of E-selectin and P-selectin [21]. Additionally, it has been shown to have anti-angiogenic effects. EMAP II induces endothelial cell apoptosis and inhibits endothelial cell proliferation, tube formation, and angiogenesis in the rat aortic ring assay [5]. One group has demonstrated that in vivo injection of EMAP II in a flank tumor model results in decreased tumor vascularization, increased rate of vascular thrombosis, and delayed tumor growth [39]. The mechanisms by which EMAP II carries out these functions are still being elucidated, however work by the Kang laboratory indicates that cell surface ATP synthase is involved in the anti-angiogenic effects. His research group identified the α subunit of ATP synthase as one of the receptors for EMAP II. Incubation of endothelial cells with EMAP II decreases proliferation by 40%, and this growth suppression can be inhibited by approximately 30% via addition of soluble α subunit of ATP synthase. The anti-proliferative activity can also be achieved by incubation with anti-α-ATP synthase antibody in a dose-dependent manner [12].

Under acidic conditions, the interaction between EMAP II and the α subunit is regulated by cell surface heparan sulfate, which binds directly to both proteins via heparin binding motifs. Heparan sulfate does not appear to affect the interaction at neutral pH, however at pH 6.5 increased binding occurs between EMAP II and α-ATP synthase. This enhanced binding can be inhibited by exogenous heparin, indicating that the effect is likely due to interaction with heparan sulfate. The authors suggest that at low pH, cross-linking by heparan sulfate enhances EMAP II-α subunit binding. This idea is supported by the increased inhibitory effect of EMAP II on endothelial cell proliferation at pH 6.5 relative to 7.5, which can be prevented by disrupting interactions with heparan sulfate by the addition of exogenous heparin or heparinase [11].

Binding of EMAP II to the α subunit of ATP synthase accounts for only part of the anti-angiogenic effects of this protein. In the initial study demonstrating EMAP II binding to α-ATP synthase, incubation of cells with excess soluble α subunit only partly reverses EMAP II-induced inhibition of endothelial cell proliferation indicating that significant anti-proliferative effects are also transmitted through other receptors [12]. EMAP II also disrupts fibronectin matrix assembly by endothelial cells via a direct interaction with the integrins α5β1 and αVβ3, likely playing a role in its effects on angiogenesis [38]. Vascular endothelial growth factor (VEGF) receptors I and II are a third target of EMAP II on the endothelial cell surface. EMAP II inhibits VEGF binding to these receptors, preventing downstream signaling in the VEGF pathway and abrogating VEGF-induced endothelial cell proliferation and migration [3]. At this time, it is not known whether any of the other effects of EMAP II on endothelial cells are also mediated in part by the interaction with α-ATP synthase.

9.6 Remaining Questions

Although many functions for cell surface ATP synthase have been identified, the precise mechanism by which it reaches the plasma membrane remains to be elucidated. Other components of the electron transport chain, normally present in the inner mitochondrial membrane, have also been demonstrated to exist on the cell surface of eukaryotic cells. Some of the subunits for these complexes are translated on cytoplasmic ribosomes and targeted to mitochondria, whereas others are encoded by mitochondrial DNA and synthesized in mitochondria. This suggests that the complexes are most likely assembled in mitochondria and then transported to the cell surface.

Yonally and Capaldi propose two potential mechanisms for this trafficking to the plasma membrane: fusion with the mitochondrial membrane or transport by a shuttling mechanism, e.g. lipid rafts [53]. This latter hypothesis is supported by the observation that ATP synthase and other mitochondrial proteins are distributed in caveolae—specialized lipid rafts [40, 46, 49, 53]. However, Yamamoto et al. demonstrated that ATP synthase levels on the cell surface are not diminished when caveolae formation is disrupted by caveolin-1 siRNA or methyl-β-cyclodextrin (MβCD). Although these findings indicate that transport to the cell surface does not require lipid rafts, ATP synthase function in shear stress-mediated ATP release is suppressed when the complex is not localized to caveolae. The authors suggest that the role of ATP synthase in this pathway might be signal transduction through P2X and/or P2Y receptors, concentrated in proximity to ATP synthase in caveolae [49]. A recent publication purports that ATP synthase is in fact not localized in lipid rafts/caveolae, given that these authors did not observe sensitivity of surface ATP synthase expression to cholesterol disruption via MβCD [55]. However, both Yamamoto et al. and Wang et al. demonstrated that increasing and depleting plasma membrane cholesterol levels in endothelial cells affects cell surface ATP synthase expression [46, 49]. These different findings have not yet been resolved, but could be due to the different cell types examined.

Collaborators of our laboratory studying ATP synthase in neural cells propose that at least the α subunit is transported to the cell surface through the secretory pathway, i.e. moving from the rough endoplasmic reticulum (ER) to the Golgi apparatus to the plasma membrane via secretory vesicles. Utilizing a neuroblastoma cell line (B103), they show that inhibiting trafficking from the ER to the Golgi with brefeldin A prevented the α-subunit from reaching the cell surface. Further supporting this hypothesis, they provide evidence that surface α-ATP synthase is N-glycosylated during the secretory process, unlike the mitochondrial form of this protein [37]. Although this is possible for the α subunit, which is encoded in nucleic DNA, this transport process would not account for trafficking of ATP synthase subunits 6 and 8, which are synthesized in mitochondria, to the cell surface. Given the conflicting evidence provided by various research groups, the question of how the various components of ATP synthase, as well as other members of the electron transport chain traditionally considered mitochondrial proteins, remains unanswered.

Aside from this trafficking mystery, the precise mechanisms by which ATP synthase carries out its other functions in endothelial cells continue to be unresolved (Fig. 9.8). With regards to its role in angiogenesis, it is unclear whether it promotes endothelial cell survival and proliferation through pH_i regulation, ATP signaling, and/or another mechanism. If ATP signaling is involved, the purinergic receptors

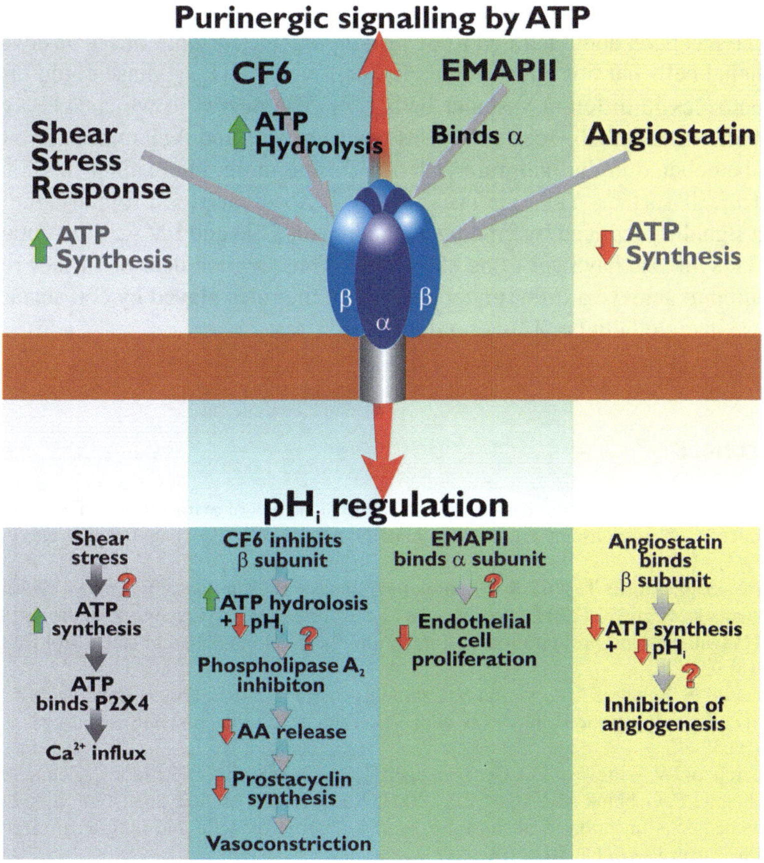

Fig. 9.8 Summary of the roles of ATP synthase on the surface of endothelial cells. Through its dual roles of proton transport and ATP synthesis/hydrolysis, cell surface ATP synthase can affect both the extracellular and intracellular environment. This enzyme complex influences the extracellular levels of ATP and ADP, leading to potential downstream effects via purinergic signaling. ATP synthase is also implicated in pH_i regulation through the bidirectional movement of protons through the transmembrane F_0 portion of the complex. The enzymatic activity of ATP synthase on endothelial cells is regulated by several factors. Shear stress stimulates increased ATP generation by ATP synthase in HPAEC through a mechanism not yet understood. Coupling factor 6 (CF6) binds to the β subunit and stimulates ATP hydrolysis. This activity is associated with inhibition of phospholipase A_2, leading to decreased synthesis of the vasoconstrictor prostacyclin. However, the connection between ATP hydrolysis by ATP synthase and phospholipase A_2 activity has not been elucidated. The α subunit of ATP synthase also acts as a receptor for endothelial monocyte-activating polypeptide II (EMAPII), playing a partial role in the effect of this tumor-derived cytokine on endothelial cell proliferation. The β subunit of ATP synthase can also affect endothelial cell proliferation through its activity as a receptor for angiostatin. Binding of this endogenous angiogenesis inhibitor inhibits ATP synthesis and pH_i regulation by endothelial cells under acidic conditions. It is unclear whether the anti-angiogenic effect is achieved through the alteration in extracellular ATP levels, pH_i, or a combination of both

and downstream cascade(s) activated by ATP also remain to be determined. Given that ATP has been demonstrated to be pro-angiogenic for pulmonary vasa vasorum endothelial cells but not other endothelial cell types, ATP synthase could also play different roles in different vascular beds [18]. The degree to which ATP synthase contributes to endothelial cell responses to shear stress and ATP release also continues to be debated in the literature [19, 52]. Furthermore, the manner in which flow activates cell surface ATP synthase activity is not understood. Finally, the downstream signaling induced by binding of the ligands CF6 and EMAP II to endothelial cell ATP synthase have yet to be elucidated. Thus, a great deal of further research is required to achieve a fuller understanding of the roles played by cell surface ATP synthase in endothelial cell functioning.

References

1. Ando J, Ohtsuka A, Korenaga R, Kamiya A. (1991) Effect of extracellular ATP level on flow-induced Ca^{++} response in cultured vascular endothelial cells. Biochem Biophys Res Commun 179(3):1192–9.
2. Arakaki N, Nagao T, Niki R, Toyofuku A, Tanaka H, Kuramoto Y, Emoto Y, Shibata H, Magota K, Higuti T. (2003) Possible role of cell surface H^+-ATP synthase in the extracellular ATP synthesis and proliferation of human umbilical vein endothelial cells. Mol Cancer Res 1(13):931–9.
3. Awasthi N, Schwarz MA, Verma V, Cappiello C, Schwarz RE. (2009) Endothelial monocyte activating polypeptide II interferes with VEGF-induced proangiogenic signaling. Lab Invest 89(1):38–46.
4. Beerepoot LV, Witteveen EO, Groenewegen G, Fogler WE, Sim BK, Sidor C, Zonnenberg BA, Schramel F, Gebbink MF, Voest EE. (2003) Recombinant human angiostatin by twice-daily subcutaneous injection in advanced cancer: a pharmacokinetic and long-term safety study. Clin Cancer Res 9(11):4025–33.
5. Berger AC, Alexander HR, Tang G, Wu PS, Hewitt SM, Turner E, Kruger E, Figg WD, Grove A, Kohn E, Stern D, Libutti SK. (2000) Endothelial monocyte activating polypeptide II induces endothelial cell apoptosis and may inhibit tumor angiogenesis. Microvasc Res 60(1):70–80.
6. Bodin P, Bailey D, Burnstock G. (1991) Increased flow-induced ATP release from isolated vascular endothelial cells but not smooth muscle cells. Br J Pharmacol 103(1):1203–5.
7. Bodin P, Burnstock G. (2001) Evidence that release of adenosine triphosphate from endothelial cells during increased shear stress is vesicular. J Cardiovasc Pharmacol 38(6):900–8.
8. Boyer PD. (1997) The ATP synthase–a splendid molecular machine. Annu Rev Biochem 66:717–49.
9. Boyer PD. (2000) Catalytic site forms and controls in ATP synthase catalysis. Biochim Biophys Acta 1458(2–3):252–62.
10. Capaldi RA, Aggeler R. (2002) Mechanism of the F(1)F(0)-type ATP synthase, a biological rotary motor. Trends Biochem Sci 27(3):154–60.
11. Chang SY, Ko HJ, Heo TH, Kang CY. (2005) Heparan sulfate regulates the antiangiogenic activity of endothelial monocyte-activating polypeptide-II at acidic pH. Mol Pharmacol 67(5):1534–43.
12. Chang SY, Park SG, Kim S, Kang CY. (2002) Interaction of the C-terminal domain of p43 and the alpha subunit of ATP synthase. Its functional implication in endothelial cell proliferation. J Biol Chem 277(10):8388–94.

13. Chi SL, Pizzo SV. (2006) Angiostatin is directly cytotoxic to tumor cells at low extracellular pH: a mechanism dependent on cell surface-associated ATP synthase. Cancer Res 66(2): 875–82.
14. Chi SL, Pizzo SV. (2006) Cell surface F1Fo ATP synthase: a new paradigm? Ann Med 38(6):429–38.
15. Chi SL, Wahl ML, Mowery YM, Shan S, Mukhopadhyay S, Hilderbrand SC, Kenan DJ, Lipes BD, Johnson CE, Marusich MF, Capaldi RA, Dewhirst MW, Pizzo SV. (2007) Angiostatin-like activity of a monoclonal antibody to the catalytic subunit of F1F0 ATP synthase. Cancer Res 67(10):4716–24.
16. Das B, Mondragon MO, Sadeghian M, Hatcher VB, Norin AJ. (1994) A novel ligand in lymphocyte-mediated cytotoxicity: expression of the beta subunit of H^+ transporting ATP synthase on the surface of tumor cell lines. J Exp Med 180(1):273–81.
17. Fessenden-Raden JM. (1972) Purification and properties of a new coupling factor required for oxidative phosphorylation in silicotungstate-treated submitochondrial particles. J Biol Chem 247(8):2351–7.
18. Gerasimovskaya EV, Woodward HN, Tucker DA, Stenmark KR. (2008) Extracellular ATP is a pro-angiogenic factor for pulmonary artery vasa vasorum endothelial cells. Angiogenesis 11(2):169–82.
19. Hahn C, Schwartz MA. (2009) Mechanotransduction in vascular physiology and atherogenesis. Nat Rev Mol Cell Biol 10(1):53–62.
20. Hassessian H, Bodin P, Burnstock G. (1993) Blockade by glibenclamide of the flow-evoked endothelial release of ATP that contributes to vasodilatation in the pulmonary vascular bed of the rat. Br J Pharmacol 109(2):466–72.
21. Kao J, Houck K, Fan Y, Haehnel I, Libutti SK, Kayton ML, Grikscheit T, Chabot J, Nowygrod R, Greenberg S, et al. (1994) Characterization of a novel tumor-derived cytokine. Endothelial-monocyte activating polypeptide II. J Biol Chem 269(40):25106–19.
22. Komi Y, Ohno O, Suzuki Y, Shimamura M, Shimokado K, Umezawa K, Kojima S. (2007) Inhibition of tumor angiogenesis by targeting endothelial surface ATP synthase with sangivamycin. Jpn J Clin Oncol 37(11):867–73.
23. Loomis CR, Bell RM. (1988) Sangivamycin, a nucleoside analogue, is a potent inhibitor of protein kinase C. J Biol Chem 263(4):1682–92.
24. Martinez LO, Jacquet S, Esteve JP, Rolland C, Cabezon E, Champagne E, Pineau T, Georgeaud V, Walker JE, Terce F, Collet X, Perret B, Barbaras R. (2003) Ectopic beta-chain of ATP synthase is an apolipoprotein A-I receptor in hepatic HDL endocytosis. Nature 421(6918):75–9.
25. Moser TL, Kenan DJ, Ashley TA, Roy JA, Goodman MD, Misra UK, Cheek DJ, Pizzo SV. (2001) Endothelial cell surface F1-F0 ATP synthase is active in ATP synthesis and is inhibited by angiostatin. Proc Natl Acad Sci USA 98(12):6656–61.
26. Moser TL, Stack MS, Asplin I, Enghild JJ, Hojrup P, Everitt L, Hubchak S, Schnaper HW, Pizzo SV. (1999) Angiostatin binds ATP synthase on the surface of human endothelial cells. Proc Natl Acad Sci USA 96(6):2811–6.
27. O'Reilly MS, Holmgren L, Chen C, Folkman J. (1996) Angiostatin induces and sustains dormancy of human primary tumors in mice. Nat Med 2(6):689–92.
28. O'Reilly MS, Holmgren L, Shing Y, Chen C, Rosenthal RA, Moses M, Lane WS, Cao Y, Sage EH, Folkman J. (1994) Angiostatin: a novel angiogenesis inhibitor that mediates the suppression of metastases by a Lewis lung carcinoma. Cell 79(2):315–28.
29. Ohno O, Shima Y, Ikeda Y, Kondo SI, Kato K, Toi M, Umezawa K. (2001) Selective growth inhibition by sangivamycin of human umbilical vein endothelial cells. Int J Oncol 18(5): 1009–15.
30. Osada H, Sonoda T, Tsunoda K, Isono K. (1989) A new biological role of sangivamycin; inhibition of protein kinases. J Antibiot (Tokyo) 42(1):102–6.
31. Osanai T, Kamada T, Fujiwara N, Katoh T, Takahashi K, Kimura M, Satoh K, Magota K, Kodama S, Tanaka T, Okumura K. (1998) A novel inhibitory effect on prostacyclin synthesis

of coupling factor 6 extracted from the heart of spontaneously hypertensive rats. J Biol Chem 273(48):31778–83.
32. Osanai T, Magota K, Tanaka M, Shimada M, Murakami R, Sasaki S, Tomita H, Maeda N, Okumura K. (2005) Intracellular signaling for vasoconstrictor coupling factor 6: novel function of beta-subunit of ATP synthase as receptor. Hypertension 46(5):1140–6.
33. Osanai T, Tanaka M, Kamada T, Nakano T, Takahashi K, Okada S, Sirato K, Magota K, Kodama S, Okumura K. (2001) Mitochondrial coupling factor 6 as a potent endogenous vasoconstrictor. J Clin Invest 108(7):1023–30.
34. Palczewski K, Kahn N, Hargrave PA. (1990) Nucleoside inhibitors of rhodopsin kinase. Biochemistry 29(26):6276–82.
35. Quevillon S, Agou F, Robinson JC, Mirande M. (1997) The p43 component of the mammalian multi-synthetase complex is likely to be the precursor of the endothelial monocyte-activating polypeptide II cytokine. J Biol Chem 272(51):32573–9.
36. Saffer JD, Glazer RI. (1981) Inhibition of histone H1 phosphorylation by sangivamycin and other pyrrolopyrimidine analogues. Mol Pharmacol 20(1):211–7.
37. Schmidt C, Lepsverdize E, Chi SL, Das AM, Pizzo SV, Dityatev A, Schachner M. (2008) Amyloid precursor protein and amyloid beta-peptide bind to ATP synthase and regulate its activity at the surface of neural cells. Mol Psychiatry 13(10):953–69.
38. Schwarz MA, Zheng H, Liu J, Corbett S, Schwarz RE. (2005) Endothelial-monocyte activating polypeptide II alters fibronectin based endothelial cell adhesion and matrix assembly via alpha5 beta1 integrin. Exp Cell Res 311(2):229–39.
39. Schwarz RE, Schwarz MA. (2004) In vivo therapy of local tumor progression by targeting vascular endothelium with EMAP-II. J Surg Res 120(1):64–72.
40. Sprenger RR, Speijer D, Back JW, De Koster CG, Pannekoek H, Horrevoets AJ. (2004) Comparative proteomics of human endothelial cell caveolae and rafts using two-dimensional gel electrophoresis and mass spectrometry. Electrophoresis 25(1):156–72.
41. Veitonmaki N, Cao R, Wu LH, Moser TL, Li B, Pizzo SV, Zhivotovsky B, Cao Y. (2004) Endothelial cell surface ATP synthase-triggered caspase-apoptotic pathway is essential for k1-5-induced antiangiogenesis. Cancer Res 64(10):3679–86.
42. Wahl ML, Grant DS. (2000) Effects of microenvironmental extracellular pH and extracellular matrix proteins on angiostatin's activity and on intracellular pH. Gen Pharmacol 35(5): 277–85.
43. Wahl ML, Kenan DJ, Gonzalez-Gronow M, Pizzo SV. (2005) Angiostatin's molecular mechanism: aspects of specificity and regulation elucidated. J Cell Biochem 96(2):242–61.
44. Wahl ML, Owen CS, Grant DS. (2002) Angiostatin induces intracellular acidosis and anoikis in endothelial cells at a tumor-like low pH. Endothelium 9(3):205–16.
45. Wang J, Han Y, Liang J, Cheng X, Yan L, Wang Y, Liu J, Luo G, Chen X, Zhao L, Zhou X, Wu K, Fan D. (2008) Effect of a novel inhibitory mAb against beta-subunit of F1F0 ATPase on HCC. Cancer Biol Ther 7(11):1829–35.
46. Wang T, Chen Z, Wang X, Shyy JY, Zhu Y. (2006) Cholesterol loading increases the translocation of ATP synthase beta chain into membrane caveolae in vascular endothelial cells. Biochim Biophys Acta 1761(10):1182–90.
47. Watts SW. (2005) Vasoconstriction caused by the ATP synthase subunit-coupling factor 6: a new function for a historical enzyme. Hypertension 46(5):1100–2.
48. Yamamoto K, Korenaga R, Kamiya A, Qi Z, Sokabe M, Ando J. (2000) P2X(4) receptors mediate ATP-induced calcium influx in human vascular endothelial cells. Am J Physiol Heart Circ Physiol 279(1):H285–92.
49. Yamamoto K, Shimizu N, Obi S, Kumagaya S, Taketani Y, Kamiya A, Ando J. (2007) Involvement of cell surface ATP synthase in flow-induced ATP release by vascular endothelial cells. Am J Physiol Heart Circ Physiol 293(3):H1646–53.
50. Yamamoto K, Sokabe T, Matsumoto T, Yoshimura K, Shibata M, Ohura N, Fukuda T, Sato T, Sekine K, Kato S, Isshiki M, Fujita T, Kobayashi M, Kawamura K, Masuda H, Kamiya A,

Ando J. (2006) Impaired flow-dependent control of vascular tone and remodeling in P2X4-deficient mice. Nat Med 12(1):133–7.

51. Yamamoto K, Sokabe T, Ohura N, Nakatsuka H, Kamiya A, Ando J. (2003) Endogenously released ATP mediates shear stress-induced Ca^{2+} influx into pulmonary artery endothelial cells. Am J Physiol Heart Circ Physiol 285(2):H793–803.
52. Yegutkin GG. (2008) Nucleotide- and nucleoside-converting ectoenzymes: Important modulators of purinergic signalling cascade. Biochim Biophys Acta 1783(5):673–694.
53. Yonally SK, Capaldi RA. (2006) The F(1)F(0) ATP synthase and mitochondrial respiratory chain complexes are present on the plasma membrane of an osteosarcoma cell line: An immunocytochemical study. Mitochondrion 6(6):305–314.
54. Zhang X, Gao F, Yu LL, Peng Y, Liu HH, Liu JY, Yin M, Ni J. (2008) Dual functions of a monoclonal antibody against cell surface F1F0 ATP synthase on both HUVEC and tumor cells. Acta Pharmacol Sin 29(8):942–50.
55. Zheng YZ, Berg KB, Foster LJ. (2009) Mitochondria do not contain lipid rafts and lipid rafts do not contain mitochondrial proteins. J Lipid Res 50(5):988–98.

Chapter 10
ATP Release Via Connexin Hemichannels Controls Intercellular Propagation of Ca^{2+} Waves in Corneal Endothelial Cells

Raf Ponsaerts, Catheleyne D'hondt, Priya Gomes, Geert Bultynck, Sangly P. Srinivas, Johan Vereecke, and Bernard Himpens

Abstract Intercellular communication (IC) is essential for coordinating cellular activity in multi-cellular organisms, and plays a role in many physiological and pathological processes. Direct IC between adjacent cells is mainly mediated by gap junctions, whereas paracrine signaling provides an indirect pathway for IC between neighboring cells. Purinergic mediators are important messengers in this paracrine signaling, and recent evidence shows that connexin and pannexin hemichannels form an important release pathway for ATP involved in purinergic IC.

We have investigated the properties of IC via Ca^{2+} waves evoked by mechanical stimulation of a single cell in monolayers of bovine corneal endothelial cells (BCEC). We have demonstrated that both gap junctions and paracrine IC are involved in the Ca^{2+} wave propagation, and that purinergic paracrine signaling is the main component of IC in BCEC. Furthermore, our results demonstrate that connexin43 hemichannels provide the pathway for ATP release. Our experiments show that the paracrine IC is strongly reduced in the presence of inflammatory agents, such as histamine and thrombin, that enhance actomyosin contraction. The effect of these agents is due to inhibition of hemichannel-mediated ATP release. This effect of thrombin can be precluded by preincubation of the cells with adenosine. Possible roles of paracrine IC to "bystander" and "good Samaritan" effects are discussed.

Keywords Intercellular communication · Cornea · Endothelial cells · Calcium wave · Mechanical stimulation · Connexin · Gap junction · Paracrine signaling · Hemichannels · Purinergic · ATP release · Thrombin · Histamine · Adenosine · Rho kinase · Phosphorylation · Myosin light chain · Myosin ATPase activity · Contractility · Cytoskeleton · Blebbistatin · Bystander effect · Good Samaritan effect

---

J. Vereecke (✉)
Laboratory of Molecular and Cellular Signaling, Department of Molecular Cell Biology, KULeuven, Campus Gasthuisberg O&N1, B-3000 Leuven, Belgium
e-mail: Johan.Vereecke@med.kuleuven.be

E. Gerasimovskaya, E. Kaczmarek (eds.), *Extracellular ATP and Adenosine as Regulators of Endothelial Cell Function*, DOI 10.1007/978-90-481-3435-9_10,

10.1 Introduction

This chapter discusses the mechanism and role of ATP-mediated intercellular communication in coordinating the activity of endothelial cells of the cornea.

10.1.1 Corneal Endothelium

The corneal endothelium is a non-regenerative monolayer at the posterior surface of the cornea. It separates the anterior chamber of the eye from the corneal stroma. While the corneal endothelial monolayer has a barrier function similar to vascular endothelium, and the cells express endothelial cell markers such as von Willebrand factor (vWF) (see Table 10.1), most of the corneal endothelial cells are derived from neural crest, instead of having a mesodermal origin [41]. However, it has been demonstrated that neural stem cells can differentiate to the endothelial lineage [125].

As in vascular endothelium, corneal endothelial cells are interconnected by a complex set of interendothelial junction proteins that comprise tight junctions, adherens junctions and gap junctions. Whereas tight junctions and adherens junctions form zipperlike pericellular structures along the cell border through their homophilic adhesions of their transmembrane proteins, gap junctions form plaques of transmembrane channels between adjacent cells.

The main physiological role of the corneal endothelium is to maintain the transparency of the cornea by controlling stromal hydration. Decompensation of the endothelium leads to stromal edema, and hence loss of visual acuity, and is a common indication for corneal transplantation. Unlike most tissues in the body, the cornea is avascular; it contains no blood vessels to nourish it or to protect it against infection. Instead, aqueous humor that circulates in the anterior chamber is largely responsible for supplying nutrients to the cornea, while oxygen is supplied via the tears as well as by the aqueous humor.

Because of its discontinuous focal tight junctions and the sinuous interdigitation of lateral membranes of adjacent cells, the endothelium forms a leaky barrier, which permits paracellular percolation of aqueous humor from the anterior chamber into the cornea. The same discontinuous tight junctions and interdigitations of the endothelium also function as a barrier by restraining excessive fluid leak into the stroma. Furthermore, the fluid leak is counterbalanced by an active fluid transport of the endothelial cells from the stroma into the anterior chamber of the eye. Thereby, the endothelium carries out hydration control of the corneal stroma, which is essential for maintaining corneal transparency (for review see [11]).

10.1.2 Intercellular Signaling Coordinates Function of the Corneal Endothelium

In order to sustain these physiological functions in the absence of regeneration, the corneal endothelium possesses a number of mechanisms to withstand extracellular

Table 10.1 mRNA expression of a number of purinergic receptors, nucleotidases, connexin isoforms, histamine-, bradykinin-, LPA-, and S1P-receptors, and cell surface markers in BCEC

mRNA	Presence	Accession N°	5′ Forward primer	5′ Reverse primer
P1A1	+	NM_174497	ACATGGAGGTCTTCTATCTGATCC	TCCAGATCTTAAGGAAGGTCACTC
P1A2B	+	XM_608217	GGTCATCTACGTCAAGATCTTCCT	GGATGTACCTGGAAATGATTTTGT
P2X1	–	XM_592814	ACCCATCTACGAGTTCCACG	GGAGCAGAAGGAGGTCACAG
P2X2B	–	XM_605891	GAGACACCCAAGGTGATCGT	CTCCGGTGGTTTCACGTACT
P2X3	+	XM_608941	TGTCTGACTACGTGACCCCA	AGTGAAGTTCTCGGCTTCCA
P2X4	+	BT020829	TGACACAGAAGTGCCAAAGC	GCCTTTCCAAACACGATGAT
P2X5	+	XM_594012	TGTGCTATGAGGACAGCGAC	TGGAGAAGTTGAATTTGGGG
P2X6	–	XM_870597	TTCCAGTGGAGGAGGAGATG	GGGCGTATCTGTCTTGGAGA
P2X7	+	XM_591410	TGCACCTTTCACAAGACTCG	GTGGCCAAACCAAAGTAGGA
P2Y1	+	NM_174410	CCTCTTCTACTCTGGCACCG	AAGCACACATTTCTGGGGTC
P2Y2	+	XM_612432	TGGTCTGTTACGTACTCATGGC	AGCGGAAGGAGTAGTAGAGGGT
P2Y4	+	XM_592537	GCAAGTTCATCCGCTTTCTC	GGTGGTGACGAAGAACAGGT
P2Y6	+	XM_588870	CTATGCCCAAGGTGACCACT	GATTCCTGTGGAGGCAAAGA
P2Y8	–	XM_867568	TCATGATCAACCTGAGCGTC	ACGAAGAGCGTGAAGAGGAA
P2Y10	+	XM_593145	ACGCAACCACCTACATCCTC	TCTGAAGGCTGATGCATGTC
P2Y12	+	AJ623293	GCCGAGTGACAAGACTGTGA	GGCATGCATTTAAGGACGTT
CD39	+	AF005940	CTGGCACCTACATCCTGGTT	GGATCAGGGAGAAGAGGACC
CD73	+	NM_174129	CTGATGATGGACGGAAGGTT	ACTGGACCAGGTCAAAGGTG
E-NPP2	+	XM_001255819	CATCGAAAAGCTGAGGTCTTG	CCACTGCAGGATGGTCAGTAT
E-NPP3	+	BC122742	TCTACATGTTTGAAGGACCCG	GGGATGGTTCATAAAAAGGGA
E-NPP6	+	NM_001040558	CCATCAGATGATCGGGAACTA	GACAGCATTGGCAAAATTGAT
Cx26	+	XM_592125	TACGACCACTACTTCCCCATCT	AACTCGGTCTTTATCTCTCCCC
Cx30	–	NM_001015546	GCGAGACGAGGAGTGAGTTC	AGATGAAGCAATCCACCAGG

Table 10.1 (continued)

mRNA	Presence	Accession N°	5′ Forward primer	5′ Reverse primer
Cx30.3	+	XM_580595	CTCTACATCTTCCATCGCCTCT	GGTCATTTCAGCCTTCATTAGG
Cx31	+	XM_590411	CACCAAGCTCTACGACGACA	GGTGAAGAGCTTCTTCTCAGTAGG
Cx31.1	–	XM_602348	CCGAGGAAAAGAACATCTTCAC	GTCTTGAGGGTGGTCTGGTAAG
Cx32	+	BC111646	GAGTATGGCTCTCTGTCATCTTCA	GTAGCCAGGGTAGAGCAGATAAAA
Cx35.4	–	XM_606006	GAGAGAAAAGGCACAGGAAGAA	TACTTGAGGTTGTGGCATTCAC
Cx36	+	AY150575	GACCGCTACCCCTGTATCAA	CTGACTTTCTCTTGGCCTGG
Cx37	–	XM_583597	ATCTTCATCATCTTCATGCTGGT	TTGTTTCTTGGAAGCAGAGCTAC
Cx40.1	–	XM_600510	TCCACTGGAGACAGAAGAACAA	CTGCAGAGTATTGCAGACGAAC
Cx43	+	NM_174068	GGACATGCACTTGAAGCAGA	TCTTTCCCTTCACACGATCC
Cx45	+	XM_588395	CAGTATTTCCTGTACGGCTTCC	TCGGTGTACTGGATTTGATCTG
Cx46	+	XM_600743	GCAGATCATTTTCGTGTCCA	TCATCCCCTGTTTCAGCTTC
Cx46.6	+	XM_582393	AAGATGAAGAGGCTGAGGACAC	CATACAGAAGTACTGGCCCAC
Cx50	+	XM_876144	CGAGAAAACCATCTTCATCCTC	ACCTCCGTCAAAGGGAAATAAT
Panx1	+	XM_606581	AATTAAAGCCGCAAAGAGCA	AACTGAGCAGCCGGAAGATA
Panx2	+	XM_584601	AGGAGCCCATTTACTGCTACAC	TCCAGGTACTTCTCGAACAGGT
Panx3	+	XM_584064	TTTTGTCCCAGTAACTTCAGCAT	ATCAGTGTCTGAGCTCTCCTGTC
PAR-1	+	XM_604958	GTGGTGTACCCATCCAGTC	CATGGGATCAGAGGAAAGGA
PAR-2	+	XM_604897	CACGATCGTATTCGTGGTTG	ATGAGCAGCCATATTCCCAG
PAR-4	–	XM_583942	GTGCTATGTCACGACGTGCT	CGTACAGGTGGCCGTAGAGT

Table 10.1 (continued)

mRNA	Presence	Accession N°	5′ Forward primer	5′ Reverse primer
H1	+	NM_174083	GCTCATGCTCTGGTTCTATGC	AGAATTTGCCAGTTCTCCAT
H2	+	XM_599517	AACTTGGTGTACGGTCTGGTAGA	GTAAACGAACACGGTGAAGTA
H3	–	XM_867027	CTCATCAGCTATGACCGCTTC	CAGTTGTAGAAGAACTCGGCG
H4	–	XM_001251983	ATAACATTTGTGTCTTTTGGCTCA	TCTTTTTAAGGAATTCAGGTTTGC
BK1	+	XM_610629	AACTTCTTTGCTTTCGTCAACAG	CTCTGTCCTATGAGACGAGGCTA
BK2	+	XM_583508	GCGTGGTGAACACCATACTCTA	CTTGAACTTCTGCATCTCGTTG
LPA1 (edg2)	+	NM_174047	TACCTTATGGCGAATCTGGC	GTTGAAAATGGCCAGAAGA
LPA2 (edg4)	+	XM_586777	CTGCTCATGGTGGCTGTCTA	TGTAGCGAACAGACTCGTGG
LPA3 (edg7)	+	XM_612024	CATGTCGATCATGAGGATGC	GCTGATGGACCCACTCGTAT
S1P1 (edg1)	+	NM_001013585	ACAATGGGAGCAACAGGTTC	GAAGACGCCCAGGACAATAA
S1P2 (edg5)	+	XM_593074	TGGCCTTCATAGCCAATACC	TGCGGATGTACAGAGCTACG
S1P4 (edg6)	+	XM_587940	GTCTCCTCCCGCTCTACTCC	AGACGTTGGAGCCGAAGATA
vWF	+	XM_584169	AGAGAGGAGAGTTCATTTGGGAG	GTCCACTTCATCTTCAGATTTGC
CD31	+	NM_174571	CTGGAGTCTTCAGCCACACA	GCTTCTCTTGAACACCCTGC
CD44	+	BC119907	AAAGGCGAATACAGAACTAACCC	CTACTGTTGACAGCAATGCAAAC
Keratocan	+	BC120375	CAAGGATCTGGTACCTTTATCTTGA	TCTGAAAGGCATTGTCTAGTAGCTT
Lumican	+	BC102271	AACAAGATTAGCAACATCCCTGA	ATCCAGCTCCAACAAAGATGATA

+: mRNA for the protein was detected by reverse-transcriptase PCR with forward and reverse primers given in the table. –: mRNA for the protein could not be detected stresses. Intercellular communication (IC), which enables the cells in the monolayer to react with a coordinated response, possibly helps in resilience of the endothelium against extracellular stresses, such as exposure to inflammatory mediators, chemical and mechanical stimuli or infection.

Purinergic signaling has been shown to be an important factor in the coordination of cellular activity and in the repertoire of defense mechanisms. For example, adenosine is able to enhance integrity of the corneal endothelial barrier via myosin light chain (MLC) dephosphorylation and by stimulation of fluid transport via activation of cAMP-activated Cl^- channels [87, 105]. It was also shown that ATP is able to prevent or rescue the loss of integrity of the corneal endothelial barrier by thrombin, which is secondary to increased MLC phosphorylation [93].

10.1.3 Mechanisms of Intercellular Signaling

Intercellular signaling can take many forms. In many cell types, including endothelial cells, Ca^{2+} waves coordinate the activity of neighboring cells within a tissue. Global Ca^{2+} signals are produced by coordinated activity of elementary events (influx of Ca^{2+} through channels permeable to Ca^{2+}, or release of Ca^{2+} from intracellular stores through e.g. IP3R channels) to generate an intracellular Ca^{2+} wave, which spreads throughout the cell. Ca^{2+} signals then propagate into neighboring cells, generating intercellular Ca^{2+} waves. In non-excitable cells, the mechanism of intercellular Ca^{2+} communication can occur via two pathways. The first pathway, the gap junctional intercellular communication (GJIC), occurs via the diffusion of Ca^{2+} or inositol-1,4,5-trisphosphate (IP_3) through gap junctions that form an intercellular hydrophilic pathway coupling adjacent cells. The second pathway, the paracrine intercellular communication (PIC), involves release of a diffusible extracellular messenger (such as ATP) that can elicit a Ca^{2+} transient in neighboring cells via Ca^{2+} influx or via Ca^{2+} release from intracellular stores (Fig. 10.1A). Both pathways can function in the same cell type, but the two pathways can also provide communication between different cell types.

10.1.4 Mechanisms of GJIC and PIC

10.1.4.1 Cx Gap Junctions Mediate GJIC

Gap junctions are plaques of proteinaceous channels that interconnect the cytoplasm of adjacent cells. Gap junction channels have a unitary conductance of 15–300 pS and a diameter of approximately 1–2 nm, depending on the connexin (Cx) isoform present in the gap junction channel [123]. Gap junction channels are large enough to permit a direct diffusion of ions and molecules smaller than about 1.2 kDa between the cytoplasms of adjacent cells, and therefore are permeable to a number of signaling molecules (e.g., Ca^{2+}, cAMP, IP_3) and metabolites.

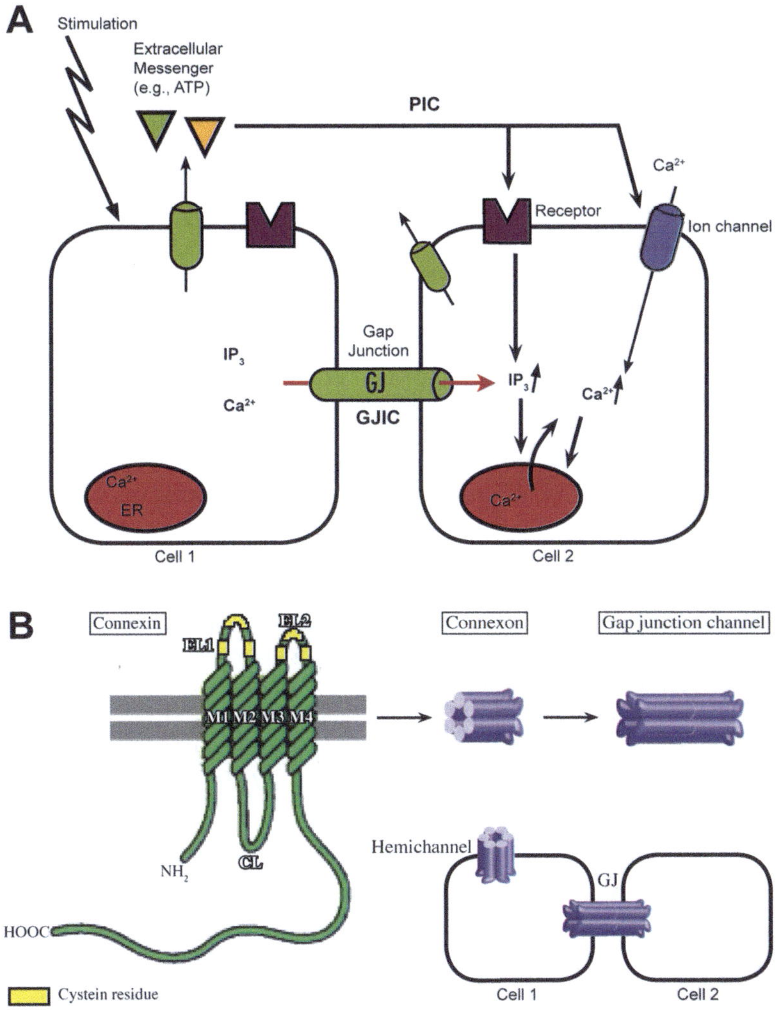

Fig. 10.1 (**a**) Mechanism of the Ca^{2+} wave propagation between cells. A specific stimulus induces a Ca^{2+} transient in the stimulated cell (cell 1) via Ca^{2+} influx and/or Ca^{2+} release from intracellular stores (ER). Diffusion of Ca^{2+} or IP_3 through the gap junctions (GJ) or secretion of an agonist into the extracellular space that diffuses to neighboring cells can induce a Ca^{2+} increase in the neighboring cells, resulting in gap junctional intercellular communication (GJIC) or paracrine intercellular communication (PIC), respectively. (**b**) Structure of connexin, connexon hemichannels and gap junction channels. *Top*: One connexin molecule, showing the four typical transmembrane domains (M1–M4), together with the one cytoplasmic (CL) and the two extracellular (EL1 and EL2) loops. The N- and C-terminal domains face the cytoplasm and are very divergent between connexins. Six connexins form a connexon, which can be homomeric or heteromeric, depending on the connexin isoform composition of the connexon. Connexons from adjacent cells can dock to form gap junction channels. Each cell can contribute the same or a different type of connexon, giving rise to either homotypic or heterotypic intercellular channels. Connexons can also be present in non-junctional membranes as connexon hemichannels

Gap junction channels are formed by the docking of two hemichannels (also called connexons) of adjacent cells. Each connexon comprises six integral transmembrane proteins called connexins (Cxs), which are radially arranged around a central pore (Fig. 10.1B).

Cxs belong to a superfamily that also includes innexins, which form gap junctions in invertebrates, and pannexins (Panx), which are found in many species, including mammals (for review see [129]). Cx isoforms are members of the highly conserved multigenic family of transmembrane proteins consisting of about 20 members, and are named on the basis of their molecular weight (between 26 and 60 kDa) [25]. Corneal endothelial cells express several Cx subtypes, including Cx26, Cx31, Cx32, Cx36, Cx43, Cx45, and Cx50. As in many other cell types, the most abundant Cx subtype is Cx43.

Cxs are folded in the membrane in the approximate shape of an "M". They contain four membrane-spanning domains (M1 to M4), two extracellular loops (named EL1 and EL2), which are involved in the docking of two connexons and which connect M1 with M2 and M3 with M4, respectively, one cytoplasmic loop (CL; connecting M2 and M3), and the cytoplasmic N- and C-terminal domains. The M3 membrane-spanning domain of each Cx contains a high proportion of hydrophilic amino acids, and is the major pore-lining helix of the gap junction channel. M1 contributes to the lining of the pore, while M2 appears to contribute to the mouth of the pore. The sequences of the N- and C- terminal domains in the cytoplasm are very diverse between different Cx subtypes. Both the highly variable cytoplasmic loop and C-terminus may be involved in the cellular regulation of gap junction formation and channel permeability. The activity of the channel is closely regulated by phosphorylation or via protein-protein interactions [46, 90, 42, 60, 92, 47].

10.1.4.2 Hemichannel-Mediated ATP Release can Produce PIC

PIC takes place via the release of signaling molecules in the extracellular space that induce a response in neighbouring target cells expressing receptors for the signaling molecules. The released messengers can act on the receptors on target cells in close proximity (PIC) or on distant target cells (endocrine signaling), and they can also eventually act on the releasing cell (autocrine effect). While some messengers, such as arachidonic acid and nitric oxide (NO), are lipophilic, penetrating the membrane and acting on intracellular receptors, many paracrine messengers are hydrophilic and bind as ligands to receptor proteins that are present in the plasma membrane.

A major form of PIC evoked by hydrophilic messengers is mediated by nucleotides (for review see [19]). Extracellular nucleotides, such as ADP and ATP, as well as adenosine, are important signaling molecules involved in paracrine intercellular communication.

ATP is known to be released from many cell types, and several mechanisms can be involved in the release of ATP and other nucleotides, depending on the conditions and cell types. The contribution of the different pathways to ATP release under various conditions in different cell types is still very controversial.

Several studies provided evidence for release of nucleotides by exocytosis. Other mechanisms include the ATP binding cassette (i.e., ABC transporters, such as cystic fibrosis transmembrane conductance regulator (CFTR)), or stretch- and voltage-gated channels, such as plasmalemmal voltage-dependent anion channel (pl-VDAC; a murine splicing variant of the mitochondrial ATP porin VDAC-1), swelling-activated Gd^{3+}-sensitive anion channel, the stretch-activated cation channels, maxi-anion channels and P2X7 channels, and carriers (for review see [19, 81]).

Recent findings have provided evidence that connexons (the constituent of gap junctions; Fig. 10.1B) are also present in non-junctional parts of the plasma membrane, where they form functional Cx hemichannels that can eventually open under physiological and pathophysiological conditions (for review see [34, 45, 51, 90–92, 38]). Furthermore, evidence has been presented that these hemichannels are involved in ATP release, providing a pathway for PIC [108, 15, 29, 53].

Because gap junction channels and hemichannels are permeable to molecules with molecular weight up to 1.2 kDa, it was thought that maintenance of gradients across the cell membrane requires that hemichannels should have very low open probability under physiological conditions. However, a growing amount of evidence indicates that opening of hemichannels does occur under both physiological and pathological conditions in many cell types [24, 5, 108, 15, 91, 43, 74, 29, 3, 53, 89]. Electrophysiological studies indicate that hemichannels have a low open probability under physiological conditions, but may have a much higher open probability under certain pathological conditions. Cx hemichannels tend to be closed by negative membrane potentials, high concentrations of extracellular Ca^{2+} and intracellular H^+ ions, gap junction blockers and protein phosphorylation. They tend to be opened by positive membrane potentials and low extracellular Ca^{2+}, and possibly by as-yet-unidentified cytoplasmic signaling molecules [92].

The permeability of hemichannels has been evaluated using a number of fluorescent tracers. Cx43 hemichannels are shown to be permeable to lucifer yellow, ethidium bromide, propidium iodide, carboxyfluorescein, 7-hydroxycoumarin-3-carboxylic acid and fura-2 [61, 58, 78, 108]. Furthermore, Cx43 hemichannels are permeable to small, physiologically significant molecules, including NAD^+, glutamate, ATP and IP_3 [18, 88, 108, 53] (for review see [90]), and may mediate paracrine as well as autocrine signaling.

Low extracellular [Ca^{2+}] or application of quinine, two conditions known to open hemichannels, elicit local and propagating Ca^{2+} signals [108]. Mechanical stimulation triggers ATP release through Cx43 hemichannels, and thus initiates propagation of Ca^{2+} waves in astrocytes and other electrically non-excitable cells [5, 108]. When mechanically stimulated or exposed to extracellular medium with low [Ca^{2+}], astrocytes showed membrane currents and uptake of small fluorescent dyes that were inhibited by flufenamic acid or Gd^{3+}, providing evidence for opening of Cx hemichannels [5, 108].

Panx channels and the Panx-P2X7 complex have also been suggested to be mechanisms for ATP release. Recent data support the concept that Panxs form non-junctional channels that play paracrine roles by releasing ATP [110, 96, 104, 95].

While ATP can be released into the extracellular space, it is not the only purinergic nucleotide that can be present in the extracellular medium after stimulation of cells. ATP can be hydrolyzed by ecto-nucleotidases to ADP, AMP and adenosine (for review see [128]). Ecto-nucleotidases belong to different subfamilies. E-NPPs hydrolyze ATP to AMP, E-NTPDases can hydrolyze ATP to ADP and ADP to AMP, while ecto-5′ nucleotidase converts AMP to adenosine. Adenosine can be further converted to inosine by adenosine deaminase. The ecto-nucleotidases are important for the relation of purinergic IC. Since they regulate the local levels of the nucleotides, they can influence the magnitude and duration of purinergic signaling, and because of the presence of different nucleotide receptors on the extracellular surface of the membranes, they can affect the type of response of cells to purinergic signaling [65].

The purinergic messengers involved in IC can act on a large variety of receptors belonging to two main families: P1 or adenosine receptors, and P2 receptors for which ATP (or ADP) are physiological agonists. The P1 receptor family is further subdivided into four subtypes: A1, A2A, A2B and A3. All adenosine receptors have seven putative transmembrane domains with an extracellular N-terminus and a cytoplasmic C-terminus, and they couple to G proteins. The A2A and A2B receptors preferably interact with members of the Gs family of G proteins, stimulating adenylyl cyclase, and the A1 and A3 receptors interact with Gi/o proteins, inhibiting adenylyl cyclase.

The P2 receptors are divided into ionotropic P2X and metabotropic (G protein-coupled) P2Y receptors, respectively. Mammalian P2X receptors belong to a family of at least seven proteins (P2X1-P2X7) of ligand-gated ion channels, which allow Ca^{2+} influx from the extracellular space upon activation. P2Y receptors form a large subfamily of purine and pyrimidine nucleotide receptors that are coupled to G proteins. They stimulate a release of intracellular Ca^{2+} from stores through phospholipase C (PLC)-mediated PIP_2 hydrolysis and activation of the IP_3 pathway. There are at least eight subtypes of P2Y receptors, namely P2Y1, P2Y2, P2Y4, P2Y6, P2Y11, P2Y12, P2Y13 and P2Y14, but purine nucleotides are full agonists only for P2Y1, P2Y2, P2Y11, P2Y12 and P2Y13. P2Y1 and P2Y12 receptors are preferentially activated by ADP, P2Y2 and P2Y4 by UTP, P2Y6 by UDP, and P2Y2 and P2Y11 are mainly activated by ATP (IUPHAR Database).

10.2 AIM

In several cell types, IC is mediated by both GJIC and purinergic PIC. Corneal endothelial cells contain gap junctions and also express purinergic receptors. It is, therefore, likely that both types of IC could be present in the corneal endothelial monolayer, and play a role in homeostasis and in the coordination of cellular activity. The importance of GJIC and PIC in normal conditions and in response to extracellular stress remains poorly understood. However, it is well known that senescence, injury and inflammation can cause intercellular gaps in the corneal endothelial monolayer, which not only affect barrier integrity, but could also hamper GJIC. In such situations, functional capacity and restoration of integrity of the corneal monolayer could depend to a large extent on PIC. Therefore, our aim was to investigate the contribution, role and mechanism of PIC and GJIC in corneal endothelial cells. In this chapter we describe some of our experimental data about IC in corneal endothelial cells, with emphasis on the mechanism, regulation and possible role of hemichannel-mediated purinergic PIC.

10.3 PIC in BCEC

10.3.1 IC in BCEC Consists of GJIC and PIC

BCEC subjected to mechanical stimulation of a single cell in a confluent monolayer show a transient $[Ca^{2+}]_i$ rise. This Ca^{2+} transient propagates as a decremental intercellular Ca^{2+} wave, which activates cells up to a distance of four to six neighboring cells away from the mechanically stimulated cell (Fig. 10.2). The amplitude of the Ca^{2+} rise and percentage of responsive cells decrease with increasing distance from the mechanically stimulated cell [44]. We investigated the mechanism of the Ca^{2+} wave propagation in BCEC to identify whether the propagation is mediated by GJIC and/or PIC.

In scrape-loading experiments with the membrane-impermeable dye lucifer yellow, and in fluorescence recovery after photobleaching (FRAP) experiments using carboxyfluorescein, significant transfer of the hydrophilic dyes between different cells was observed. These results provide evidence for the presence of functional GJIC in BCEC. However, Ca^{2+} wave propagation was also found to activate cells without cell-cell contacts in non-confluent monolayers obtained by sparse seeding of cells. Furthermore, the Ca^{2+} wave was also able to cross cell-free lanes obtained by deliberate scratching of confluent monolayers [43, 44]. These experiments clearly demonstrate the contribution of an extracellular messenger to the intercellular Ca^{2+} wave propagation. Together these experiments provided evidence that both GJIC and PIC contribute to IC in BCEC.

10.3.2 Mechanism of PIC in BCEC

10.3.2.1 PIC is Mediated Via Purinergic Signaling

ATP is a prime candidate for eliciting PIC-induced Ca^{2+} wave propagation through its action on the P2 family of purinergic receptors, which can either cause Ca^{2+} influx via P2X receptors or evoke IP_3-induced Ca^{2+} release from intracellular stores by activating PLC via P2Y receptors.

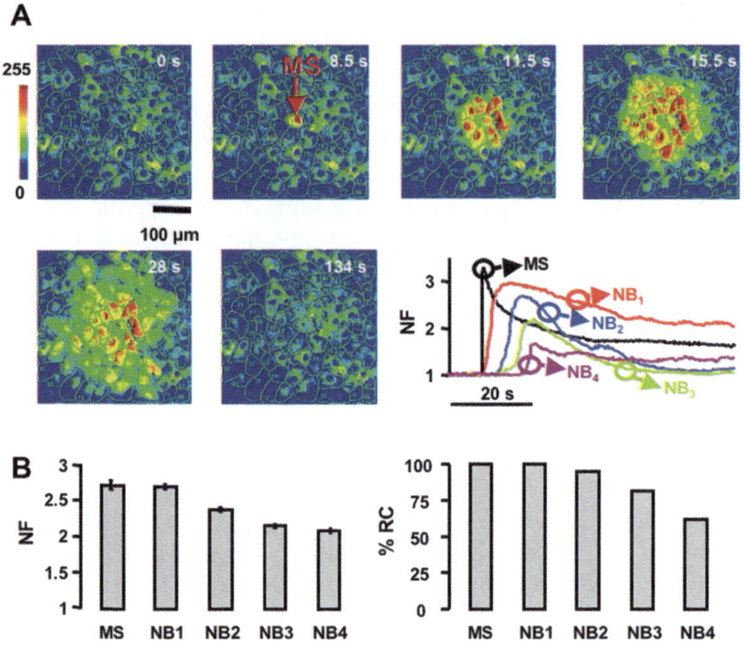

Fig. 10.2 Example of intercellular Ca^{2+} wave propagation elicited by point mechanical stimulation of a single BCEC cell in cells loaded with Fluo-4 AM. (**Panel A**) Fluorescence images at different time points following mechanical stimulation of a single cell. The intensity scale at the *top left* represents the fluorescence changes induced by the $[Ca^{2+}]_i$ changes. The first image shows the fluorescence intensities before stimulation. The arrow in the second image (time = 8.5 s) identifies the mechanically stimulated (MS) cell. The time course of $[Ca^{2+}]_i$-sensitive fluorescence of MS and a typical cell from different neighboring (NB) cell layers is shown at the lower right of the panel. (**Panel B**) The bar graph at the left of the panel shows the maximal normalized fluorescence (NF) in the MS cell and in responsive NB cells. The bar graph of the lower right panel shows the percentage of responsive cells (%RC) for each of the NB cell layers. From [44] with permission. Copyright: *Association for Research in Vision and Ophthalmology*

We first examined the presence of purinergic receptors in BCEC. RT-PCR demonstrated expression in BCEC of mRNA for P2Y1, P2Y2, P2Y4, P2Y6, P2Y12 and for P2X4, P2X5 and P2X7 (Table 10.1). It was previously shown that exogenous application of purinergic agonist results in Ca^{2+} mobilization in BCEC [107]. Using the fluorescent Ca^{2+} indicator Fluo-4-AM, or measuring intracellular Ca^{2+} using a high throughput Fura-2 assay, we demonstrated that application of extracellular ATP induces a Ca^{2+} rise in BCEC, with an EC_{50} for ATP of 1.5 μM [44].

To examine whether ATP is the messenger in the intercellular propagation of the Ca^{2+} wave in BCEC, we used P2 receptor antagonists. Pretreatment of cells with the non-selective purinergic receptor antagonist suramin (200 μM) indeed inhibited Ca^{2+} wave propagation. Also, exogenous application of apyrases (10 U/ml), which hydrolyze extracellular nucleotides, decreased the active area of the Ca^{2+} wave, which is the total area of cells that show a rise in $[Ca^{2+}]_i$ [44] (Fig. 10.3). Application

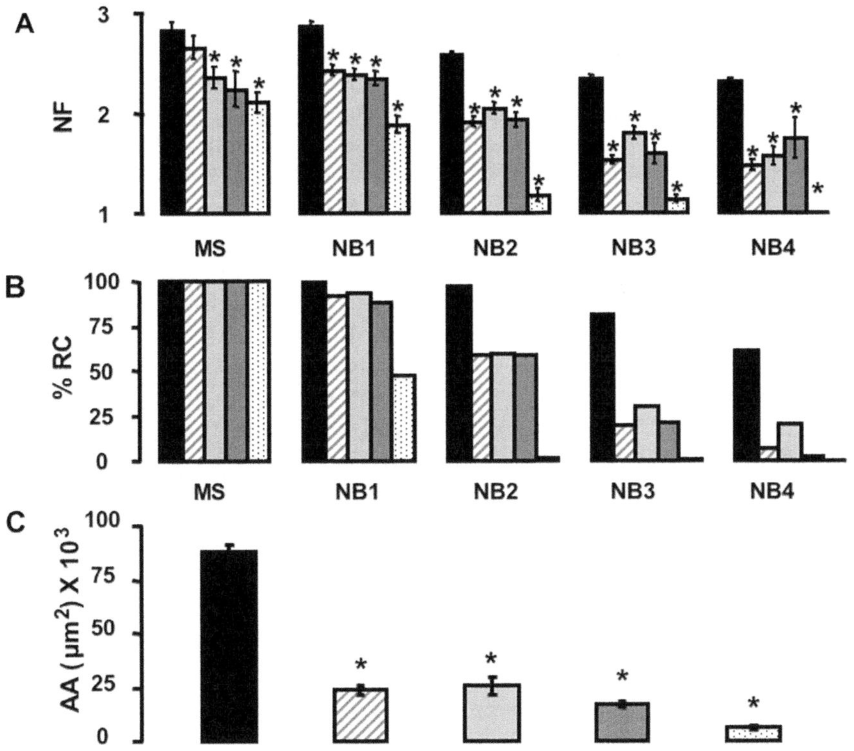

Fig. 10.3 Reduced Ca^{2+} wave propagation in the presence of purinergic modulators: Cells were treated with suramin (200 μM), apyrase VI (10 U/ml), apyrase VII (10 U/ml), or apyrase VI+VII (5 U/ml each) for 30 min. The Ca^{2+}wave propagation in response to mechanical stimulation is represented in control (*black bars*) (N=53), suramin (*striped bars*) (N=42), apyrase VI (*light grey bars*) (N=25), apyrase VII (*dark grey bars*) (N=17), and apyrase VI+VII (*dotted bars*) (N=21) treated cells. Panel A shows the bar graph of the maximal normalized fluorescence (NF) and panel B the bar graph of the percentage responsive cells (%RC) for MS and neighboring (NB) cell layers NB1 to NB4. Panel C shows the active areas (AA) for the different experimental conditions. $^{*}p <$ 0.05 vs. control. From [44] with permission. Copyright: *Association for Research in Vision and Ophthalmology*

of the ectonucleotidase inhibitor ARL-67156 (100 μM) significantly enhanced the active area of the Ca^{2+} wave. These data provide evidence that PIC via an agonist of P2 receptors is involved in the IC [44].

Brilliant Blue G (BBG), a selective antagonist for P2X7-receptors [50], which at higher concentrations also inhibits other P2X receptors such as P2X4 and P2X5, had no significant influence on the propagation of the Ca^{2+} wave. In contrast, the PLC inhibitor U-73122 (10 μM) caused a marked reduction of the wave propagation, providing evidence that the P2 receptors involved belong to the class of PLCβ-coupled P2Y receptors. We can therefore conclude that the purinergic receptors that are mainly responsible for the paracrine intercellular Ca^{2+} wave propagation in BCEC are of the P2Y subtype.

Since P2Y4 and P2Y6 are pyrimidine receptors with very low sensitivity to ATP, these receptors are unlikely to have an important contribution to the Ca^{2+} wave propagation. To investigate the involvement of P2Y1 receptors (which can be activated by ATP and ADP) and P2Y2 receptors (which are activated by ATP but only very weakly sensitive to ADP), we also examined the effect of nucleotide hydrolysis by exogenous apyrase VI (which has a high ATPase/ADPase ratio) and apyrase VII (which preferentially hydrolyses ADP). Apyrase VII caused a somewhat stronger inhibition in the outermost cell layers when compared to apyrase VI. The combination of 5 U/ml apyrase VI and 5 U/ml apyrase VII had a cumulative effect, causing a more pronounced inhibition of the Ca^{2+} wave than either 10 U/ml apyrase VI or 10 U/ml apyrase VII (Fig. 10.3). These experiments provide evidence that both ATP and ADP are involved [44], and suggest that P2Y1 and P2Y2 receptors are involved in PIC in BCEC. Furthermore, these results indicate that ATP is released in response to mechanical stimulation. Direct evidence for the involvement of ATP release in the intercellular Ca^{2+} wave propagation was obtained by ATP-dependent luciferin-luciferase bioluminescence measurements, showing that mechanical stimulation increases the amount of ATP in the extracellular medium [43].

10.3.2.2 Mechanism of ATP Release

Around the time of our findings, Stout et al. [108] and Braet et al. [15] provided evidence for the involvement of hemichannels in ATP release during Ca^{2+} wave propagation in astrocytes. We therefore examined Cx hemichannels as candidate pathways for ATP release involved in Ca^{2+} wave propagation in BCECs.

Using RT-PCR, we showed the presence of mRNA for Cx26, Cx43, Cx45 and Cx50 in BCEC (Table 10.1) and we demonstrated expression of Cx43 protein in the membrane, using Western blot and immunofluorescence (Fig. 10.4).

Next, we investigated the contribution of Cx hemichannels to PIC in BCECs using the hemichannel blockers flufenamic acid (FFA) and ^{43}Gap26 (VCYDKSFPISHVR) [37]. ^{43}Gap26 is a Cx mimetic peptide with a sequence identical to a part of the EL1 sub-domain (first extracellular loop sequence) of Cx43, which constitutes a significant portion of the pore-lining. This peptide has been shown to inhibit Cx43-mediated hemichannel opening [15]. Application of FFA or ^{43}Gap26 resulted in a pronounced inhibition of the propagation of the Ca^{2+} wave evoked by mechanical stimulation (Fig. 10.5).

Since Stout et al. [108] have shown that Cx hemichannels open upon exposure to low extracellular Ca^{2+} and show permeability for a number of hydrophilic dyes such as lucifer yellow, we performed dye uptake experiments in Ca^{2+}-free solution using lucifer yellow as a membrane-impermeable hemichannel-permeable reporter dye. Removal of extracellular Ca^{2+}-induced uptake of lucifer yellow, which was completely inhibited by FFA and ^{43}Gap26 [43]. Our observations provide evidence that hemichannels are present in BCECs and can be blocked by FFA and ^{43}Gap26.

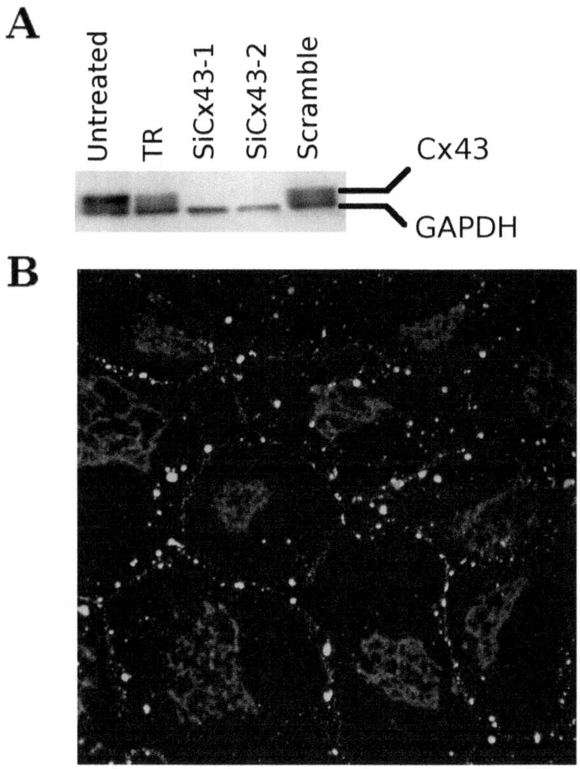

Fig. 10.4 Presence of Cx43 in BCEC. **A**: Western blotting of Cx43 (43 kDa) in BCEC, with knock-down of endogenous Cx43 by RNA silencing (100 nM duplex). TR represents treatment with transfection reagent only. Scrambled siRNA duplex of Cx43-1 was used as negative control. GAPDH (37 kDa) levels are visualized as a loading control. Cx43-1 (region 322-344): 5'-GAAGGAGGAGGAACUCAAAdTdT Cx43-2 (region 883-905): 5'-CAAUUCUUCCUGCCGCAAUdTdT Scramble (region 322-344): 5'-GGUAAACGGAACGAGAAGAdTdT **B**: Immunofluorescence staining of Cx43 in BCEC. A monoclonal mouse anti-Cx43 antibody (Sigma, Clone CXN-6; Catalog number C8093) was used for Cx43-detection in Western blot and immunofluorescence

This conclusion is further supported by experiments demonstrating that application of Ca^{2+}-free solution resulted in immediate marked increase in the rate of ATP release. Similar to their effects on the dye uptake, ^{43}Gap26 and FFA also inhibited the enhancement of ATP release in the absence of extracellular Ca^{2+}, indicating that both agents inhibit the release by the same mechanism [43].

Direct measurement of ATP in samples of the bathing fluid immediately after mechanical stimulation demonstrated enhanced ATP release after mechanical stimulation (Fig. 10.7A). ^{43}Gap26 and FFA inhibited this ATP release [43]. Our experiments therefore demonstrated that functional hemichannels are present in BCECs, which are opened by low extracellular Ca^{2+} and in response to mechanical stimulation, and that ATP release through hemichannels contributes to mechanical stimulation-induced intercellular Ca^{2+} wave propagation.

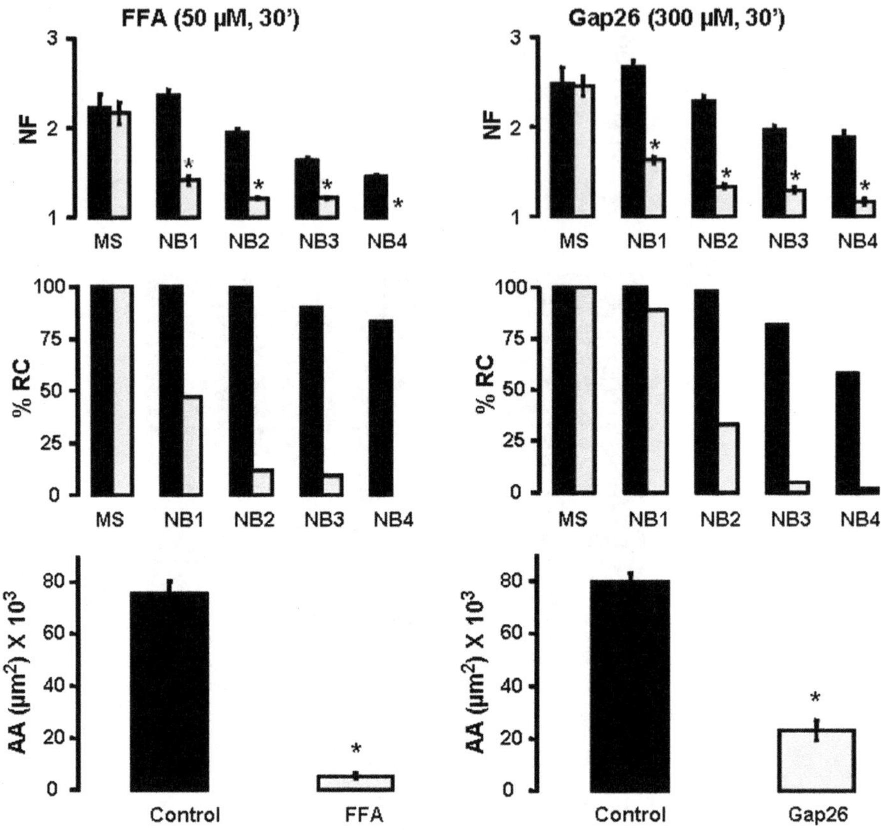

Fig. 10.5 Inhibition of Ca^{2+} wave propagation in BCEC by FFA and [43]Gap26. The bar graphs show the effect of a 30 min application of 50 μM FFA (*left*) or 300 μM [43]Gap26 (*right*) on the maximal normalized fluorescence (NF), percentage of responsive cells (%RC) and active area (AA) of the Ca^{2+} wave propagation in response to point mechanical stimulation. The black bars represent control conditions while the white bars show the values in the presence of the drug. $^*p < 0.05$ vs. control. From [43] with permission. Copyright: *Association for Research in Vision and Ophthalmology*

10.3.3 Regulation of PIC

10.3.3.1 PIC is Inhibited by Histamine and Thrombin

Histamine [106] and thrombin [94] induce actomyosin contraction, leading to breakdown of barrier integrity of the corneal endothelial cell monolayer, concomitant with formation of gaps between the cells. Since formation of intercellular gaps could influence IC, we investigated whether histamine and thrombin affect intercellular propagation of Ca^{2+} waves in BCEC.

RT-PCR showed transcripts for histamine receptors HR1 and HR2 in BCEC (Table 10.1). Histamine significantly reduced the active area of the Ca^{2+} wave

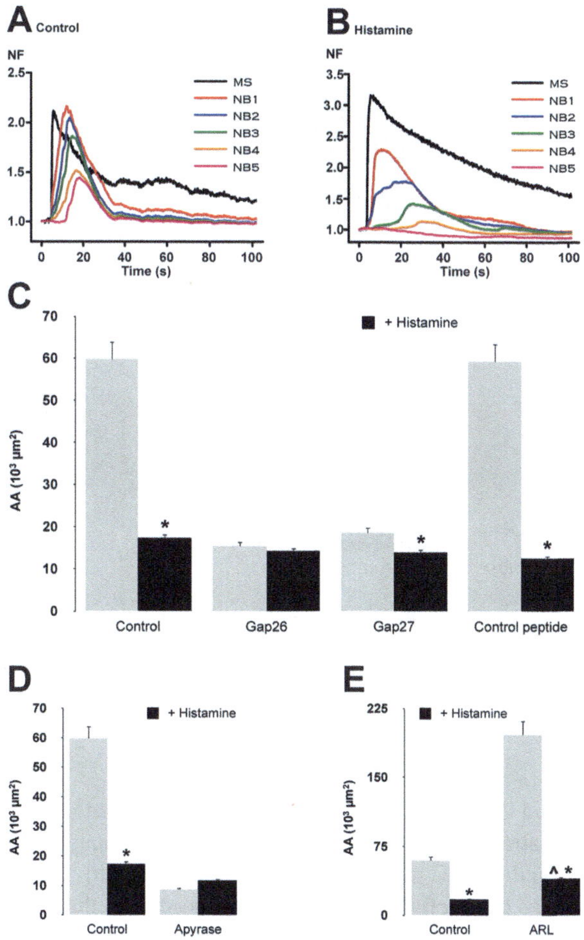

Fig. 10.6 Effect of histamine on the Ca^{2+} wave propagation. (**a and b**) Normalized Fluorescence (NF) values of the Ca^{2+} transient measured with Fluo-4AM in a typical experiment in the mechanically stimulated (MS) cell and in neighboring (NB) cell layers NB1-NB5 surrounding the mechanically stimulated cell. (**a**) in control. (**b**) in the presence of histamine (100 μM for 10 min). (**c**): Active area (AA) of the Ca^{2+} wave propagation in control conditions and in the presence of histamine, and in the presence of histamine after preincubation with ^{43}Gap26 or ^{43}Gap27 (300 μM for 30 min). In the presence of ^{43}Gap26, histamine does not cause a significant further decrease of the active area. ($N = 50$) $^{*}p < 0.001$ for presence versus absence of histamine. (**d**) Active area of the Ca^{2+} wave propagation in control conditions and in the presence of histamine, and in the presence of histamine after preincubation with apyrase VI + apyrase VII (5 U/ml each). In the presence of apyrase, histamine does not cause a further decrease of the active area. ($N = 35$). $^{*}p < 0.001$ for presence versus absence of histamine. (**e**) Active area of the Ca^{2+} wave propagation in control conditions and in the presence of histamine, and in the presence of histamine after preincubation with ARL-67156 (100 μM for 30 min). The very marked increase in active area by ARL is abolished in the presence of histamine ($N = 30$). $^{*}p < 0.001$ for presence versus absence of histamine. $\hat{}p < 0.001$ for ARL vs. control

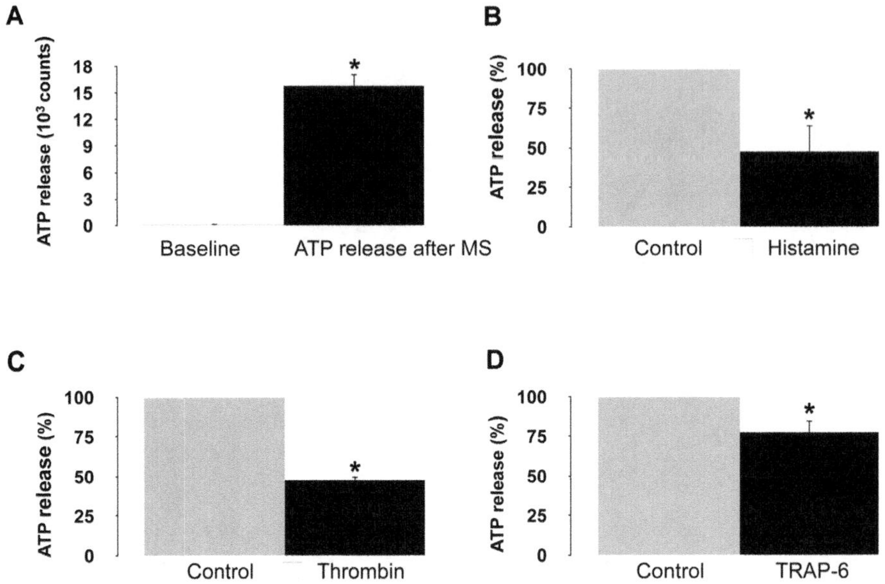

Fig. 10.7 ATP release after mechanical stimulation in control and in the presence of histamine, thrombin or TRAP-6. ATP release in basal conditions and after mechanical stimulation (MS) of BCEC ($N = 63$) (**a**). Effect of histamine (100 μM for 10 min; $N = 8$) (**b**), thrombin (2 U/ml for 5 min; $N = 55$) (**c**), and TRAP-6 (10 μM for 30 min; $N = 16$) (**d**) on ATP release after mechanical stimulation. $^{*}p < 0.05$

(Fig. 10.6), but did not significantly reduce the fluorescence recovery in FRAP experiments, indicating that the inhibition of Ca^{2+} wave propagation by histamine is not via an effect on the GJIC pathway. In cells pretreated with ^{43}Gap26 (Fig. 10.6c) or apyrases (Fig. 10.6d), activation of histamine receptors did not cause a significant further reduction of Ca^{2+} wave propagation, providing evidence that histamine exerts its effect by inhibiting PIC. In consistence with this conclusion, histamine precluded enhancement of Ca^{2+} wave propagation by the ecto-nucleotidase inhibitor ARL-67156 (Fig. 10.6e). Exposure of BCEC to histamine in Ca^{2+}-free medium led to complete inhibition of lucifer yellow uptake. Histamine also markedly reduced ATP release upon mechanical stimulation (Fig. 10.7b). These experiments provide evidence that histamine inhibits PIC by inhibiting the hemichannel-mediated ATP release.

To investigate whether thrombin also affects IC, we first examined expression of PAR receptors. RT-PCR showed transcripts for the PAR receptors PAR-1 and PAR-2, but not for PAR-4 (Table 10.1). Immunocytochemistry showed thrombin-sensitive PAR receptors as well as trypsin-sensitive PAR-2 receptors. Both thrombin and the selective PAR-1 agonist TRAP-6 strongly reduced the active area of the Ca^{2+} wave (Fig. 10.8) [26]. The effect of thrombin and TRAP-6 on the Ca^{2+} wave was inhibited by a peptide antagonist of PAR-1, but not by the indazole derivative YD-3, a selective PAR-4 antagonist. While thrombin and TRAP-6 reduced the fluorescence

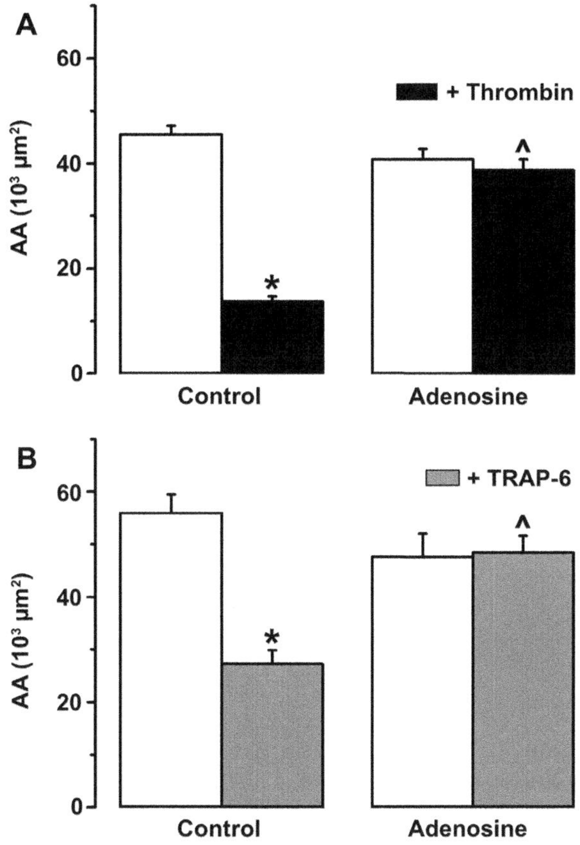

Fig. 10.8 Effect of PAR-1 receptor agonists on the active area of the Ca^{2+} wave in control and in cells pretreated with adenosine. The *bar graphs* in the *left panels* show the reduction of the active area (AA) of the Ca^{2+} wave propagation by thrombin (2 U/ml for 5 min; $N = 145$) (**a**) and by TRAP-6 (10 µM for 30 min; $N = 25$) (**b**): The *bar graphs* in the *right panels* show the inhibition of the effect of thrombin and TRAP-6 by pretreatment of the cells with adenosine (200 µM for 30 min) $^{*}p < 0.001$ for comparison between aa in the presence vs. absence of the PAR-1 agonist (i.e., comparison of *black or grey bars* with *white bars* in each condition). $^{\wedge}p < 0.001$ for comparison between AA in the presence vs. absence of adenosine (i.e., comparison of filled (*black* respectively *grey*) bar with corresponding *filled bar* in control condition. From [27] with permission. Copyright: *Association for Research in Vision and Ophthalmology*

recovery in FRAP experiments, indicating an effect on GJIC, the main effect of these agents was a reduction of PIC. Activation of PAR-1 did not significantly affect the Ca^{2+} wave propagation in cells pretreated with the hemichannel blocker ^{43}Gap26 or in the presence of exogenous apyrases. Thrombin abolished enhancement of the Ca^{2+} wave propagation by the ectonucleotidase inhibitor ARL-67156 [26]. These experiments demonstrate that the effect of thrombin on Ca^{2+} wave propagation is mainly due to inhibition of purinergic PIC.

10.3.3.2 Effect of Thrombin on PIC is Inhibited by Adenosine

Adenosine was previously shown to decrease the permeability of the corneal endothelial barrier [87]. We therefore investigated the effect of adenosine on IC. Pretreatment with adenosine prevented the inhibitory effect of thrombin on intercellular Ca^{2+} wave propagation, while it had no significant effect on Ca^{2+} wave propagation in control conditions (Fig. 10.8). NECA (a potent A2B agonist) and forskolin, agents known to elevate cAMP in BCEC, also suppressed the effect of thrombin. The A1 receptor agonist CPA (N6-cyclopentyladenosine) failed to inhibit the effect of thrombin [27].

Similarly to the effects on Ca^{2+} wave propagation, adenosine prevented the thrombin-induced reduction in the fluorescence recovery during photobleaching experiments, providing evidence for inhibition by adenosine of the effect of thrombin on GJIC. Pretreatment with adenosine also prevented both thrombin and TRAP-6 from blocking the uptake of lucifer yellow in a Ca^{2+}-free medium. Adenosine was ineffective in overcoming the inhibition of lucifer yellow by ^{43}Gap26 [27]. These experiments provide evidence that the main mechanism of the effect of adenosine in overcoming the inhibition of IC by thrombin involves the hemichannel-mediated paracrine pathway. In consistence with this conclusion, the thrombin-induced inhibition of ATP release was also overcome by pretreatment with adenosine.

10.3.3.3 Mechanism of Effect of Effects of Thrombin and Adenosine on Intercellular Communication

Signaling Pathways Initiated by Thrombin and Adenosine

It has been demonstrated that the thrombin-induced decrease in barrier integrity of BCEC is mediated via myosin light chain (MLC) phosphorylation [94]. MLC phosphorylation is driven by MLC kinase (MLCK), while its dephosphorylation is brought about by myosin light chain phosphatase (MLCP). MLCK can be activated by thrombin, since PAR-1 is coupled to $G_{\alpha q/11}$, which results in Ca^{2+} release, and hence Ca^{2+}-calmodulin binding to MLCK, which activates the kinase. PAR-1 receptors also activate PKC via $G_{\alpha q/11}$ and Rho kinase via $G_{\alpha 12/13}$; both pathways contribute to inactivation of MLCP. The transduction of the thrombin effect on MLC phosphorylation can thus occur via two pathways, namely inhibition of the MLCP or stimulation of the MLCK (Fig. 10.9). We performed experiments to investigate whether thrombin exerts its effect on the Ca^{2+} wave propagation through the same transduction cascade that decreases the barrier integrity.

Pretreatment with MLCK inhibitor ML-7, with the Rho kinase inhibitor Y-27632 or with the PKC inhibitor chelerythrine caused a significant suppression of the effect of thrombin or TRAP-6 on the propagation of the Ca^{2+} wave. Upon pretreatment of the cells with a combination of Y-27632 plus chelerythrine, or with the combination of ML-7 plus Y-27632 plus chelerythrine, application of thrombin did not significantly decrease the active area. The lack of a significant effect of thrombin in the

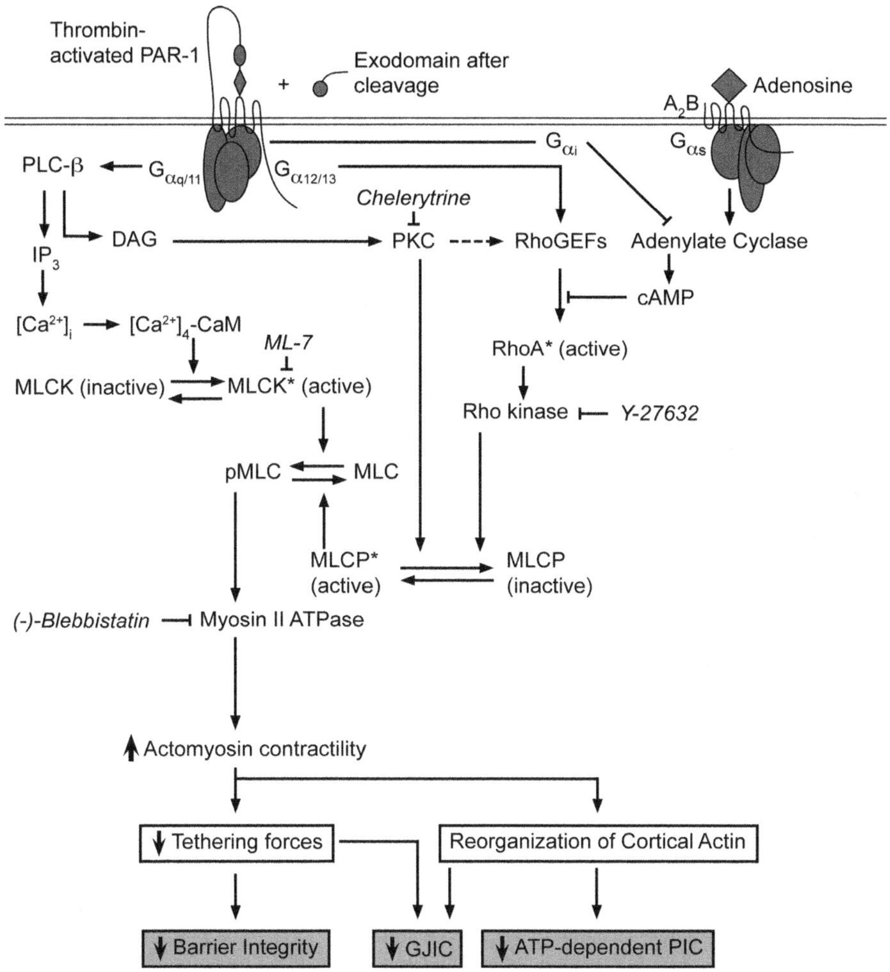

Fig. 10.9 Transduction of the effect of thrombin and adenosine on intercellular communication in BCEC

presence of the combination of these three protein kinase inhibitors provides evidence that the major pathways for the effect of thrombin on the active area are via MLCK, Rho kinase and PKC [26].

Our experiments showed that adenosine, which elevates cAMP levels through activation of $G_{\alpha s}$-coupled A2B, inhibits the effect of thrombin on IC. Since PKA inhibits RhoA activation, elevated cAMP causes MLC dephosphorylation. Pretreatment of BCEC with adenosine reduced the thrombin-induced MLC phosphorylation [80]. Therefore, these experiments suggest that adenosine decreases the inhibitory effect of thrombin on ATP release and PIC by inhibition of RhoA through elevated cAMP, and thereby reduces MLC phosphorylation [27].

Involvement of Cytoskeleton Contraction in Thrombin-Induced Effects on IC

Thrombin induces actomyosin contraction in corneal endothelial cells, leading to significant changes in the organization of the cortical actin cytoskeleton. Such changes in the cytoskeleton have been implicated in loss of barrier integrity of corneal endothelial monolayers [94]. Specifically, thrombin, which can be generated in the cornea [6], may lead to onset of centripetal forces by the peri-junctional actomyosin ring (PAMR), a dense band of actin at the apical junctional complex. These forces are opposed by intercellular tethering forces, which are essential to establish interactions of the transmembrane proteins that make up the tight junctions. Since the formation of intercellular gaps due to actomyosin contraction could also influence GJIC, we investigated a possible role of contractility in the propagation of intercellular Ca^{2+} waves by using blebbistatin, a selective myosin II ATPase inhibitor [109, 62, 73].

Pretreatment with blebbistatin (5 μM for 20 min) or its nitro derivative [64] prevented the thrombin-induced inhibition of the Ca^{2+} wave [79]. Neither photo-inactivated blebbistatin nor the inactive enantiomers prevented the thrombin effect. Blebbistatin also prevented thrombin-induced inhibition of lucifer yellow uptake, ATP release and FRAP, indicating that it prevented the thrombin effect on PIC and GJIC. In the absence of thrombin, blebbistatin had no significant effect on PIC or GJIC. The drug had no influence on MLC phosphorylation or on $[Ca^{2+}]_i$ transients in response to thrombin or ATP [79]. Our findings demonstrate that myosin II-mediated actomyosin contractility plays a central role in thrombin-induced inhibition of GJIC and of hemichannel-mediated PIC.

10.4 Discussion

10.4.1 Contribution of Cx Hemichannels Towards IC in BCEC

Our experiments demonstrate that intercellular Ca^{2+} wave propagation in BCEC consists of both PIC and GJIC, and that PIC is the main contributor to IC. Furthermore, our experiments show that ATP and ADP are the mediators of this PIC, acting via activation of P2Y receptors, and that ATP is released via hemichannels. Our results also provide evidence that the hemichannels involved in PIC in corneal endothelial cells are of Cx43 isoform.

Recent experiments from other laboratories have provided strong evidence for Panx hemichannels [17] as pathways for ATP release in many different cell types (for review see [95]). The Panx subtype Panx1 is widely distributed among tissues that exhibit IC via Ca^{2+} waves [17]. The channel formed by Panx1 can be opened by mechanical perturbation at the resting membrane potential [7]. The channel is permeable to ATP and can be opened at physiological Ca^{2+} levels [17]. Since high concentrations of Cx mimetic peptides were shown to reduce the current through Panx hemichannels, and since opening of Cx hemichannels has not been demonstrated by electrophysiological methods under physiological conditions, it has been argued that the hemichannels involved in ATP release in different preparations are Panx hemichannels (see [95]). There is, however, compelling evidence for the presence and opening of Cx hemichannels in a number of different conditions. Direct electrophysiological evidence for Cx43 hemichannel opening was demonstrated in Cx43 overexpressing HeLa cells and C6 cells [23, 22, 53]. Romanov et al. [89] provided evidence that Cx hemichannels are responsible for ATP release in type II taste cells, based on sensitivity to Cx- and Panx-mimetic peptides, effect of carbenoxolone and kinetics of voltage-dependence. Anselmi et al. [3] provided evidence for propagation of Ca^{2+} signals via Cx hemichannel-mediated ATP release in the inner ear. Single channel experiments in astrocytes have also demonstrated Cx43-hemichannel-mediated ATP release [53]. Cx32 hemichannel-mediated ATP release upon rise in intracellular Ca^{2+} concentration has been established in a number of different cell types [29, 30, 3].

Since Panx hemichannels do not open in Ca^{2+}-free conditions [16], our experiments provide evidence that the ATP release in BCEC is mediated via Cx43 hemichannels. Furthermore, Gap26 almost completely blocked Ca^{2+} wave propagation, lucifer yellow uptake and ATP release [43]. In addition, the Ca^{2+} wave propagation was not significantly inhibited by 10 μM carbenoxolone, which is known to inhibit Panx hemichannels with an IC_{50} of ~5 μM[16]. Furthermore, the Ca^{2+} wave propagation in BCEC was almost completely inhibited by 50 μM flufenamic acid [43], which has only limited effect on Panx hemichannels [16]. In addition, preliminary evidence from experiments in our laboratory shows that cell-permeable TAT-peptides corresponding to the L2 sequence of Cx43 (DGANVDMHLKQIEIKKFKYGIEEHGK) significantly inhibit the Ca^{2+} wave propagation in BCEC by more than 50%, while the H126K/I130N mutant used as control peptide [100, 101] only causes a limited reduction of the active area, which was, however, not significant (unpublished findings). These experiments provide evidence that the ATP release in BCEC is mediated by hemichannels formed by Cx43.

10.4.2 Effect of Histamine and Thrombin on PIC in BCEC

Our experiments demonstrate that histamine and thrombin, two agents that decrease integrity of the endothelial barrier [94, 106], markedly reduce the purinergic signaling by inhibiting ATP release via Cx43 hemichannels. We demonstrated that thrombin enhances MLC phosphorylation through activation of PAR-1 receptors [26]. As in smooth muscle cells, thrombin regulates the level of MLC phosphorylation in the endothelial cells by two opposing pathways: activation of myosin light chain kinase (MLCK)-driven phosphorylation, and myosin light chain phosphatase (MLCP)-driven dephosphorylation of MLC. Specifically, by increasing $[Ca^{2+}]_i$, thrombin activates MLCK via Ca^{2+}-Calmodulin, while it causes inactivation of MLC phosphatase by mobilizing the RhoA-Rho kinase axis (Fig. 10.9). Since phosphorylation of MLC regulates myosin II activity in BCEC, we investigated the role of contractility of the cytoskeleton in the effect of thrombin. Inhibition of the myosin II ATPase activity by blebbistatin inhibited the effect of thrombin, suggesting that contractility of the actin cytoskeleton, including that of PAMR, plays an important regulatory role in the activity of the hemichannels (Fig. 10.9).

Our findings on thrombin-mediated inhibition of hemichannel-mediated IC help us understand the role of IC in corneal endothelial cells under physiological and pathological situations. Thus, our results suggest that thrombin, besides enhancing the permeability of the corneal endothelial monolayer, decreases the IC in corneal endothelial cells, which may relate to pathological ophthalmic conditions that result in inflammation, e.g., uveitis and infection. Endotoxin (LPS), a component of the cell wall of gram-negative bacteria, triggers a cascade for macrophagal/endothelial cells to secrete pro-inflammatory cytokines, such as TNFα, IL-1, IL-6 and IL-8. These proinflammatory cytokines can induce the expression of tissue factor on monocytes [68] and endothelial cells [118] (for review see [63]) via Toll-like receptors (TLRs) [33] leading to thrombin formation. LPS can also cause activation of factor XII, which is involved in the formation of thrombin. Since it has been shown that all extrinsic and intrinsic tenase factors for generation of thrombin are present in the cornea, thrombin can be formed in the cornea in situations of infection and inflammation, where it causes permeabilization of the endothelial monolayer [6].

Our experiments demonstrate that thrombin also inhibits intercellular propagation of Ca^{2+} waves, and that the inhibition is mainly due to a reduction of PIC, but that also GJIC is reduced. The role of the reduction of IC between corneal endothelial cells in situations of inflammation is not yet clear. However, IC may control the spread and release of inflammatory mediators, thereby coordinating the corneal inflammatory responses. At least in the lung capillary bed, Cx43-dependent intercellular Ca^{2+} wave propagation conveys the spread of inflammation by controlling expression of inflammatory mediators (P-selectin) [72]. Inhibition of Cx43 by the Cx43-derived Cx-mimetic peptides ^{43}Gap26 or ^{43}Gap27 blocked the increase of lung micro-vascular permeability by thrombin. It therefore seems plausible that Ca^{2+} wave propagation in corneal endothelial cells via PIC and/or GJIC may also coordinate the inflammatory response of cells in the cornea.

Our experiments revealed that increased MLC phosphorylation inhibits IC in the corneal endothelium, thereby suggesting a role of contractility of the actin cytoskeleton in the regulation of IC. The role of the actin cytoskeleton and tethering forces for gap junctions is reminiscent of interactions of the cytoskeleton with tight junctions and adherens junctions in the regulation of barrier integrity [94, 105, 93]. However, the block of hemichannels involved in PIC in response to increased actomyosin contraction suggests that the disposition of cortical actin also plays an important role in the functioning and regulation of endothelial hemichannels. In situations of inflammatory stress, which often result in MLC phosphorylation [117, 121, 8, 103], block of hemichannels could be thought of as being a defense mechanism. Thus, such a block could prevent potential "bystander effects" [2, 102] by limiting the release of signaling molecules. While IC could contribute to the spread of inflammatory mediators, thrombin might act as a negative feedback mechanism, limiting the spreading of the response of the cells to inflammation by reducing

IC. In addition, it is plausible that hemichannel block can have a beneficial effect by preventing disturbances of intracellular homeostasis caused by solute entry/loss through the channels.

The influence of the organization and/or contractility of the cortical actin cytoskeleton on hemichannels suggests altered activity of hemichannels in response to hypoxia and aging, since under these conditions, the corneal endothelial cells show significant alterations in the actin cytoskeleton, as suggested by polymegathism and pleomorphism [48, 127].

Our experiments also demonstrated that exposure of BCEC to adenosine, which increases intracellular cAMP through activation of A2B receptors resulting in inhibition of RhoA activation [105], opposes the thrombin effect on IC. Our observations on the effect of adenosine add yet another mechanism to the growing repertoire of defense mechanisms by which adenosine may protect and enhance the physiological functions of corneal endothelium. While adenosine is known to enhance fluid transport [86], to activate cAMP-activated Cl^- channels [12, 112] and to rescue the loss of barrier integrity by reducing MLC phosphorylation [105, 93], our experiments demonstrate that adenosine can influence the coordination of cellular responses to stimuli. Despite the focus of this study on the thrombin-induced inhibition of the IC, our observations have more general implications in that our experiments demonstrated that the action of adenosine in preventing the inhibition of IC by thrombine was mediated via a pathway involving MLC. Since pro-inflammatory molecules can cause MLC phosphorylation, it is plausible that by inducing MLC dephosphorylation, adenosine could be beneficial in overcoming the potential threat to loss of IC concomitant with its ability to rescue barrier integrity. Furthermore, our finding that GJIC is enhanced in response to increase levels of cAMP is supported by a previous finding that HCO_3^-, which activates soluble adenylate cyclase [111], enhances dye coupling in the corneal endothelium [124]. Our finding that adenosine not only affects GJIC, but also overcomes the thrombin-induced inhibition of hemichannels was surprising. However, it is in line with our findings that the major effect of thrombin on IC is through inhibition of PIC. Improving PIC could be helpful especially for cells that have lost direct contact with neighboring cells.

10.4.3 IC Via ATP in Different Cell Types

Together with the major progress in the field of purinergic receptors and characterization of ectonucleotidases, IC via ATP release and its potential physiological relevance, have been examined in a large number of cell types/tissues. It has become clear that, in addition to other mechanisms, hemichannels are an important pathway for ATP release, playing an important role in PIC in a large number of physiological and pathophysiological processes. A brief summary is provided in the Table 10.2.

Table 10.2 ATP release through hemichannels in other cell types

Tissue/cell type	Types of stimulation	Signaling, pathway of release and physiological and pathological influence	Study
Airway epithelial cells	Cell swelling	Carbenoxolone and probenecid, but not FFA, inhibited ATP release; attributed to Panx1 hemichannels. Suppression of Panx1 expression by shRNA inhibited release	[82]
Neurosensory epithelium inner ear	Photostimulation with caged IP_3	ATP release through connexin Cx26/Cx30 hemichannels propagates Ca^{2+} signals across inner ear	[3]
T Cells	Cell swelling T-cell receptor stimulation	Release stimulates Ca^{2+} influx via P2X7 receptors leading to T-cell activation. Panx1 channels are implicated	[97]
Human microvascular endothelial cells	Basal release	Hypoxia reduced release through Cx43-serine368 phosphorylation, which switches Cx43 hemichannels to closed state.	[39]
C6 cells expressing Cx43		Patch clamp studies show that Cx43 hemichannels are permeable to ATP	[53]
Cultured neurons	Membrane depolarization	Depolarization led to release and ischemic tolerance; inhibited by Cx hemichannel blockers and by siRNA knockdown of Cx36	[99]
Ovarian granulosa cells	Mechanical stimulation or EGTA	Cells expressing hemichannels showed dye uptake and ATP release, although Cx43 hemichannels did not support folliculogenesis,	[115]
Receptor cells in mouse taste buds	A tastant mix	ATP release mediated through Panx1 hemichannels contributes to taste transduction	[49]
Neutrophils	fMLP	Release stimulated by fMLP-dependent dephosphorylation Cx43 and subsequent opening Cx43 hemichannels.	[36]
Articular chondrocytes	Mechanical stimulation or EGTA	Cx43 hemichannels and P2 receptors form putative mechanoreceptor complex in cilia	[57]
Cardiac myocytes	Simulated ischemia	^{43}Gap26 inhibited ATP release. A peptide with antiarrhythmic properties stimulates ATP release under ischemic conditions.	[21]

10.4.4 Physiological and Pathological Significance of PIC in Corneal Endothelium and Other Cell Types

In humans, the corneal endothelium is a non-regenerative monolayer [52]. In a number of situations, including mechanical injury (e.g., during intraocular surgery such as phaco-emulsification), aging, Fuch's dystrophy, inflammation and hypoxia, corneal endothelial cells manifest a significant change in morphology and apoptosis, leading to loss of functional integrity of the corneal endothelial monolayer [35, 11, 14, 13]. Such changes not only affect the integrity of the corneal barrier, but can also decrease IC, thereby hampering the coordinated response of a monolayer of cells to extracellular stimuli. The role of the two modalities of IC (GJIC and PIC) has thus far not been fully elucidated, but our experiments show effects on GJIC and PIC and a possible role of IC in response to mechanical stimuli.

GJIC can be involved in intracellular homeostasis by exchange of solutes through the gap junction channels, thereby regulating cell growth, proliferation, differentiation and apoptosis [1, 28, 32]. Timely exchange of molecules via gap junctions can protect neighboring cells from cytotoxic substances. This effect has been called "metabolic cooperation", "good Samaritan effect" or "kiss of life" [2, 20, 59, 69]. Exchange of molecules via gap junctions can also affect neighboring cells by spreading factors that inhibit cell proliferation, or by transferring toxic factors or stimuli of apoptosis to adjacent cells. Spreading of such factors has been called the "bystander effect" or "kiss of death" [2, 20, 59, 69].

Cx-mediated GJIC is involved in the bystander effect [40, 120, 59]. In situations of injury, decrease of GJIC may attenuate the spread of toxic metabolites from injured to healthy cells, or reduce the loss of important cellular metabolites [28]. Cxs are known to function as tumor suppressors and numerous studies have explored restoration of GJIC as a potential therapy against cancer [116]. Gap junctions are down-regulated in many types of cancer, including gliomas, breast carcinoma, and prostate cancers [67, 70]. Transfection of these cells with Cx43 leads to restoration of gap junctions, inhibited tumor growth [130, 71].

Also PIC plays an important role in cell growth, differentiation and migration and injury repair. PIC, via extracellular release of ATP, controls the propagation of injury-induced Ca^{2+} waves in corneal epithelial cells. This effect was due to activation of P2Y receptors, thereby affecting cell migration and proliferation [55, 56], and activating MAPK [126]. Cx-based hemichannels also play an important role in cell death (reviewed in [32]). Hemichannels in the plasma membrane could affect cell death or survival by acting as (i) a pathway for loss of essential metabolites or for uptake of toxic or survival enhancing substances, (ii) a transmembrane transduction signaling pathway and (iii) as a intercellular signaling pathway (see [31]). Recently, hemichannel-mediated PIC was suggested to regulate the propagation of the cytochrome C-induced apoptotic cell death signal in concert with gap junctions [31]. In C6 glioma cells stably transfected with Cx43, gap junctions mediate the spread of cytochrome C-induced apoptosis in a zone next to where apoptosis was triggered, whereas hemichannels also promoted cell death beyond this area [31].

In addition, Cx43 hemichannels are regulated by the redox potential and oxidative stress. On the one hand, reducing the intracellular redox potential, either by chemical reducing agents like DTT or by intracellular physiological reducing molecules like GSH enhanced Cx43-hemichannel activity [85]. This effect of reducing agents on the opening of Cx43 hemichannels is likely mediated by reduction of intracellular cysteines that are located in the C-terminal tail of Cx43. On the other hand, opening of Cx43 hemichannels is induced by metabolic inhibition or ischemic conditions, which leads to intracellular accumulation of NO, and S-nitrosylation of the three intracellular cysteines located in the C-terminal tail of Cx43 hemichannels [84]. The exact mechanism by which the redox potential and oxidative stress regulates hemichannel opening still remains to be elucidated. It is not definitely clear how reducing agents inhibit the increase in hemichannel permeability caused by oxidative stress during metabolic inhibition, but enhance hemichannel opening under normoxic conditions. This may suggest that the same cysteine residues are substrates of different redox reactions, including formation and reduction of disulfide bonds, cysteine S-nitrosylation, and/or glutathionation [4]. Alternatively, depending on the Cx43-phosphorylation states, or on the presence of Cx43-interaction partners, the same modifications may lead to different conformational changes or modulation of different cysteine residues. Therefore, it will be essential to identify the physiological function for each of the three cysteine residues in the intracellular tail of Cx43 by site-directed mutagenesis approaches.

Not only Cx channels play a role in the propagation of factors that inhibit cell proliferation, but also Panx channels, which mainly form hemichannels [9, 10], control inflammatory responses and cell death. Panx1 levels are elevated upon exposure to diverse pro-inflammatory stimuli (e.g., TNF-α, IFN-α, IFN-γ, lipopolysaccharide, cold and systemic inflammation) [104]. Panx1 hemichannels represent the non-selective pore that opens upon P2X7 activation, thereby facilitating the entry of pro-inflammatory molecules into the cytosol required for activation of cryopyrin-dependent inflammosome and caspase-1 cleavage [98, 122, 119, 76, 54, 77]. In addition, it has been shown that a component of IL-1 cytokine release through ATP-dependent activation of P2X7 is dependent on Panx1 [75]. Furthermore, the P2X7/Panx1-protein complex controls the expression of several proteins and participates in wound healing [66]. The link of Panx1 opening with ATP-induced stimulation of P2X7 also contributed to ATP-induced cell death [76]. In addition, in acute glaucoma conditions, Panx hemichannels mediate increased ATP release during retinal pressure elevation, leading to the death of ganglion cells [83].

Furthermore, Panx1 activation causes neuronal excitotoxicity during stroke, leading to swelling, Ca^{2+} dysregulation and ischemic neuronal death in pyramidal neurons, which express endogenous Panx1 [114]. Opening of Panx1 hemichannels at postsynaptic sites [131], is triggered by NMDAR activation, resulting in rhythmic epileptiform-like bursting, which correlates to the bursting observed in epilepsy patients [113]. Finally, the opening of large numbers of hemichannels following ischemia or inflammatory injury may also be involved in pathophysiological cascades leading to cell depolarization, collapse of ionic gradients, loss of small metabolites and elevation of intracellular Ca^{2+} [104].

In conclusion, Cx and Panx hemichannels form an important pathway for release of ATP. This release contributes to IC in addition to the well-known gap junctional IC. Thus, hemichannels may contribute to bystander and/or good samaritan effects in pathological conditions, even between cells that are disjointed and lack gap junctionial coupling. Further research is needed to investigate factors that regulate the open probability of hemichannels and the role of hemichannels in physiological and pathological conditions in different cell types. A clear distinction should be made between effects that can be attributed to the Cx or Panx hemichannel-mediated IC versus effects of gap junctional IC or effects due to expression and distribution of these proteins in intracellular compartments, such as mitochondria or endoplasmic reticulum. Availability of Panx and Cx isoform-specific agonists and antagonists, and specific hemichannel blockers, in addition to gene abblation, silencing, mutation and protein expression studies, would be very helpful to clarify the role of the hemichannels in physiological and pathological conditions in different cell types and may provide therapeutic tools for treatment of pathological conditions associated with abberant IC.

Acknowledgments Supported by FWO-Vlaanderen G.0218.03, GOA/2004/07, IAP program 5/05 (BH and JV) and NIH grant R21-EY019119 and Faculty Research Grant, VP of Research, IU Bloomington, IN (SPS).

References

1. Abbracchio MP, Burnstock G. (1998) Purinergic signalling: pathophysiological roles. Jpn J Pharmacol 78:113–45.
2. Andrade-Rozental AF, Rozental R, Hopperstad MG, Wu JK, Vrionis FD, Spray DC. (2000) Gap junctions: the "kiss of death" and the "kiss of life". Brain Res Brain Res Rev 32: 308–15.
3. Anselmi F, Hernandez VH, Crispino G, Seydel A, Ortolano S, Roper SD, Kessaris N, Richardson W, Rickheit G, Filippov MA, Monyer H, Mammano F. (2008) ATP release through connexin hemichannels and gap junction transfer of second messengers propagate Ca^{2+} signals across the inner ear. Proc Natl Acad Sci USA 105:18770–5.
4. Aracena P, Sanchez G, Donoso P, Hamilton SL, Hidalgo C. (2003) S-glutathionylation decreases Mg^{2+} inhibition and S-nitrosylation enhances Ca^{2+} activation of RyR1 channels. J Biol Chem 278:42927–35.
5. Arcuino G, Lin JH, Takano T, Liu C, Jiang L, Gao Q, Kang J, Nedergaard M. (2002) Intercellular calcium signaling mediated by point-source burst release of ATP. Proc Natl Acad Sci USA 99:9840–5.
6. Ayala A, Warejcka DJ, Olague-Marchan M, Twining SS. (2007) Corneal activation of prothrombin to form thrombin, independent of vascular injury. Invest Ophthalmol Vis Sci 48:134–43.
7. Bao L, Locovei S, Dahl G. (2004) Pannexin membrane channels are mechanosensitive conduits for ATP. FEBS Lett 572:65–8.
8. Blair SA, Kane SV, Clayburgh DR, Turner JR. (2006) Epithelial myosin light chain kinase expression and activity are upregulated in inflammatory bowel disease. Lab Invest 86:191–201.
9. Boassa D, Ambrosi C, Qiu F, Dahl G, Gaietta G, Sosinsky G. (2007) Pannexin1 channels contain a glycosylation site that targets the hexamer to the plasma membrane. J Biol Chem 282:31733–43.

10. Boassa D, Qiu F, Dahl G, Sosinsky G. (2008) Trafficking dynamics of glycosylated pannexin 1 proteins. Cell Commun Adhes 15:119–32.
11. Bonanno JA. (2003) Identity and regulation of ion transport mechanisms in the corneal endothelium. Prog Retin Eye Res 22:69–94.
12. Bonanno JA, Srinivas SP. (1997) Cyclic AMP activates anion channels in cultured bovine corneal endothelial cells. Exp Eye Res 64:953–62.
13. Bourne RR, Minassian DC, Dart JK, Rosen P, Kaushal S, Wingate N. (2004) Effect of cataract surgery on the corneal endothelium: modern phacoemulsification compared with extracapsular cataract surgery. Ophthalmology 111:679–85.
14. Bourne WM. (2003) Biology of the corneal endothelium in health and disease. Eye 17: 912–8.
15. Braet K, Vandamme W, Martin PE, Evans WH, Leybaert L. (2003) Photoliberating inositol-1,4,5-trisphosphate triggers ATP release that is blocked by the connexin mimetic peptide gap 26. Cell Calcium 33:37–48.
16. Bruzzone R, Barbe MT, Jakob NJ, Monyer H. (2005) Pharmacological properties of homomeric and heteromeric pannexin hemichannels expressed in Xenopus oocytes. J Neurochem 92:1033–43.
17. Bruzzone R, Hormuzdi SG, Barbe MT, Herb A, Monyer H. (2003) Pannexins, a family of gap junction proteins expressed in brain. Proc Natl Acad Sci USA 100:13644–9.
18. Bruzzone S, Guida L, Zocchi E, Franco L, De Flora A. (2001) Connexin 43 hemi channels mediate Ca^{2+}-regulated transmembrane NAD^{+} fluxes in intact cells. Faseb J 15:10–2.
19. Burnstock G. (2007) Physiology and pathophysiology of purinergic neurotransmission. Physiol Rev 87:659–797.
20. Chipman JK, Mally A, Edwards GO. (2003) Disruption of gap junctions in toxicity and carcinogenicity. Toxicol Sci 71:146–53.
21. Clarke TC, Williams OJ, Martin PE, Evans WH. (2009) ATP release by cardiac myocytes in a simulated ischaemia model: inhibition by a connexin mimetic and enhancement by an antiarrhythmic peptide. Eur J Pharmacol 605:9–14.
22. Contreras JE, Saez JC, Bukauskas FF, Bennett MV. (2003) Functioning of cx43 hemichannels demonstrated by single channel properties. Cell Commun Adhes 10: 245–9.
23. Contreras JE, Saez JC, Bukauskas FF, Bennett MV. (2003) Gating and regulation of connexin 43. (Cx43) hemichannels. Proc Natl Acad Sci USA 100:11388–93.
24. Cotrina ML, Lin JH, Alves-Rodrigues A, Liu S, Li J, Azmi-Ghadimi H, Kang J, Naus CC, Nedergaard M. (1998) Connexins regulate calcium signaling by controlling ATP release. Proc Natl Acad Sci USA 95:15735–40.
25. Cruciani V, Mikalsen SO. (2006) The vertebrate connexin family. Cell Mol Life Sci 63: 1125–40.
26. D'hondt C, Ponsaerts R, Srinivas SP, Vereecke J, Himpens B. (2007) Thrombin inhibits intercellular calcium wave propagation in corneal endothelial cells by modulation of hemichannels and gap junctions. Invest Ophthalmol Vis Sci 48:120–33.
27. D'hondt C, Srinivas SP, Vereecke J, Himpens B. (2007) Adenosine opposes thrombin-induced inhibition of intercellular calcium wave in corneal endothelial cells. Invest Ophthalmol Vis Sci 48:1518–27.
28. De Maio A, Vega VL, Contreras JE. (2002) Gap junctions, homeostasis, and injury. J Cell Physiol 191:269–82.
29. De Vuyst E, Decrock E, Cabooter L, Dubyak GR, Naus CC, Evans WH, Leybaert L. (2006) Intracellular calcium changes trigger connexin 32 hemichannel opening. Embo J 25: 34–44.
30. De Vuyst E, Decrock E, De Bock M, Yamasaki H, Naus CC, Evans WH, Leybaert L. (2007) Connexin hemichannels and gap junction channels are differentially influenced by lipopolysaccharide and basic fibroblast growth factor. Mol Biol Cell 18:34–46.
31. Decrock E, De Vuyst E, Vinken M, Van Moorhem M, Vranckx K, Wang N, Van Laeken L, De Bock M, D'Herde K, Lai CP, Rogiers V, Evans WH, Naus CC, Leybaert L. (2009)

Connexin 43 hemichannels contribute to the propagation of apoptotic cell death in a rat C6 glioma cell model. Cell Death Differ 16: 151–63.
32. Decrock E, Vinken M, De Vuyst E, Krysko DV, D'Herde K, Vanhaecke T, Vandenabeele P, Rogiers V, Leybaert L. (2009) Connexin-related signaling in cell death: to live or let die? Cell Death Differ 16: 524–36.
33. Dunzendorfer S, Lee HK, Soldau K, Tobias PS. (2004) Toll-like receptor 4 functions intracellularly in human coronary artery endothelial cells: roles of LBP and sCD14 in mediating LPS responses. Faseb J 18:1117–9.
34. Ebihara L. (2003) New roles for connexons. News Physiol Sci 18:100–3.
35. Edelhauser HF. (2000) The resiliency of the corneal endothelium to refractive and intraocular surgery. Cornea 19: 263–73.
36. Eltzschig HK, Eckle T, Mager A, Kuper N, Karcher C, Weissmuller T, Boengler K, Schulz R, Robson SC, Colgan SP. (2006) ATP release from activated neutrophils occurs via connexin 43 and modulates adenosine-dependent endothelial cell function. Circ Res 99:1100–8.
37. Evans WH, Boitano S. (2001) Connexin mimetic peptides: specific inhibitors of gap-junctional intercellular communication. Biochem Soc Trans 29: 606–12.
38. Evans WH, De Vuyst E, Leybaert L. (2006) The gap junction cellular internet: connexin hemichannels enter the signalling limelight. Biochem J 397:1–14.
39. Faigle M, Seessle J, Zug S, El Kasmi KC, Eltzschig HK. (2008) ATP release from vascular endothelia occurs across Cx43 hemichannels and is attenuated during hypoxia. PLoS ONE 3: e2801.
40. Fick J, Barker FG, 2nd, Dazin P, Westphale EM, Beyer EC, Israel MA. (1995) The extent of heterocellular communication mediated by gap junctions is predictive of bystander tumor cytotoxicity in vitro. Proc Natl Acad Sci USA 92:11071–5.
41. Gage PJ, Rhoades W, Prucka SK, Hjalt T. (2005) Fate maps of neural crest and mesoderm in the mammalian eye. Invest Ophthalmol Vis Sci 46:4200–8.
42. Giepmans BN. (2004) Gap junctions and connexin-interacting proteins. Cardiovasc Res 62: 233–45.
43. Gomes P, Srinivas SP, Van Driessche W, Vereecke J, Himpens B. (2005) ATP release through connexin hemichannels in corneal endothelial cells. Invest Ophthalmol Vis Sci 46: 1208–18.
44. Gomes P, Srinivas SP, Vereecke J, Himpens B. (2005) ATP-dependent paracrine intercellular communication in cultured bovine corneal endothelial cells. Invest Ophthalmol Vis Sci 46:104–13.
45. Goodenough DA, Paul DL. (2003) Beyond the gap: functions of unpaired connexon channels. Nat Rev Mol Cell Biol 4:285–94.
46. Harris AL. (2001) Emerging issues of connexin channels: biophysics fills the gap. Q Rev Biophys 34: 325–472.
47. Herve JC, Bourmeyster N, Sarrouilhe D, Duffy HS. (2007) Gap junctional complexes: from partners to functions. Prog Biophys Mol Biol 94:29–65.
48. Holden BA, Sweeney DF, Vannas A, Nilsson KT, Efron N. (1985) Effects of long-term extended contact lens wear on the human cornea. Invest Ophthalmol Vis Sci 26:1489–501.
49. Huang YJ, Maruyama Y, Dvoryanchikov G, Pereira E, Chaudhari N, Roper SD. (2007) The role of pannexin 1 hemichannels in ATP release and cell-cell communication in mouse taste buds. Proc Natl Acad Sci USA 104:6436–41.
50. Jiang LH, Mackenzie AB, North RA, Surprenant A. (2000) Brilliant blue G selectively blocks ATP-gated rat P2X(7) receptors. Mol Pharmacol 58:82–8.
51. John S, Cesario D, Weiss JN. (2003) Gap junctional hemichannels in the heart. Acta Physiol Scand 179:23–31.
52. Joyce NC. (2003) Proliferative capacity of the corneal endothelium. Prog Retin Eye Res 22:359–89.
53. Kang J, Kang N, Lovatt D, Torres A, Zhao Z, Lin J, Nedergaard M. (2008) Connexin 43 hemichannels are permeable to ATP. J Neurosci 28:4702–11.

54. Kanneganti TD, Lamkanfi M, Kim YG, Chen G, Park JH, Franchi L, Vandenabeele P, Nunez G. (2007) Pannexin-1-mediated recognition of bacterial molecules activates the cryopyrin inflammasome independent of Toll-like receptor signaling. Immunity 26:433–43.
55. Klepeis VE, Cornell-Bell A, Trinkaus-Randall V. (2001) Growth factors but not gap junctions play a role in injury-induced Ca^{2+} waves in epithelial cells. J Cell Sci 114:4185–95.
56. Klepeis VE, Weinger I, Kaczmarek E, Trinkaus-Randall V. (2004) P2Y receptors play a critical role in epithelial cell communication and migration. J Cell Biochem 93:1115–33.
57. Knight MM, McGlashan SR, Garcia M, Jensen CG, Poole CA. (2009) Articular chondrocytes express connexin 43 hemichannels and P2 receptors – a putative mechanoreceptor complex involving the primary cilium? J Anat 214: 275–83.
58. Kondo RP, Wang SY, John SA, Weiss JN, Goldhaber JI. (2000) Metabolic inhibition activates a non-selective current through connexin hemichannels in isolated ventricular myocytes. J Mol Cell Cardiol 32:1859–72.
59. Krysko DV, Mussche S, Leybaert L, D'Herde K. (2004) Gap junctional communication and connexin43 expression in relation to apoptotic cell death and survival of granulosa cells. J Histochem Cytochem 52: 1199–207.
60. Lampe PD, Lau AF. (2004) The effects of connexin phosphorylation on gap junctional communication. Int J Biochem Cell Biol 36:1171–86.
61. Li H, Liu TF, Lazrak A, Peracchia C, Goldberg GS, Lampe PD, Johnson RG. (1996) Properties and regulation of gap junctional hemichannels in the plasma membranes of cultured cells. J Cell Biol 134:1019–30.
62. Limouze J, Straight AF, Mitchison T, Sellers JR. (2004) Specificity of blebbistatin, an inhibitor of myosin II. J Muscle Res Cell Motil 25:337–41.
63. Lopes-Bezerra LM, Filler SG. (2003) Endothelial cells, tissue factor and infectious diseases. Braz J Med Biol Res 36:987–91.
64. Lucas-Lopez C, Patterson S, Blum T, Straight AF, Toth J, Slawin AMZ, Mitchison TJ, Sellers JR, Westwood NJ. (2005) Absolute Stereochemical Assignment and Fluorescence Tuning of the Small Molecule Tool (-)-Blebbistatin. Eur J Organic Chem 2005:1736–40.
65. Matsuoka I, Ohkubo S. (2004) ATP- and adenosine-mediated signaling in the central nervous system: adenosine receptor activation by ATP through rapid and localized generation of adenosine by ecto-nucleotidases. J Pharmacol Sci 94:95–9.
66. Mayo C, Ren R, Rich C, Stepp MA, Trinkaus-Randall V. (2008) Regulation by P2X7: Epithelial Migration and Stromal Organization in the Cornea. Invest Ophthalmol Vis Sci 49(10):4384–91.
67. Mesnil M. (2002) Connexins and cancer. Biol Cell 94:493–500.
68. Meszaros K, Aberle S, Dedrick R, Machovich R, Horwitz A, Birr C, Theofan G, Parent JB. (1994) Monocyte tissue factor induction by lipopolysaccharide. (LPS): dependence on LPS-binding protein and CD14, and inhibition by a recombinant fragment of bactericidal/permeability-increasing protein. Blood 83:2516–25.
69. Mitchell SA, Randers-Pehrson G, Brenner DJ, Hall EJ. (2004) The bystander response in C3H 10T1/2 cells: the influence of cell-to-cell contact. Radiat Res 161:397–401.
70. Naus CC. (2002) Gap junctions and tumour progression. Can J Physiol Pharmacol 80: 136–141.
71. Naus CC, Elisevich K, Zhu D, Belliveau DJ, Del Maestro RF. (1992) In vivo growth of C6 glioma cells transfected with connexin43 cDNA. Cancer Res 52:4208–13.
72. Parthasarathi K, Ichimura H, Monma E, Lindert J, Quadri S, Issekutz A, Bhattacharya J. (2006) Connexin 43 mediates spread of Ca^{2+}-dependent proinflammatory responses in lung capillaries. J Clin Invest 116:2193–200.
73. Patterson S, Lucas-Lopez C, Westwood NJ. (2005) Selective chemical intervention in biological systems: the small molecule tool (S)-(-)-Blebbistatin. In: Hicks MG, Kettner C (eds) The Chemical Theatre of Biological Systems.
74. Pearson RA, Dale N, Llaudet E, Mobbs P. (2005) ATP released via gap junction hemichannels from the pigment epithelium regulates neural retinal progenitor proliferation. Neuron 46:731–44.

75. Pelegrin P, Barroso-Gutierrez C, Surprenant A. (2008) P2X7 receptor differentially couples to distinct release pathways for IL-1beta in mouse macrophage. J Immunol 180:7147–57.
76. Pelegrin P, Surprenant A. (2006) Pannexin-1 mediates large pore formation and interleukin-1beta release by the ATP-gated P2X7 receptor. Embo J 25: 5071–82.
77. Pelegrin P, Surprenant A. (2007) Pannexin-1 couples to maitotoxin- and nigericin-induced interleukin-1beta release through a dye uptake-independent pathway. J Biol Chem 282: 2386–94.
78. Plotkin LI, Manolagas SC, Bellido T. (2002) Transduction of cell survival signals by connexin-43 hemichannels. J Biol Chem 277:8648–57.
79. Ponsaerts R, D'Hondt C, Bultynck G, Srinivas SP, Vereecke J, Himpens B. (2008) The myosin II ATPase inhibitor blebbistatin prevents thrombin-induced inhibition of intercellular calcium wave propagation in corneal endothelial cells. Invest Ophthalmol Vis Sci 49:4816–27.
80. Ponsaerts R, D'Hondt C, Srinivas SP, Vereecke J, Himpens B. (2008) Effects of thrombin and adenosine on MLC phosphorylation in bovine corneal endothelial cells studied by confocal immunofluorescence microscopy. Acta Physiologica 192:P0–12.
81. Praetorius HA, Leipziger J. (2009) ATP release from non-excitable cells. Purinergic Signal, DOI 10.1007/s11302-009-9146-2.
82. Ransford GA, Fregien N, Qiu F, Dahl G, Conner GE, Salathe M. (2009) Pannexin 1 Contributes to ATP Release in Airway Epithelia. Am J Respir Cell Mol Biol 41(5):525–34.
83. Reigada D, Lu W, Zhang M, Mitchell CH. (2008) Elevated pressure triggers a physiological release of ATP from the retina: Possible role for pannexin hemichannels. Neuroscience 157:396–404.
84. Retamal MA, Cortes CJ, Reuss L, Bennett MV, Saez JC. (2006) S-nitrosylation and permeation through connexin 43 hemichannels in astrocytes: induction by oxidant stress and reversal by reducing agents. Proc Natl Acad Sci USA 103:4475–80.
85. Retamal MA, Schalper KA, Shoji KF, Bennett MV, Saez JC. (2007) Opening of connexin 43 hemichannels is increased by lowering intracellular redox potential. Proc Natl Acad Sci USA 104:8322–7.
86. Riley MV, Winkler BS, Starnes CA, Peters MI. (1996) Adenosine promotes regulation of corneal hydration through cyclic adenosine monophosphate. Invest Ophthalmol Vis Sci 37:1–10.
87. Riley MV, Winkler BS, Starnes CA, Peters MI, Dang L. (1998) Regulation of corneal endothelial barrier function by adenosine, cyclic AMP, and protein kinases. Invest Ophthalmol Vis Sci 39:2076–84.
88. Romanello M, Pani B, Bicego M, D'Andrea P. (2001) Mechanically induced ATP release from human osteoblastic cells. Biochem Biophys Res Commun 289:1275–81.
89. Romanov RA, Rogachevskaja OA, Khokhlov AA, Kolesnikov SS. (2008) Voltage dependence of ATP secretion in mammalian taste cells. J Gen Physiol 132:731–44.
90. Saez JC, Berthoud VM, Branes MC, Martinez AD, Beyer EC. (2003) Plasma membrane channels formed by connexins: their regulation and functions. Physiol Rev 83:1359–400.
91. Saez JC, Contreras JE, Bukauskas FF, Retamal MA, Bennett MV. (2003) Gap junction hemichannels in astrocytes of the CNS. Acta Physiol Scand 179:9–22.
92. Saez JC, Retamal MA, Basilio D, Bukauskas FF, Bennett MV. (2005) Connexin-based gap junction hemichannels: gating mechanisms. Biochim Biophys Acta 1711: 215–24.
93. Satpathy M, Gallagher P, Jin Y, Srinivas SP. (2005) Extracellular ATP opposes thrombin-induced myosin light chain phosphorylation and loss of barrier integrity in corneal endothelial cells. Exp Eye Res 81:183–92.
94. Satpathy M, Gallagher P, Waniewski ML, Srinivas SP. (2004) Thrombin-induced phosphorylation of the regulatory light chain of myosin II in cultured bovine corneal endothelial cells. Exp Eye Res 79:477–86.
95. Scemes E, Spray DC, Meda P. (2009) Connexins, pannexins, innexins: novel roles of "hemi-channels". Pflugers Arch 457:1207–26.

96. Scemes E, Suadicani SO, Dahl G, Spray DC. (2007) Connexin and pannexin mediated cell-cell communication. Neuron Glia Biol 3:199–208.
97. Schenk U, Westendorf AM, Radaelli E, Casati A, Ferro M, Fumagalli M, Verderio C, Buer J, Scanziani E, Grassi F. (2008) Purinergic control of T cell activation by ATP released through pannexin-1 hemichannels. Sci Signal 1:ra6.
98. Schilling WP, Wasylyna T, Dubyak GR, Humphreys BD, Sinkins WG. (1999) Maitotoxin and P2Z/P2X7 purinergic receptor stimulation activate a common cytolytic pore. Am J Physiol 277:C766–76.
99. Schock SC, Leblanc D, Hakim AM, Thompson CS. (2008) ATP release by way of connexin 36 hemichannels mediates ischemic tolerance in vitro. Biochem Biophys Res Commun 368:138–44.
100. Seki A, Coombs W, Taffet SM, Delmar M. (2004) Loss of electrical communication, but not plaque formation, after mutations in the cytoplasmic loop of connexin43. Heart Rhythm 1:227–33.
101. Seki A, Duffy HS, Coombs W, Spray DC, Taffet SM, Delmar M. (2004) Modifications in the biophysical properties of connexin43 channels by a peptide of the cytoplasmic loop region. Circ Res 95:e22–8.
102. Shao C, Furusawa Y, Aoki M, Ando K. (2003) Role of gap junctional intercellular communication in radiation-induced bystander effects in human fibroblasts. Radiat Res 160:318–23.
103. Shen L, Black ED, Witkowski ED, Lencer WI, Guerriero V, Schneeberger EE, Turner JR. (2006) Myosin light chain phosphorylation regulates barrier function by remodeling tight junction structure. J Cell Sci 119:2095–106.
104. Shestopalov VI, Panchin Y. (2008) Pannexins and gap junction protein diversity. Cell Mol Life Sci 65:376–94.
105. Srinivas SP, Satpathy M, Gallagher P, Larivière E, Van Driessche W. (2004) Adenosine induces dephosphorylation of myosin II regulatory light chain in cultured bovine corneal endothelial cells. Exp Eye Res 79:543–51.
106. Srinivas SP, Satpathy M, Guo Y, Anandan V. (2006) Histamine-induced phosphorylation of the regulatory light chain of myosin II disrupts the barrier integrity of corneal endothelial cells. Invest Ophthalmol Vis Sci 47:4011–18.
107. Srinivas SP, Yeh JC, Ong A, Bonanno JA. (1998) Ca^{2+} mobilization in bovine corneal endothelial cells by P2 purinergic receptors. Curr Eye Res 17:994–1004.
108. Stout CE, Costantin JL, Naus CC, Charles AC. (2002) Intercellular calcium signaling in astrocytes via ATP release through connexin hemichannels. J Biol Chem 277:10482–8.
109. Straight AF, Cheung A, Limouze J, Chen I, Westwood NJ, Sellers JR, Mitchison TJ. (2003) Dissecting temporal and spatial control of cytokinesis with a myosin II Inhibitor. Science 299:1743–7.
110. Suadicani SO, Brosnan CF, Scemes E. (2006) P2X7 receptors mediate ATP release and amplification of astrocytic intercellular Ca^{2+} signaling. J Neurosci 26: 1378–85.
111. Sun XC, Cui M, Bonanno JA. (2004) [HCO3-]-regulated expression and activity of soluble adenylyl cyclase in corneal endothelial and Calu-3 cells. BMC Physiol 4:8.
112. Tan-Allen KY, Sun XC, Bonanno JA. (2005) Characterization of adenosine receptors in bovine corneal endothelium. Exp Eye Res 80:687–96.
113. Thompson RJ, Jackson MF, Olah ME, Rungta RL, Hines DJ, Beazely MA, MacDonald JF, MacVicarr BA. (2008) Activation of pannexin-1 hemichannels augments aberrant bursting in the hippocampus. Science 322.
114. Thompson RJ, Zhou N, MacVicar BA. (2006) Ischemia opens neuronal gap junction hemichannels. Science 312:924–7.
115. Tong D, Li TY, Naus KE, Bai D, Kidder GM. (2007) In vivo analysis of undocked connexin43 gap junction hemichannels in ovarian granulosa cells. J Cell Sci 120:4016–24.
116. Trosko JE, Ruch RJ. (2002) Gap junctions as targets for cancer chemoprevention and chemotherapy. Curr Drug Targets 3:465–82.

117. Utech M, Ivanov AI, Samarin SN, Bruewer M, Turner JR, Mrsny RJ, Parkos CA, Nusrat A. (2005) Mechanism of IFN-gamma-induced endocytosis of tight junction proteins: myosin II-dependent vacuolarization of the apical plasma membrane. Mol Biol Cell 16:5040–52.
118. Veltrop MH, Beekhuizen H, Thompson J. (1999) Bacterial species- and strain-dependent induction of tissue factor in human vascular endothelial cells. Infect Immun 67:6130–8.
119. Verhoef PA, Kertesy SB, Estacion M, Schilling WP, Dubyak GR. (2004) Maitotoxin induces biphasic interleukin-1beta secretion and membrane blebbing in murine macrophages. Mol Pharmacol 66:909–20.
120. Vrionis FD, Wu JK, Qi P, Waltzman M, Cherington V, Spray DC. (1997) The bystander effect exerted by tumor cells expressing the herpes simplex virus thymidine kinase. (HSVtk) gene is dependent on connexin expression and cell communication via gap junctions. Gene Ther 4:577–85.
121. Wang F, Graham WV, Wang Y, Witkowski ED, Schwarz BT, Turner JR. (2005) Interferon-gamma and tumor necrosis factor-alpha synergize to induce intestinal epithelial barrier dysfunction by up-regulating myosin light chain kinase expression. Am J Pathol 166:409–19.
122. Warny M, Kelly CP. (1999) Monocytic cell necrosis is mediated by potassium depletion and caspase-like proteases. Am J Physiol 276: C717–24.
123. Weber PA, Chang HC, Spaeth KE, Nitsche JM, Nicholson BJ. (2004) The permeability of gap junction channels to probes of different size is dependent on connexin composition and permeant-pore affinities. Biophys J 87:958–73.
124. Williams KK, Watsky MA. (2004) Bicarbonate promotes dye coupling in the epithelium and endothelium of the rabbit cornea. Curr Eye Res 28:109–20.
125. Wurmser AE, Nakashima K, Summers RG, Toni N, D′Amour KA, Lie DC, Gage FH. (2004) Cell fusion-independent differentiation of neural stem cells to the endothelial lineage. Nature 430:350–6.
126. Yang L, Cranson D, Trinkaus-Randall V. (2004) Cellular injury induces activation of MAPK via P2Y receptors. J Cell Biochem 91:938–50.
127. Yee RW, Matsuda M, Schultz RO, Edelhauser HF. (1985) Changes in the normal corneal endothelial cellular pattern as a function of age. Curr Eye Res 4:671–8.
128. Yegutkin GG. (2008) Nucleotide- and nucleoside-converting ectoenzymes: Important modulators of purinergic signalling cascade. Biochim Biophys Acta 1783:673–94.
129. Yen MR, Saier MH, Jr.. (2007) Gap junctional proteins of animals: the innexin/pannexin superfamily. Prog Biophys Mol Biol 94:5–14.
130. Zhu D, Caveney S, Kidder GM, Naus CC. (1991) Transfection of C6 glioma cells with connexin 43 cDNA: analysis of expression, intercellular coupling, and cell proliferation. Proc Natl Acad Sci USA 88:1883–7.
131. Zoidl G, Petrasch-Parwez E, Ray A, Meier C, Bunse S, Habbes HW, Dahl G, Dermietzel R. (2007) Localization of the pannexin1 protein at postsynaptic sites in the cerebral cortex and hippocampus. Neuroscience 146:9–16.

Chapter 11
Pregnancy Induced Reprogramming of Endothelial Function in Response to ATP: Evidence for Post Receptor Ca^{2+} Signaling Plasticity

FuXian Yi, Derek S. Boeldt, and Ian M. Bird

Abstract Studies of uterine artery endothelial cells in primary culture (UAEC) have shown that ATP stimulates eNOS activation in cells obtained during late pregnancy (P-UAEC) to a greater extent that in cells obtained in the nonpregnant state (NP-UAEC). While ATP-stimulated intracellular free Ca^{2+} ($[Ca^{2+}]i$) in the initial peak is similar, P-UAEC show a greater sustained phase that underlies a greater and more prolonged eNOS activation in uterine artery endothelium. While P2X and P2Y receptors are both present in UAEC, a number of studies suggest P2Y2 mediates the Ca^{2+} and eNOS responses to ATP. After the initial $[Ca^{2+}]i$ peak, the sustained phase is a form of Capacitative Ca^{2+} Entry (CCE) which is dependent on extracellular Ca^{2+} and fully blocked (along with the initial peak) by blockers of IP3 generation (U73122) or IP3 action (2-APB). One form of channel associated with CCE in an IP3 dependent fashion is the transient receptor potential cation channels (TRPC) family, and particularly TRPC3 and 6. While TRPC3 and 6, and many other proteins involved in the responses (P2Y2, Gq, IP3R1/2/3 and PLC beta3) are all expressed in UAEC, none are expressed at different levels in NP- vs. P-UAEC. Nonetheless, immunoprecipitation studies still show that functional association of TRPC3 with IP3R2 in response to ATP in P-UAEC far exceeds that observed in NP-UAEC. We have found that the pregnancy enhanced sustained $[Ca^{2+}]i$ response requires a pregnancy enhanced cell-cell communication via connexin 43 (CX43) Gap junctions. We have also found that the prolonged $[Ca^{2+}]i$ response observed in UAEC is buffered by the mitochondria in a manner further enhanced by pregnancy. We conclude that pregnancy adaptation of UAEC function involves reprogramming at multiple levels of cell signaling to prolong $[Ca^{2+}]i$ responses and cell-cell synchronization, thereby greatly enhancing eNOS activation.

Keywords Uterine artery · Endothelial · UAEC · Programming · Adaptation · Pregnancy · Ca^{2+} · eNOS · NO · P2Y2 · IP3 (inositol 1,4,5-trisphosphate) · IP3-R

I.M. Bird (✉)
Perinatal Research Laboratories, Department OBGyn, University Wisconsin School Medicine and Public Health, 7E Meriter Hospital, 202 South Park Street, Madison, WI 53715, USA
e-mail: imbird@wisc.edu

E. Gerasimovskaya, E. Kaczmarek (eds.), *Extracellular ATP and Adenosine as Regulators of Endothelial Cell Function*, DOI 10.1007/978-90-481-3435-9_11,

(IP3 receptor) · Phospholipase C · TRPC · CX43 · Gap junction · Capacitative entry · Mitochondria · U73122 · 2-APB

11.1 Background

In the nonpregnant state the blood flow to the uterus is low, but as the fetus grows, uterine blood flow must increase dramatically to match the progressively increasing demand for nutrients and oxygen [1]. Although growth of new vessels as well as remodeling of existing vessels during early pregnancy contributes to the increased flow, the period of greatest increase in flow occurs after the completion of new vessel growth, so maintenance of vasodilation in existing or newly-developed vessels is probably crucial. It is now clear from studies of both vessels and isolated cells that uterine artery (UA) endothelial nitric oxide (NO) and indeed prostacyclin production by endothelial cells changes in response to growth factors, bradykinin, ATP, and angiotensin II (AII) which elicit a rise in $[Ca^{2+}]i$ and/or activate kinases which in turn activate eNOS and indeed cytosolic phospholipase A2 [8, 18]. The physiologic mechanisms underlying these adaptive responses to pregnancy in UA are clearly complex and although the subject of intense investigation, still not fully understood. Nonetheless, the importance is clear given the mounting evidence that dysfunctional endocrine regulation of endothelial vasodilator production during pregnancy may directly result in preeclampsia and possibly associated growth retardation of the fetus [7, 28].

Most early studies of the effects of agonists on UA endothelial function were assayed indirectly through the use of inhibitors such at L-NAME on vessel myography and/or indirect assay of NO action i.e. assay of cGMP (reviewed in [3]). Before the methods for the isolation of UA endothelial cells and their maintenance in culture were developed, it was generally assumed that pregnancy-specific increases in NO production in response to this broad array of agonists related directly to the increased expression of eNOS observed at this time [20]. Nonetheless evidence for the ability to remap cell signaling within UA endothelium was implied by further in vivo studies performed in different physiologic states. While studies of expression levels of eNOS protein in UA endothelium showed that during the ovarian cycle, the follicular phase is associated with a marginal increase in eNOS expression and systemic NO while pregnancy is associated with a far greater increase in both eNOS protein and systemic NO [21], further analysis of the UA endothelial eNOS levels and systemic NO* levels (by luminescent measurement of NO after conversion from nitrate and nitrite) in pregnancies with singleton, twins and triplets showed that UA endothelial eNOS does not undergo any substantial further increase with number of offspring while NO* clearly does [21]. Also, in ovarectomized ewes in response to exogenous steroid, changes in eNOS expression in response to estrogen (E2B) and progesterone (P4) alone or together are not uniformly paralleled by changes in systemic NO [27]. While these studies suggest dissociation between eNOS protein and function in UA endothelium, the use of systemic NO measurements assumes that uterine NO production is the major determinant of systemic NO*, which may not always be the case. Other studies by Xiao et al [34], however, were more clear cut, since both eNOS level and NO* were measured pair wise ex vivo. While pregnancy increased eNOS protein levels in the UA endothelium, and hypoxia increased the expression further still, increases in NO* were far greater than the associated increases in eNOS protein.

The development of the in vitro UAEC model in 1996 gave the clearest direct evidence for the ability to remap cell signaling in UA endothelium was largely responsible for enhancement of NO production in vivo. In maintaining NP-UAEC or P-UAEC to the fourth passage the different expression levels of eNOS are greatly reduced [3] and yet the ability of the cells to produce NO* in response to a number of agonists was still clearly different, with P-UAEC production of NO* being greater than NP-UAEC [3, 6]. Initial studies also showed that ATP in particular could elicit a strong initial $[Ca^{2+}]i$ response (the first peak within the first minute of a 5 min observation time) in both NP-UAEC and P-UAEC but also that the response in P-UAEC was thereafter generally more sustained above basal (Fig. 11.1), apparently mediated by a G-protein coupled receptor and coupled to

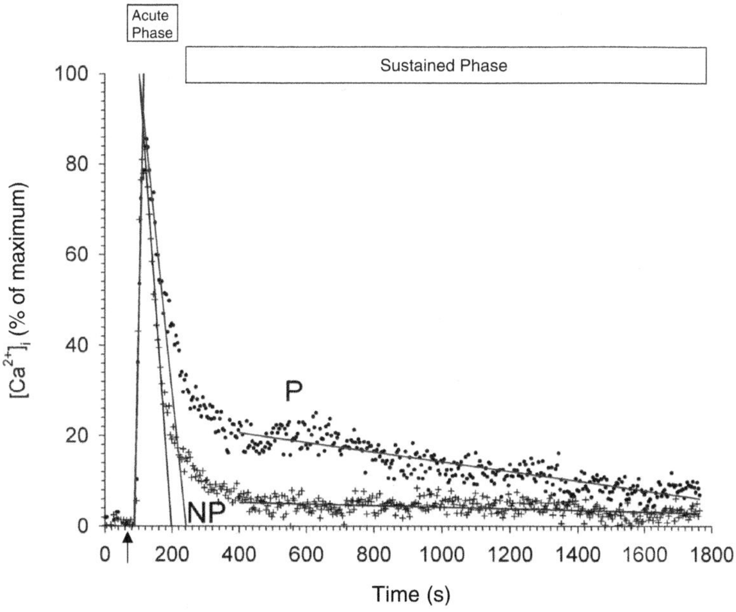

Fig. 11.1 Representative average $[Ca^{2+}]i$ responses to ATP (100 μM) in NP- and P-UAEC. Dishes of cells at 70% confluency were imaged 'full field' so an average response was obtained for each dish of cells, rather than an individual cell response. Comparison of P-UAEC with NP-UAEC showed quite clearly the difference in duration of the overall responses over a 30 min period. For clarity, data are normalized to the maximum in the initial peak and each data point shown is every 10th data point collected from $n = 8$ dishes, each for both NP- and P-UAEC. Lines indicate regression fits for the acute and sustained phases. Note that the up slopes of the initial peaks are virtually identical. The regression lines for the down slopes of the initial peaks begin to disassociate and the regression lines for the sustained phase are completely different, with the sustained phase of the $[Ca^{2+}]i$ response far less pronounced in NP-UAEC. Periods of time referred to as 'acute' phase and 'sustained' phase are indicated by bars. The arrow indicates the time of addition of ATP to the dish (90 s)

PLC Beta. While the initial peak was largely independent of extracellular Ca^{2+}, the subsequent sustained phase (beyond 2 min) was entirely dependent on extracellular Ca^{2+}. Moreover, P-UAEC but not NP-UAEC were capable of rapid (within seconds) $[Ca^{2+}]i$ oscillations superimposed upon the down slope of the initial $[Ca^{2+}]i$ peak. In addition to these observations on the $[Ca^{2+}]i$ response itself, a causal relationship between $[Ca^{2+}]i$ and NO was also implied by a strong correlation between the dose-dependency for $[Ca^{2+}]i$ responses and NO* production in NP- and P-UAEC [6]. A previous review describes these initial studies in more detail [3] and we will not repeat that here. Instead we will now move on to focus on the more recent studies which have gone on to examine the molecular basis of the $[Ca^{2+}]i$ responses in single cells as well as in groups of cells over a longer period of time (30 min of stimulation). We will also relate this to recent data from direct imaging of NO and $[Ca^{2+}]i$ simultaneously in freshly isolated endothelium. Such additional studies have allowed us to identify the molecular signaling components underlying these events and how pregnancy has altered the Ca^{2+} signaling pathway to function in a different way to achieve dramatically enhanced eNOS activation during pregnancy.

11.2 Investigating Differences in P2Y Receptors, G Protein Alpha Subunits and Phospholipase C B3 Expression in P-UAEC Vs. NP-UAEC

Early studies had suggested a P2Y class receptor was most likely responsible for the $[Ca^{2+}]i$ response [6]. More recent studies have since confirmed that multiple subtypes of P2Y and P2X receptors are present in UAEC [11], but further that the expression levels of those receptor classes coupled to PLC in particular are not significantly different in P-UAEC when compared to NP-UAEC. In spite of the wide range of P2X and P2Y receptor subtypes in UAEC, comparison of $[Ca^{2+}]i$ responses to a wide range of selective purinergic agonists with differential affinities for members of the P2X and P2Y receptor families have now clearly established that the primary receptor mediating the stimulation of $[Ca^{2+}]i$ itself is indeed P2Y2 [11].

At a post receptor level, additional western blot analysis of all G protein alpha subtypes known to couple to the PLC beta signaling pathway further confirmed that Gq, the isoform most likely to mediate this P2Y2 receptor response, is present in equal amounts in both NP- and P-UAEC [11]. Alpha subunit isoforms for Gi, Go, Gs and Gz were also identified, again in amounts that were not different in NP- vs. P-UAEC. At the level of phospholipase C itself, microarray analysis has confirmed that PLC beta 3 isoform is highly abundant and again expressed at similar levels in NP- and P-UAEC [9]. Thus enhanced $[Ca^{2+}]i$ signaling in P-UAEC is not due to simple changes in P2X or P2Y receptor expression, and the same P2Y2 receptor mediates the effect in each case. We must conclude, therefore, that while pregnancy adaptation of signaling must occur at a *post* receptor level, it is not apparently due to simple changes in expression of individual G proteins or PLC itself.

11.3 IP3 as a Regulator of the Sustained Phase [Ca^{2+}]i Response to ATP

In a preliminary screen, the IP3 receptor antagonist 2-APB was found to not only block the initial [Ca^{2+}]i response that was not dependent on extracellular Ca^{2+}, but also the sustained [Ca^{2+}]i response to ATP which was dependent on extracellular Ca^{2+}. Similar results were seen using the PLC inhibitor U73122, a compound structurally distinct from 2-APB which prevents IP3 generation at the source rather than at its site of action. Given the complete dependence of all aspects of the [Ca^{2+}]i response to IP3 (or more specifically Ins(1,4,5)P3), P2X involvement in the sustained [Ca^{2+}]i phase could be eliminated altogether [30], and the conclusion had to be made that either IP3 receptor was in some way directly involved in controlling the sustained Ca^{2+} influx or, more likely, the influx from outside was being channeled in a manner sensitive to the IP3-dependent emptying of the ER, possibly through a Store Operated Channel Entry (SOCE)/Capacitative Ca^{2+} Entry (CCE) type mechanism. Further examination of NP-and P-UAEC [Ca^{2+}]i responses to both Thapsigargin [11] and to ATP [12] in the absence of extracellular Ca^{2+} and then with subsequent re-addition of Ca^{2+} back to the extracellular medium confirmed that NP- and P-UAEC were both capable of CCE; removal of Ca^{2+} from the medium before stimulation with ATP resulted in the complete loss of the sustained phase in UAEC, and subsequent replacement of the Ca^{2+} only after the initial peak had gone and [Ca^{2+}]i was down to basal resulted in both a recovery and even some overshoot in the sustained phase [12]. Of note the magnitude of the CCE response to ATP in P-UAEC and NP-UAEC was similar under these non-physiologic conditions [12], but the experiment using Thapsigargin [11] revealed P-UAEC had a greater capacity to respond than NP-UAEC. Thus responses to ATP only utilize a part of the capacity of the cell to respond through CCE type mechanisms. The ability of Thapsigargin to trigger CCE in UAEC also shows that the opening of the channels mediating CCE can occur independently of P2Y2 receptor occupancy and so are directly responsive to emptying of the ER Ca^{2+} pool (SOCE, or CCE), such as could occur when IP3 opens IP3-R for any length of time. In addition, studies with either ATP or Thapsigargin revealed that the CCE mediated influx of Ca^{2+} (on reapplication of Ca^{2+} to the medium) in each case could be prevented if 2-APB (an inhibitor of IP3 binding to its receptor) was already present, and could also be arrested and even reversed if 2-APB was added after the extracellular Ca^{2+} [11, 12]. The speed of this action was pretty much instantaneous and subsequent studies (see below) confirmed that indeed 2-APB may be working on a plasma membrane channel mediating CCE or, more likely, interacting with IP3-R that had docked with plasma membrane Ca^{2+} channels responsible for CCE, consistent with the findings emerging from other laboratories at that time (reviewed in [4]).

11.4 Comparison of the Dose-Dependent Effects of 2-APB and U73122 on $[Ca^{2+}]i$ and eNOS Activity

Having established that U73122 (an inhibitor of phospholipase C, so blocking IP3 generation) and 2-APB (blocking IP3 action) could completely block the elevation of $[Ca^{2+}]i$ in NP- and P-UAEC, it followed that a comparison of the dose dependency of the effects of 2-APB and U73122 on initial and sustained phase $[Ca^{2+}]i$ alongside a parallel comparison of eNOS activation may prove useful to establish more clearly the causal relationship between the two events. Given that previous imaging of $[Ca^{2+}]i$ in UAEC had often been performed at low cell density, but eNOS activity was also assessed by arginine-citrulline conversion assay in UAEC at 70–80% confluency, $[Ca^{2+}]i$ imaging was performed using cells at the same density. Since the activity assay measures the total of all the cells in a dish, $[Ca^{2+}]i$ imaging was also performed using full field measurement, i.e. recording the average $[Ca^{2+}]i$ of many cells, rather than imaging of single cells. As such the initial maximum was measured as the peak of the initial spike, and the sustained phase $[Ca^{2+}]i$ level was measured as the maximum plateau value achieved thereafter [30]. Figure 11.2 illustrates the effect of 2-APB in NP- and P-UAEC. The data shows one striking feature, that in all cases the decline in eNOS activity in the presence of antagonist only occurred at higher doses than the decline in $[Ca^{2+}]i$, regardless of whether we are referring to $[Ca^{2+}]i$ peak or plateau. This suggests that while $[Ca^{2+}]i$ may play a role in eNOS activation, it may not be the sole regulator. This finding is consistent with other work on kinase regulation of UAEC eNOS activation, and similar observations have also been made in freshly isolated UA endothelium by DAF imaging [36], namely that 2-APB will fully block ATP-stimulated elevation of $[Ca^{2+}]i$ in freshly isolated P-UA endothelium but not fully block corresponding increases in

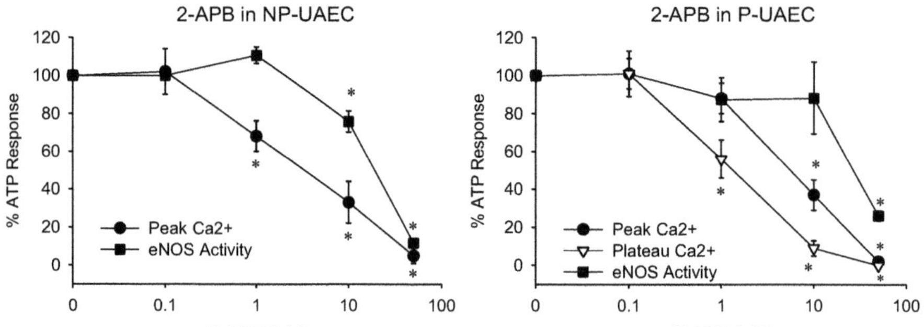

Fig. 11.2 Concentration-dependency of 2-APB inhibition of Ca^{2+} and NO production in UAEC. The effect of 2-APB on the peak $[Ca^{2+}]i$, maximum of the sustained phase $[Ca^{2+}]i$ and the eNOS activity as measured by arginine-citrulline conversion are shown. P and NP-UAEC were treated with ATP (100 μM) for 10 min with or without 5 min pretreatment with indicated concentrations of 2-APB. Data are normalized for the different assays by expression as % of ATP response alone ($n = 4$–6). Values shown are mean ± SE. (*$P < 0.05$ relative to ATP control response). Figure is modified from that published in [30]

NO (inhibition ~60%) – see Fig. 11.3. Closer examination reveals some interesting points. First, while eNOS activity in UEAC may not be directly correlated with the $[Ca^{2+}]i$ level in the cell, there is no doubt that when peak and plateau $[Ca^{2+}]i$ are no longer allowed to increase above basal there is effective inhibition of eNOS activity, and this is greater still (in fact almost complete) in NP than in P-UAEC. This is consistent with reports from other endothelial cells that sustained CCE is necessary for prolonged eNOS activation [19] but also with our previous suggestion, not reviewed here, that wortmannin – sensitive kinases in particular may play an increased but still incomplete role in Ca^{2+}-independent eNOS activation during pregnancy over that in nonpregnancy in UAEC [3, 30, 14].

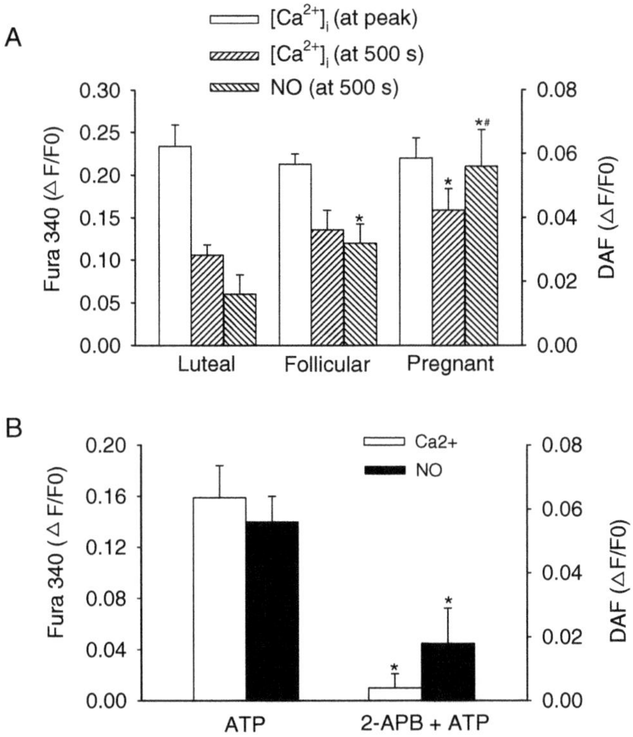

Fig. 11.3 Relationship between Ca^{2+} and NO in freshly isolated UA endothelium. (**A**) Effects of ATP (100 μM, 500 s stimulation) on $[Ca^{2+}]i$ and NO production in the freshly-isolated UA endothelium. Delta F/F0 (net increment fluorescence intensity) for Fura 2 at the initial $[Ca^{2+}]i$ peak, or for both Fura 2 and DAF signals at 500 s of treatment with ATP (100 mM) in pregnant-, follicular- and luteal- derived UA endothelium are shown. Data are means ± SE of UAE from 6 to 8 ewes. $*P < 0.05$ vs. luteal; $\#P < 0.05$ vs. follicular. (**B**) Effects of 2-APB (50 μM) on ATP- and ionomycin-induced increase of $[Ca^{2+}]i$ and NO production in the freshly-isolated pregnant UA endothelium. After pretreatment with 2-APB (50 μM) or vehicle for 10 min, UA endothelium was stimulated with ATP (100 μM) for 500 s. Data for Fura 2 and DAF signals at 500 s are means ± SE of UAE from $n = 6$ animals. $*P < 0.05$ vs. ATP alone. Figure is modified from that published in [36]

11.5 2-APB as an Effective Inhibitor of Ca^{2+} Signaling and NO Production in Endothelial Cells Freshly Isolated from the Luminal Surface of the UA Ex Vivo

The development of imaging methods to detect NO production and [Ca^{2+}]i simultaneously in freshly isolated UA endothelium provides a powerful means to monitor cell function in the very cells that perform the vasodilation response in vivo. In these studies, the pharmacological agonists ionomycin (calcium ionophore) and Thapsigargin (an endoplasmic reticulum Ca^{2+} pump inhibitor), were used to maximally elevate $[Ca^{2+}]_i$ above the physiologic range, and so fully activate eNOS in a receptor independent manner as a measure of total eNOS expression [36]. The corresponding NO production stimulated by ionomycin (5 μM) vs TG (10 μM) were similar at 1.95 /2.05 fold respectively in endothelium from late term pregnant ewes and 1.34/1.37 fold in endothelium from ewes synchronized to the follicular phase when compared to control responses determined in endothelium from ewes in the luteal phase. In contrast, the physiologic agonist, ATP (100 μM), stimulated a 3.43 fold increase in NO in endothelium from pregnant ewes, and 1.90 fold in endothelium from follicular phase ewes when expressed normalized to the response observed in the luteal phase [36]. This once again suggested that during pregnancy and in the follicular phase, eNOS activation is enhanced beyond simple changes in expression of eNOS protein in vivo. Of further relevance here, however, is the observation that the extent to which [Ca^{2+}]i rose in the sustained phase of the response to ATP was a closer indicator of the extent to which NO was increased (Fig. 11.3). In addition, studies performed in endothelium freshly isolated from pregnant ewes showed that 2-APB could totally prevented the ATP-induced $[Ca^{2+}]_i$ response, just as it had in UAEC in culture, but could not completely inhibit NO production (Fig. 11.3), consistent with findings in UAEC in primary culture. This is a particularly important observation since it strongly implies the observations made in UAEC are directly relevant to the situation observed in the intact endothelium in vivo.

11.6 The Case for TRPC Channels as Mediators of CCE in UAEC

Of further relevance to the actual mechanism of Ca^{2+} influx from outside the cell, a striking finding using 2-APB in P-UAEC is that the Ki for inhibition of the [Ca^{2+}]i plateau is almost a log unit lower than that for inhibiting the initial release from the ER. Indeed the Ki for the inhibition of the initial peak is actually that expected for IP3-R inhibition (6–7 μM) while that for inhibition of [Ca^{2+}]i plateau is closer to 1 μM. Assuming the effects to be on the IP3-R itself this can be interpreted in two possible ways. The first is that following the initial peak the IP3-R is altered in some way (phosphorylation, calmodulin binding, or binding another protein) that raises its affinity for 2-APB. The second is 2-APB binds another related protein with higher affinity and this protein is involved in Ca^{2+} influx. Of direct relevance, a newly described family of proteins (TRPC) are now increasingly implicated as mediators of Ca^{2+} influx in many cells [38] including endothelial cells (Reviewed in [24]), and interactions of IP3-R with TRPC have more recently been described in several cells [5, 17, 23, 25]. Further, a recent report proposed evidence that 2-APB binds TRPC with a higher affinity than IP3-R and inhibits Ca^{2+} influx to cells [4]; it remains unclear, however, if 2-APB binds directly on TRPC itself or, more likely, that 2-APB binds to IP3-R with higher affinity once TRPC binds IP3-R and so has in turn altered IP3-R conformation. Either mechanism would explain how ATP stimulated Ca^{2+} influx during the sustained phase was immediately and completely blocked by 2-APB. Nonetheless, studies on the effects of Thapsigargin on $[Ca^{2+}]i$ in UAEC [11] are at least supportive of the activation of TRPC by emptying of the ER stores, an event known to be sensed by IP3-R.

Thapsigargin is known to elevate $[Ca^{2+}]i$ by promoting leakage of Ca^{2+} from the ER through the inhibition of reuptake. The effects of Thapsigargin to empty the ER should be complete within just a few minutes yet the effect on $[Ca^{2+}]i$ in UAEC are particularly long lasting – as long as 60 min elevation of $[Ca^{2+}]i$ before a return to basal levels [11]. Nonetheless in the absence of extracellular Ca^{2+} this is dramatically blunted to a smaller ($\sim$30% maximum) response lasting $\sim$20 min. Thus the capacity of the ER itself is not unusually excessive, and the prolonged $[Ca^{2+}]i$ response in normal medium must involve a mechanism that is greatly dependent upon prolonged Ca^{2+} influx. Whether emptying of the ER promotes IP3-R binding to TRPC to allow direct Ca^{2+} entry to the cytosol or alternatively the two form a pipe to allow direct refilling of the ER, so sustaining the ER capacity to function, is unclear from these studies alone. Either way, however, the behavior of the plasma membrane Ca^{2+} channels activated in response to emptying of the ER in UAEC, and mediating a CCE type response in a manner sensitive to 2-APB inhibition, strongly implicates TRPC channel involvement.

11.7 Evidence for Enhanced TRPC3-IP3R-2 Functional Interaction in P-UAEC

The proposal that TRPC type channels are involved in the sustained CCE phase of $[Ca^{2+}]i$ response to ATP in UAEC is given some further support by the observation that transfection of TRPC3 into bovine pulmonary artery endothelial cells allowed the induction of a more sustained Ca^{2+} entry in response to ATP, bradykinin and intracellular IP3 where no such response occurred before [16]. A subsequent survey of NP- and P-UAEC using antisera specific to different TRPC subtypes has clearly identified the presence of TRPC3 and TRPC6 [12]. In addition the presence of IP3-R1, IP3-R2 and IP3-R3 in UAEC were also confirmed [11]. Of interest, however, there were no apparent differences in the level of expression of any of these proteins in NP- vs P-UAEC. Nonetheless it was intriguing to note that

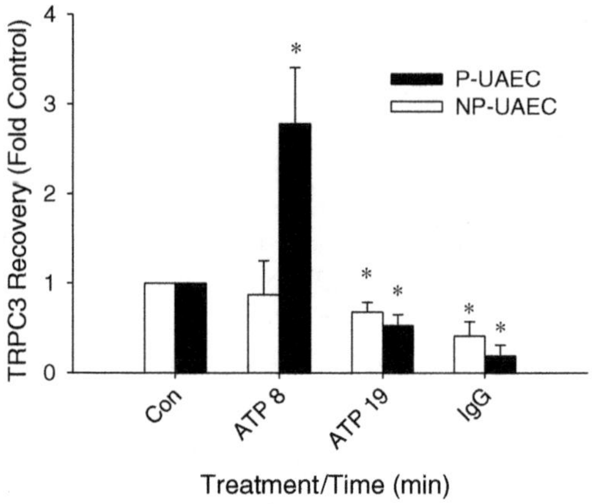

Fig. 11.4 Immunoprecipitation evidence for IP3R2 interaction with TRPC3 in NP- and P-UAEC. P- and NP-UAEC were stimulated with 100 μM ATP for 8 or 19 min, after which immunoprecipitation of IP3R2 was performed. Western blots of recovered protein were probed for TRPC3 and IP3R2. The graph depicts the ratio of TRPC3 in the elutions to TRPC3 in the whole cell lysates, representing fold enrichment. Data are expressed as mean±SE. $n = 4$. $P = 0.05$. * denotes significant difference from unstimulated control. IgG control values are also shown to verify the specificity of the IP3R2 antisera. All data are modified from that published in [12]

association between IP3R2 and TRPC3 was observed by immunoprecipitation in NP- and P-UAEC and, more importantly, this association was only increased after ATP stimulation of P-UAEC but NOT in the case of NP-UAEC (Fig. 11.4) [12]. These observations confirmed for the first time the direct interaction of IP3R2 with TRPC3 in UAEC and also showed that this specific interaction was selectively enhanced by pregnancy. The key to further understanding the pregnancy specific difference in CCE was in identifying at what level this association was being facilitated in P-UAEC that was comparatively lacking in NP-UAEC. The answer to this question has since been found to be in the pregnancy specific control of cell-cell communication.

11.8 Evidence that UA Endothelial Cells Communicate and Coordinate Their Function Via Gap Junctions

The first study that suggested cell-cell communication occurred in UAEC came from a comparison of the signal responses of UAEC at passage 4 to those of small groups of freshly isolated cells (initially performed to ensure the pregnancy specific changes in cell signaling we observed were a true reflection of events in vivo [9]). Studies of the enhancement of ERK phosphorylation and the more sustained nature of the $[Ca^{2+}]i$ responses in cells freshly isolated from pregnant ewes were all consistently observed as previously reported for cultured cells [3, 6]. However, the surprise was in the nature of the $[Ca^{2+}]i$ responses observed in groups of cells. Previous studies had reported rapid $[Ca^{2+}]i$ oscillations in single cells but observations of groups of cells had not been made prior to that time. Since the prior imaging technique was photometry (i.e. simply measures total light output from the view field without actually 'imaging' pixel by pixel) then it was assumed any oscillations of single cells in a crowd would be lost in the noise of unsynchronized responses. However when groups of freshly isolated endothelial cells from P-UA comprising patches of 12–25 endothelial cells at a time were imaged there was indeed evidence of synchronous peaks of $[Ca^{2+}]i$ and a pregnancy related difference in their likelihood. At doses of ATP from 30 to 300 μM almost half the observations of cells from pregnant ewes showed synchronous and regular periodic oscillations similar to those reported previously in P-UAEC, while in fresh cells from NP ewes only 1/22 recordings showed periodic oscillatory behavior [9]. This was consistent with observations of single cells at passage 4 and suggested that cells may communicate and synchronize their oscillations in vivo. There are several profound physiologic implications of this phenomenon, including the recruitment of increasing numbers of cells to provide vasodilator synthesis, the fact such events may be agonist-specific. Furthermore it raised the intriguing possibility of 'digital' communication between cells and further that distributing vessels of the uterus may be able to communicate with resistance level vessels and vice versa, independently of the direction of blood flow.

In order to further address the possibility of cell-cell communication and synchronization, further studies using digital video imaging system to examine UAEC responses to ATP at low (20%), intermediate (70%, our normal density in the past) and high density (100%) have been performed, and recording times extended to 30 min [37]. Such studies have clearly shown that increased cell density is associated with a greater number of cells responding to ATP, and further that cells treated at high density with ATP can oscillate synchronously in groups with P-UAEC > NP-UAEC. A more subtle difference for the response to ATP was the ability of the oscillatory response to show sharply localized $[Ca^{2+}]i$ bursts separated by a return to basal level (Fig. 11.5a). The number of 'bursts' in a given 30 min observation period in response to ATP was found to be greater in pregnancy. In particular, while the number of cells responding with the initial spike of $[Ca^{2+}]i$ that is already dependent on release of Ca^{2+} from the ER is only marginally density dependent for ATP in NP and P-UAEC, the ability to show higher numbers of subsequent bursts in the CCE phase of the remaining 30 min is far greater in P- than in NP-UAEC. Indeed cells from pregnant ewes have *twice* the likelihood of showing 5 bursts in 30 min than the nonpregnants at the 100% density seen in vivo [37] (Fig. 11.5b). This is perhaps the most physiologically relevant observation in this field made to date since (1) it relates to cells at the most physiologic density, i.e. confluent at 100%, (2) even if NO secretion per second was not greater in P than in NP UA endothelium, a more sustained level of NO production would maintain chronic vasodilation more effectively, and (3) a greater frequency of 'bursts' over a longer period of time could sustain signal propagation between cells more effectively via Gap junctions.

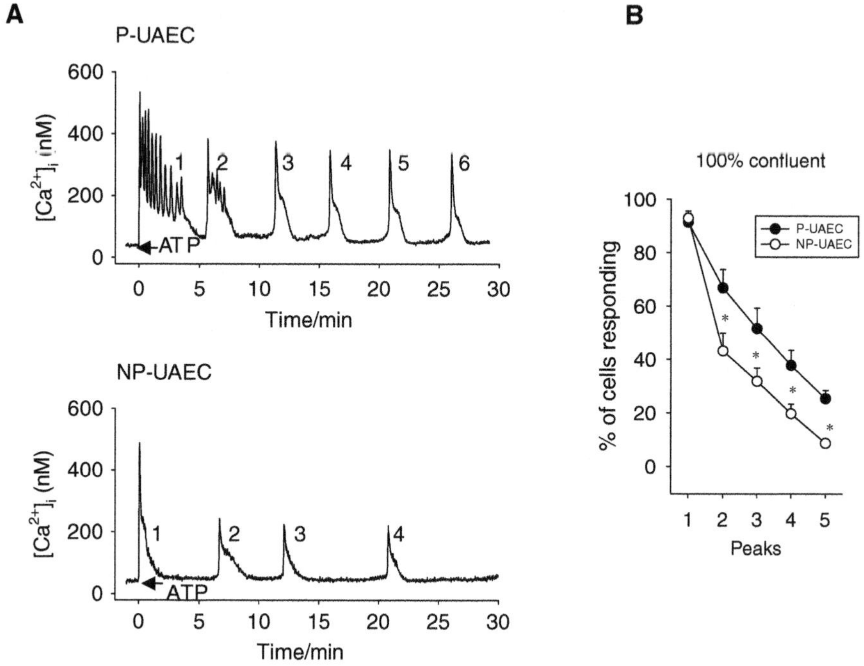

Fig. 11.5 Effects of pregnancy on sustained phase Ca^{2+} bursts in UAEC. (**A**) Responses to 100 μM ATP in individual NP- and P-UAEC at 100% confluency over 30 min. Note the presence of secondary bursts in each case and the more sustained nature of the secondary bursts in P-UAEC following the initial 5 min of the response. (**B**) Pregnancy specific changes in likelihood of $[Ca^{2+}]i$ bursts observed in NP- and P-UAEC in response to 100 μM ATP. The incidences of $[Ca^{2+}]i$ bursts (numbered as in **A**) were determined for 60 or more cells per field in data from $n = 5–7$ separate cell preparations and the percentage of cells responding with each successive burst are shown. Data are mean ± SE, with significant difference between NP- and P-UAEC response shown by $*P < 0.05$. Figure is modified from that published in [37]

11.9 Evidence for Expression of CX43 in UAEC and its Role in Mediating Cell Synchronization Through Formation of CX43 Gap Junctions

In view of the findings of cell synchronization, Yi and coworkers sought to identify the means by which this may occur. In most instances of such behavior in endothelial cells in particular, the mediator is often a Gap junction comprised of a complex of members of the Connexin family. There are several members that may form hemichannels on the plasma membrane (to act as pores to release small molecules), or they may dock with each other in macromolecular arrays joining two adjacent cells and so create a cluster of transmembrane pores (Gap junctions) that link the cytosols of the adjacent cells. Investigation of the possible involvement of the 37, 40, and 43 Kd isoforms that are most commonly seen in endothelial cells [15] revealed only the clear presence of the 43 Kd isoform by Western blot. Furthermore, CX43 was detectable at both native and elevated apparent molecular weight [37], implying the CX43 could be subject to phosphorylation and so potential regulation by kinases. Confirmation that the CX proteins expressed in UAEC were indeed forming true Gap junctions rather than hemichannels was demonstrated by the finding that scrape loading of a high density cell monolayer with Lucifer yellow dye promoted dye entry alongside the cut in the cell monolayer, but not generally across all the cells [37]. Hemichannels on the surface of every cell alone would simply have allowed uniform dye entry to the entire monolayer independent of the cut, while only true Gap junctions would give the result observed with dye entry only to cells adjacent to the cut site. Further evidence for the formation of true Gap junctions by CX43 itself was achieved using peptides of sequences encoding the extracellular loop regions of the different CX isoforms. In docking to make Gap junctions, CX proteins bind each other via extracellular loops, and peptides made from copies of these same sequences can competitively interfere with the coupling process. Communication via a true Gap junction is then lost, while function through a pore on the surface of the cell to release agents to the extracellular space remains unaltered. We have found that a peptide to the CX43 sequence (SRPTEKTIFII – termed GAP27) is completely successful in preventing cell synchronization and further has a profound effect in preventing long term $[Ca^{2+}]i$ burst responses to ATP at high cell density. In contrast, other classes of GAP peptides encoding sequences of CX40 or 37, or a scrambled version of CX43 GAP27 were ineffective [37]. Thus CX43 alone mediates inter-cell communication of relevance to $[Ca^{2+}]i$ signaling, and further that such communication in the form of true Gap junctions is critical to long term $[Ca^{2+}]i$ bursts (see Fig. 11.5b). In addition, GAP27 will inhibit the burst pattern of both P-UAEC and NP-UAEC down to a common lower level. It is not clear at this time if the CX43 Gap junctions transport Ca^{2+} or IP3 itself, but either way there are two important conclusions that can be drawn concerning pregnancy adaptation of cell signaling in UAEC, namely: (1) while such bursts of capacitative entry may be a function of TRPC/IP3R interaction, this process is in turn regulated by CX43 Gap junctions, and (2) pregnancy specific enhancement of this burst activity is not regulated at the level of CX43 expression but instead is regulated by pregnancy specific enhancement of Gap junction function. Given the additional finding that this same GAP27 peptide also inhibits the activation of eNOS in P-UAEC and NP-UAEC down to a common level [12], it is possible to conclude that while the molecular means by which pregnancy enhances NO production is indeed through an associated increase in capacitative entry in UAEC at the level of IP3R2-TRPC3 interaction, this process is only increased because of the pregnancy enhancement of CX43 GAP junction communication.

11.10 Reprogramming of the UA Endothelial Cells Also Occurs at the Level of the Mitochondria

Discovery that pregnancy control of Gap junction communication is the means by which the potential for interaction of IP3-R with TRPC3 is regulated is clearly a substantial advance, yet there are clearly other factors involved in pregnancy adaptation of UAEC cell signaling. It is apparent in a number of cells that the mitochondria can 'communicate' with the ER and influence its ability to respond to agonists, and recent evidence suggest this may occur in UAEC. Treatment of the UAEC with carbonyl cyanide m-chloro phenyl hydrozone (CCCP) to inhibit mitochondrial function limits the increase in $[Ca^{2+}]i$ in response to Thapsigargin or ATP [35]. This effect, however, was not due to a collapse of ATP levels within the cell, at least in the time period studied. This effect was also not due to a change in mitochondrial Ca^{2+} uptake. Of particular relevance here, the P-UAEC are more resistant to this effect of CCCP on $[Ca^{2+}]i$ than NP-UAEC, suggesting that pregnancy is perhaps altering mitochondrial function itself. This is intriguing and the importance of this functional adaptation to pregnancy may indeed be substantial, given that families with a heritable disorder of mitochondrial function are also seen to display preeclampsia like symptoms during pregnancy [26, 32].

11.11 Future Studies

Given the evidence that CX43 is phosphorylated in UAEC, further study of the endocrine events that result in changes in permeability of CX43 through phosphorylation are clearly warranted. It will be no surprise that these studies are ongoing in our laboratory at this time. Likewise the ability to directly image NO in *individual* cells of the intact UA endothelium while still attached to the vessel has now been achieved and the examination of $[Ca^{2+}]i$ bursts and synchronization of endothelial cell function is underway ex vivo. A third area of interest is the question of mitochondrial adaptation in pregnancy. While it is clear the mitochondria can effect the $[Ca^{2+}]i$ response to agonists, the exact mechanisms remains unknown. Nonetheless it is noteworthy that TNF-alpha, known to be elevated in preeclampsia, is capable of damaging UAEC through excess reactive oxygen species generation (Yi, unpublished) as expected. Further studies into the question of whether TNF-alpha impairs ATP-stimulated $[Ca^{2+}]i$ and, if so, if this is mediated through damage to the mitochondria is of particular relevance. It is our hope that as such combined studies further reveal the molecular components of the $[Ca^{2+}]i$ signaling pathway in UAEC and the manner by which pregnancy acts to alter their function, it will become possible to identify pharmacologic targets for clinical therapy of signaling dysfunction in pregnancy associated diseases. More specifically, it is already clear that preeclampsia is in part characterized by a failure of the endothelium to adapt its function. Observations in hand vein endothelial cells [22] and umbilical vein endothelial cells [29] suggest those subjects with preeclampsia show a blunted $[Ca^{2+}]i$ response just like that observed in NP-UAEC and those with healthy pregnancy show responses like that in P-UAEC. It is also noteworthy that in 'normal' HUVEC at least, P2Y2 receptor protein is expressed [10] and functionally coupled to $[Ca^{2+}]i$ signaling [10, 31], PLC beta 3 is present [2], and both TRPC3 [13] and CX43 [33] protein expression have been reported in HUVEC preparations. Thus all the components present in UAEC, whose interaction is adapted during pregnancy to achieve maximal $[Ca^{2+}]i$ sensitive eNOS activation, are also present in HUVEC. The translation of the studies described above on UAEC into the HUVEC culture model, as well as the parallel imaging of $[Ca^{2+}]i$ and NO in single cells still attached to the luminal surface of freshly isolated vessels from normal and preeclamptic pregnancies could yield invaluable insight into the mechanistic failures associated with this disease, and a useful model to subsequently test treatments that could restore more normal function.

Acknowledgments IMB would like to dedicate this review to all those who worked so diligently on the many studies quoted herein over the past 15 years. The authors would also like to acknowledge the support of NIH grants HL64601 and HL079020 dedicated to these Ca^{2+} imaging studies, and to NIH Program Project award HD 38843 for complimentary studies of UAEC function. DB is the recipient of T32 award HD41921 for his part in this work towards a PhD in the Endocrinology and Reproductive Physiology Graduate Training Program, at the University of Wisconsin, Madison.

References

1. Ahokas RA, Mercer BM, Sibai BM. (1991) Enhanced endothelium-derived relaxing factor activity in pregnant, spontaneously hypertensive rats. Am J Obstet Gynecol 165: 801–7.
2. Bhattacharya R, Kwon J, Li X, Wang E, Patra S, Bida JP, Bajzer Z, Claesson-Welsh L, Mukhopadhyay D. (2009) Distinct role of PLCbeta3 in VEGF-mediated directional migration and vascular sprouting. J Cell Sci 122(Pt 7):1025–34.
3. Bird IM, Sullivan JA, Di T, Cale JM, Zhang L, Zheng J, Magness RR. (2000) Pregnancy-dependent changes in cell signaling underlie changes in differential control of vasodilator production in uterine artery endothelial cells. Endocrinology 141:1107–17.
4. Bootman MD, Collins TJ, Mackenzie L, Roderick HL, Berridge MJ, Peppiatt CM. (2002) 2-aminoethoxydiphenyl borate (2-APB) is a reliable blocker of store-operated Ca^{2+} entry but an inconsistent inhibitor of InsP3-induced Ca^{2+} release. FASEB J 16(10):1145–50.
5. Boulay G, Brown DM, Qin N, Jiang M, Dietrich A, Zhu MX, Chen Z, Birnbaumer M, Mikoshiba K, Birnbaumer L. (1999) Modulation of Ca(2+) entry by polypeptides of the inositol 1,4, 5-trisphosphate receptor (IP3-R) that bind transient receptor potential (TRP): evidence for roles of TRP and IP3-R in store depletion-activated Ca(2+) entry. Proc Natl Acad Sci USA 96(26):14955–60.
6. Di T, Sullivan JA, Magness RR, Zhang L, Bird IM (2001) Pregnancy-specific enhancement of agonist- stimulated ERK-1/2 signaling in uterine artery endothelial cells (UAEC) increases Ca^{2+} – sensitivity of endothelial Nitric Oxide Synthase (eNOS) as well as cytosolic Phospholipase A2 (cPLA2). Endocrinology 142:3014–26.
7. Fitzgerald DJ, Entman SS, Mulloy K, FitzGerald GA. (1987) Decreased prostacyclin biosynthesis preceding the clinical manifestation of pregnancy-induced hypertension. Circulation 75:956–63.
8. Fleming I, Busse R. (1999) Signal Transduction of eNOS Activation. Cardiovasc Res 43: 532–41.

9. Gifford SM, Cale JM, Tsoi S, Magness RR, Bird IM (2003). Pregnancy-specific changes in uterine artery endothelial cell signaling in vivo are both programmed and retained in primary culture. Endocrinology 144(8):3639–50.
10. Gifford SM, Grummer MA, Pierre SA, Austin JL, Zheng J, Bird IM. (2004) Functional characterization of HUVEC-CS: Ca^{2+} signaling, ERK 1/2 activation, mitogenesis and vasodilator production. J Endocrinol 182(3):485–99.
11. Gifford SM, Yi FX, Bird IM. (2006) Pregnancy-enhanced Ca^{2+} responses to ATP in uterine artery endothelial cells is due to greater capacitative Ca^{2+} entry rather than altered receptor coupling. J Endocrinol 190(2):373–84.
12. Gifford SM, Yi FX, Bird IM. (2006) Pregnancy-enhanced store-operated Ca^{2+} channel function in uterine artery endothelial cells is associated with enhanced agonist-specific transient receptor potential channel 3-inositol 1,4,5-trisphosphate receptor 2 interaction. J Endocrinol 190(2):385–95.
13. Groschner K, Hingel S, Lintschinger B, Balzer M, Romanin C, Zhu X, Schreibmayer W. (1998) Trp proteins form store-operated cation channels in human vascular endothelial cells. FEBS Lett 437(1–2):101–6. Erratum in: FEBS Lett Jan 8;442(1):122.
14. Grummer MA, Sullivan JA, Magness RR, Bird IM. (2009) Vascular endothelial growth factor acts through novel, pregnancy-enhanced receptor signalling pathways to stimulate endothelial nitric oxide synthase activity in uterine artery endothelial cells. Biochem J 417(2): 501–11.
15. Hill CE, Phillips JK, Sandow SL. (2001) Heterogeneous control of blood flow amongst different vascular beds. Med Res Rev 21(1):1–60.
16. Kamouchi M, Philipp S, Flockerzi V, Wissenbach U, Mamin A, Raeymaekers L, Eggermont J, Droogmans G, Nilius B. (1999) Properties of heterologously expressed hTRP3 channels in bovine pulmonary artery endothelial cells. J Physiol 518(Pt 2):345–58.
17. Kiselyov K, Xu X, Mozhayeva G, Kuo T, Pessah I, Mignery G, Zhu X, Birnbaumer L, Muallem S. (1998) Functional interaction between InsP3 receptors and store-operated Htrp3 channels. Nature 396(6710):478–82.
18. Kramer RM, Sharp JD (1997) Structure, function and regulation of Ca^{2+}-sensitive cytosolic phospholipase A2 (cPLA2). FEBS Lett 410:49–53.
19. Lin S, Fagan KA, Li KX, Shaul PW, Cooper DM, Rodman DM. (2000) Sustained endothelial nitric-oxide synthase activation requires capacitative Ca^{2+} entry. J Biol Chem 275(24): 17979–85.
20. Magness RR, Shaw CE, Phernetton TM, Zheng J, Bird IM. (1997) Endothelial vasodilator production by uterine and systemic arteries. II. Pregnancy effects on NO synthase expression. Am J Physiol. 272(4 Pt 2):H1730–40.
21. Magness RR, Sullivan JA, Li Y, Shaw CE, Phernetton TM, Bird IM. (2001) Endothelial vasodilator production by uterine and systemic arteries. VI. Ovarian and pregnancy effects on eNOS and NOx. Am J Physiol 280: (Heart Circ. Physiol) H1692–8.
22. Mahdy Z, Otun HA, Dunlop W, Gillespie JI. (1998) The responsiveness of isolated human hand vein endothelial cells in normal pregnancy and in pre-eclampsia. J Physiol 508(Pt 2):609–17.
23. Montell C. (2001) Physiology, phylogeny, and functions of the TRP superfamily of cation channels. Sci STKE 2001(90):RE1.
24. Nilius B, Droogmans G. (2001) Ion channels and their functional role in vascular endothelium. Physiol Rev 81(4):1415–59.
25. Putney JW Jr. (1999) TRP, inositol 1,4,5-trisphosphate receptors, and capacitative calcium entry. Proc Natl Acad Sci USA 96(26):14669–71.
26. Racine AC, Blanchot G, Le Vaillant C, Boog G. (2004) Pregnancy in a patient with mitochondrial disease. J Gynecol Obstet Biol Reprod (Paris) 33:131–9.
27. Rupnow HL Phernetton TM, Shaw CE, Modrick ML, Bird IM, Magness RR. (2001) Endothelial vasodilator production by uterine and systemic arteries. VII. Estrogen and progesterone effects on eNOS. Am J Physiol 280: (Heart Circ Physiol) H1699–705.

28. Seligman SP, Buyon JP, Clancy RM, Young BK, Abramson SB. (1994) The role of nitric oxide in the pathogenesis of preeclampsia. Am J Obstet Gynecol 171:944–8.
29. Steinert Jr, Wyatt AW, Poston L, et al. (2002). Preeclampsia is associated with altered Ca^{2+} regulation and NO production in human fetal venous endothelial cells. FASEB J 16:721–3.
30. Sullivan JA, Grummer MA, Yi FX, Bird IM. (2006) Pregnancy-enhanced endothelial nitric oxide synthase (eNOS) activation in uterine artery endothelial cells shows altered sensitivity to Ca^{2+}, U0126, and wortmannin but not LY294002 – evidence that pregnancy adaptation of eNOS activation occurs at multiple levels of cell signaling. Endocrinology 147(5):2442–57.
31. Tanaka N, Kawasaki K, Nejime N, Kubota Y, Nakamura K, Kunitomo M, Takahashi K, Hashimoto M, Shinozuka K. (2004) P2Y receptor-mediated Ca(2+) signaling increases human vascular endothelial cell permeability. J Pharmacol Sci 95(2):174–80.
32. Torbergsen T, Oian P, Mathiesen E, Borud O (1989) Pre-eclampsia – a mitochondrial disease? Acta Obstet Gynecol Scand 68:145–8.
33. Van Rijen H, van Kempen MJ, Analbers LJ, Rook MB, van Ginneken AC, Gros D, Jongsma HJ. (1997) Gap junctions in human umbilical cord endothelial cells contain multiple connexins. Am J Physiol 272(1 Pt 1):C117–30.
34. Xiao D, Bird IM, Magness RR, Longo LD, Zhang L. (2001) Upregulation of eNOS in pregnant ovine uterine arteries by chronic hypoxia. Am J Physiol 280: (Heart Circ Physiol) H812–20.
35. Yi FX, Bird IM. (2005) Pregnancy-specific modulatory role of mitochondria on adenosine 5?-triphosphate-induced cytosolic [Ca^{2+}] signaling in uterine artery endothelial cells. Endocrinology 146(11):4844–50.
36. Yi FX, Magness RR, Bird IM. (2005) Simultaneous imaging of [Ca^{2+}]i and intracellular NO production in freshly isolated uterine artery endothelial cells: effects of ovarian cycle and pregnancy. Am J Physiol Regul Integr Comp Physiol 288(1):R140–8.
37. Yi FX, Boeldt DS, Gifford SM, Sullivan JA, Grummer MA, Magness RR, Bird IM. (2010) Pregnancy Enhances Sustained Ca^{2+} Bursts and Endothelial Nitric Oxide Synthase Activation in ovine Uterine Artery Endothelial Cells Through Increased Connexin 43 Function. Biology Reproduction 82(1):66–75.
38. Zhu X, Jiang M, Peyton M, Boulay G, Hurst R, Stefani E, Birnbaumer L. (1996) trp, a novel mammalian gene family essential for agonist-activated capacitative Ca^{2+} entry. Cell 85(5):661–71.

Chapter 12
Purinergic Signalling in Pancreatic Islet Endothelial Cells

Leif Jansson, Eva Grapengiesser, and Bo Hellman

Abstract Both the microvascular endothelium and the endocrine cells in the pancreatic islet can release and react upon ATP. In support for the idea that intermittently released ATP, related to exocytosis of insulin secretory granules, affects the secretory activity of the capillary endothelium, we have demonstrated that isolated endothelial cells respond to activation of $P2Y_2$ receptors with pronounced and extended rises of $[Ca^{2+}]_i$. The presence of such ATP effect is consistent with reports that β-cells regulate the blood flow within islets, where adenosine is a key mediator, and that the endothelial cell produce pro-angiogenic and angiostatic factors. In β-cells down-regulation of $P2Y_1$ receptors results in disappearance of the transients of $[Ca^{2+}]_i$ supposed to entrain these cells into a common rhythm. Since the islet endothelial cells respond to activation of $P2Y_2$ receptors with extended elevation of $[Ca^{2+}]_i$, it is likely that the accompanying release of ATP is prolonged. Accordingly, the endothelial cells may have a tonic inhibitory action on the coordination of islet release pulses.

Keywords Adenosine · Arteriole · ATP · β-cells · Ca^{2+} oscillations · Capillaries · Endothelium · Insulin · Islet blood flow · Pancreas · Pancreatic islets · Purinergic signaling · P2X receptors · P2Y receptors · UTP

12.1 Introduction

Purines and pyrimidines released to or formed within the extracellular space serve as signalling molecules in many organs of the body by acting on one or several ionotropic P2X and metabotropic P2Y receptors [11]. The purinergic receptors

L. Jansson (✉)
Department of Medical Cell Biology, Biomedical Centre, Uppsala University, Husargatan 3, Box 571, SE-751 23, Uppsala, Sweden
e-mail: leif.jansson@mcb.uu.se

E. Gerasimovskaya, E. Kaczmarek (eds.), *Extracellular ATP and Adenosine as Regulators of Endothelial Cell Function*, DOI 10.1007/978-90-481-3435-9_12,

are present on different types of vascular cells. The endothelial cells have high expression of $P2Y_1$, $P2Y_2$ and $P2X_4$ receptors, medium expression of $P2Y_6$ receptors and possible expression of $P2Y_{11}$ receptors [11]. Also vascular smooth muscle cells (VSM), and adventitial cells, have purinergic receptors that enable purines and pyrimidines to induce contraction or dilation of VSM [24, 64].

External ATP has organ-specific effects on vascular endothelium. In the present review evidence is presented that purinoceptor-mediated interactions between islet capillaries and endocrine cells are key factors for coordinating glucose metabolism of the β-cells with blood flow responses and insulin release. Recent findings emphasize the importance of the islet microvasculature for the functions of this highly complex micro-organ, which crucially depends on ATP as a signalling molecule [37].

12.2 ATP and Vascular Endothelium

12.2.1 General Aspects on Vascular ATP Signalling

External ATP affects the vascular tone through several mechanisms. Released as a co-transmitter from perivascular sympathetic nerves ATP acts on P2 receptors causing VSM contraction [12]. ATP can also be released from endothelial cells during changes in blood flow (mainly increased shear stress) or hypoxia to act on endothelial P2X and P2Y receptors which results in release of nitric oxide (NO) with vasodilation [11, 86]. A third possibility for ATP to affect local circulation is release of purines from erythrocytes and granulocytes [23]. Local production of ecto-nucleotidases helps to modulate the effects of ATP [87]. These mechanisms will be discussed further below with special emphasis on islet endothelium.

In addition to controlling the vascular tone, external ATP and its degradation products have important long-term actions on endothelial and smooth muscle cell proliferation, differentiation, and death [10, 24, 26]. Recent studies on islet vasculature suggest that interactions between endocrine cells, nerves and endothelial cells may be important for the phenotype properties of parenchyme cells [47, 48, 55].

12.2.2 External ATP Generates and Propagates Ca^{2+} Signals in Microvascular Endothelium

The responses of microvascular endothelial cells are highly heterogenous in space and time [56]. Thus, there are functional differences not only between endothelium from large blood vessels and microvascular endothelium, but also with regard to the organs involved. All types of microvascular endothelial cells studied so far respond to ATP with increase of $[Ca^{2+}]_i$ [20, 83]. The elevation of $[Ca^{2+}]_i$ is known to initiate a number of processes [83], including release of vasoactive NO, endothelium-derived hyperpolarizing factor (EDHF) and prostacyclin [6, 57, 81]. Constitutive nitric oxide synthase, activated by the Ca^{2+}-calmodulin complex [57], is present in islet capillaries and an essential regulator islet blood flow [80]. ATP is a secretory product of endothelium. Indeed, regenerative release of ATP is known to propagate stimulatory Ca^{2+} signals between the endothelial cells [13, 40]. The observation that external ATP induces oscillatory $[Ca^{2+}]_i$ in capillaries and arterioles adds further weight to the argument that microvascular endothelium has a synchronizing effect on adjacent cells. Changes in endothelial $[Ca^{2+}]_i$ have profound effects on release of vasoactive substances, and consequently on regional microcirculation. As referred to below addition of ATP increases $[Ca^{2+}]_i$ in the pancreatic islet endothelium.

12.3 ATP and the Pancreas

12.3.1 Purinergic Nerves and Pancreas

Pancreatic exocrine secretion is regulated both by neural and hormonal factors, including external ATP [66, 68]. Moreover, multiple functional P2X and P2Y receptors have been identified in pancreatic duct cells [58, 68] and found to modify the exocrine pancreatic secretion. It is not yet clear whether the nucleotides are exerting their effects through autocrine, paracrine, or neuronal mechanisms. The pancreatic islets possess a rich innervation, which comprises parasympathetic, sympathetic, purinergic as well as sensory nerves [1, 74]. It has been suggested that islets constitute paraneurons ensheathed in Schwann cells [82]. The various types of islet nerves are known to modulate the hormone secretion and blood perfusion of the islets.

12.3.2 Effects of ATP and Adenosine on Exocrine Pancreatic Arterioles

Measurement of pressure during perfusion of the isolated rat pancreas made it possible to assess the vascular dilation or constriction in the arterioles. Using this approach it was found that the effect of ATP reflected the balance between P2Y-receptor mediated vasodilation and P2X-mediated vasoconstriction [39]. On the other hand adenosine was found to be an unequivocal dilator of arterioles in the exocrine pancreas [4, 17].

12.3.3 Anatomy of the Pancreatic Islets

Pancreatic islets consist of different types of cells located in aggregates of varying sizes with a diameter up to 400 μm. In the human pancreas there is 1–2 × 10^6 islets containing insulin-producing β-cells and to a lesser extent of glucagon-producing α-cells and somatostatin-producing δ-cells. All islets possess a dense vascular network with fenestrated capillaries [72]. Each β-cell, and presumably also the other endocrine cells have contact with at least one capillary [7].

12.3.4 Islet Vascular Anatomy

There is a direct and separate arteriolar flow to the islets (Fig. 12.1), implying that the islets can regulate their blood perfusion separately from that of the exocrine parenchyma. Normally islet blood flow is 5–10 times higher than that in the exocrine parts [9]. Besides mediating transport of nutrients and hormones, the islet endothelium also affects a number of processes (Table 12.1) including the differentiation of endocrine islet cells during development [50, 54], and in adults the regulation of β-cell proliferation [48]. It is likely that the basement membrane of islet capillaries has a special role in this context, due to differences in molecular composition and integrin expression compared to exocrine capillaries [49, 65]. It was recently shown that humans have a fused basement membrane between the islet endothelial and endocrine cells, constituting a blood-islet barrier. [84]. The functional importance of this is unknown. We have preliminary results suggesting that rodent islet endothelium, like bone marrow and liver sinusoidal endothelium, produces the protein stabilin-1, a marker for fagocytotic endothelia [31]. This suggests that there is a need for protection of the islet endocrine cells against direct contact with the blood stream, although the mechanisms may differ between species.

Islet endothelial cells are difficult to study in vitro, due to dedifferentiation upon culture [59, 72, 73]. It is known from transplantation experiments, that freshly isolated islets are revascularized by endogenous endothelial cells [71]. However, cultured islets require host endothelial cells to form a new vascular system.

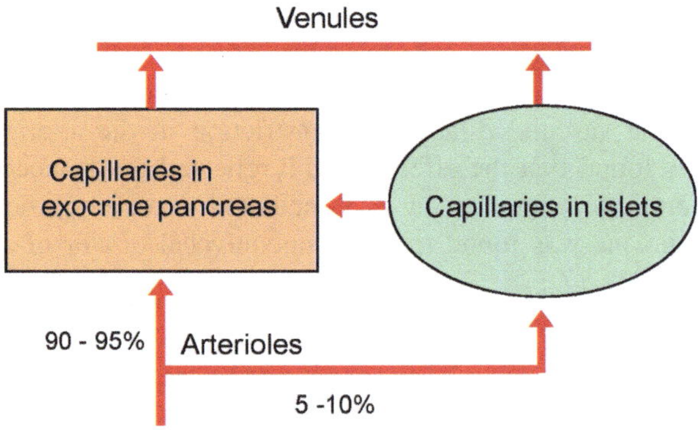

Fig. 12.1 Schematic drawing of the vasculature in the pancreas. Arterial blood flow is separate for the endocrine and exocrine parenchyma, and 5–10% is diverted to the islets. Note that the islets comprise only 1% of the pancreas, which means that their blood perfusion is 5–10 times higher than that of the exocrine gland. An unknown quantity, mainly in larger islets, flows through an insulo-acinar portal system, before being emptied into veins. The blood flow regulation occurs in the pre-capillary circulation

Table 12.1 Features of islet endothelial cells compared to those in exocrine pancreas. See text for details

Exposed to high blood flow	[9, 45]
High production of and sensitivity to endothelium-derived relaxing and constricting factors; e.g. nitric oxide and endothelin-1	[52, 80]
Different types of surface receptors	[72]
Different production of pro-angiogenic and angiostatic peptides; e.g. VEGF-A and thrombospondin-1, basic fibroblast growth factor	[55, 59, 70]
Regulates β-cell proliferation by growth factors and provides a niche for β-cell growth	[48, 50]
Transgenic animals with impaired glucose tolerance often show altered capillary morphology	[2, 55]

Both the islet vasculature and blood flow are affected in animal models of type 2 diabetes. It has been suggested that islet vascular defects, including impaired signaling between β- and islet endothelial cells, adversely affect the endocrine function in human diabetes [45, 53]. The adaptor protein Shb is associated with and relays signals from VEGF-activated VEGFR2. The resulting stimulation of the PI3-kinase with activation of FAK in concert with Src promote cytoskeletal rearrangements [41].

12.3.5 Blood Flow Direction in the Islets

Despite species differences most islets are usually supplied with arterioles separate from those to the exocrine pancreas (Fig. 12.1). Each islet receives 1–3 arterioles, which branch into fenestrated capillaries. An important issue is how the arterioles enter the islets. A summary of this debate has been provided [9]. Briefly, it was claimed that the arterioles enter through discontinuities in the islet periphery lacking β-cells and branch into capillaries. This has led to the hypothesis that the arterial blood first reaches β-cells, in the core of the islets providing high concentrations of insulin to α- and δ-cells. Glucagon from α-cells mainly affects the adjacent δ-cells. Accordingly, there is a preferential flow direction through the islets important for paracrine interactions between islet cells [78]. However, other morphological data indicate that the arterioles branch into capillaries before the entry into the islets [9]. A third theory is that the cellular order of blood perfusion varies between the species, depending upon islet cytoarchitecture [9, 33]. We recently developed a technique to perfuse large single rodent islets (Fig. 12.2), and found that the branching occurs outside the islets. However, it cannot be excluded that arterioles in some large islets can enter the central parts of the islets [69].

The absence of arterioles within the islets would exclude the possibility of islet blood flow regulation by direct actions of islet metabolites and/or hormones on VSM in the blood vessel wall. It is possible that endocrine cell metabolism, which is known to affect islet blood flow [17], must affect the endothelial cells, which then retrogradely informs the arterioles about the need for vasoconstriction or

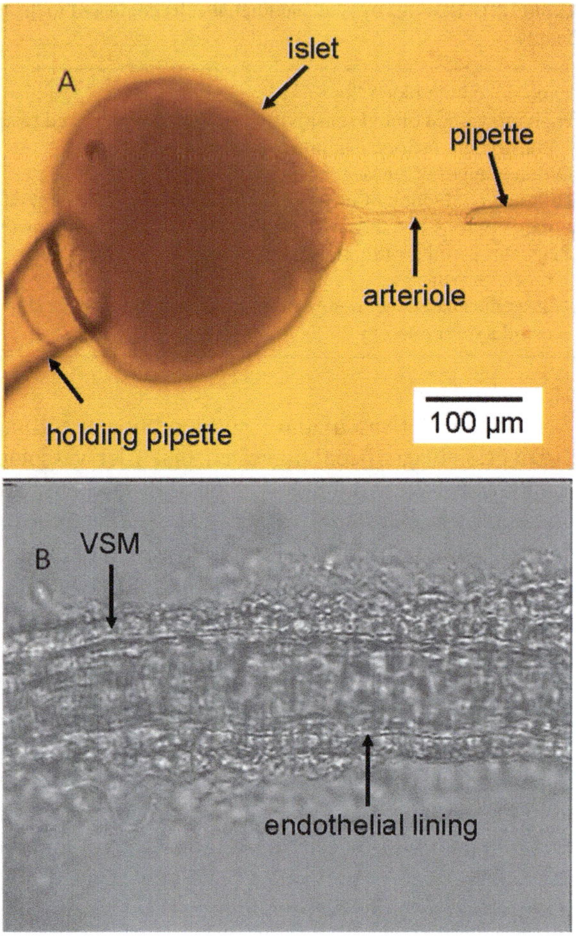

Fig. 12.2 (**a**) Isolated islet attached to a holding pipette (*left*) with an arteriole cannulated via a perfusion pipette (*right*). (**b**) Structure of the islet arteriole (diameter 25 μm). The endothelial cell lining clearly demarcates the vessel lumen. The periphery of the arteriole wall contains vascular smooth muscle (VSM)

vasodilation. In a recent set of experiments a real-time, multidimensional imaging technique allowed the study of the islet blood flow and its direction in vivo [69]. A flow from pole-to-pole was found in some islets, and centripetal flow in others. These observations support the idea of a difference in blood flow pattern between islets in the same animal. It is not known if an islet can change its flow pattern or not, even if this is likely to be the case.

The drainage of the islets is accomplished directly through veins, as seen in large islets, but also by an insulo-acinar portal system [8]. Especially the latter is the subject to large species variations, and the functional importance is unclear. Evidence has been provided that the portal system allows exposure of peri-insular acini to high concentrations of islet hormones [25, 38], and hereby resulting in an increase in their protein synthesis. Taken together available data suggest that precapillary VSM in the arterioles is the most important site for islet blood flow regulation, and that postcapillary venulae contribute little, or not.

12.3.6 Islet Blood Flow Regulation

As mentioned above the pancreatic islets possess an autonomous blood flow regulation (Fig. 12.1). Although the islets constitute only approximately 1% of the pancreas they account for 5–10% of total pancreatic blood flow [9, 45]. The mechanisms for regulation of islet blood flow are complex and multifactorial, but especially influenced by the ambient blood glucose concentration [9, 45]. Other factors of importance include metabolism- and endothelium-derived factors, such as adenosine, ATP and nitric oxide, as well as neurogenic modulation by sensors located to vascular bed of the brain, intestines and liver [17, 45, 72]. The islet blood flow is closely coupled to the β-cell glucose metabolism and the accompanying release of insulin. After food intake, the glucose receptors in the brain [46], intestine [15] and liver [16] transiently stimulate the islet blood flow, mainly through the parasympathetic nerves [46]. During hyperlipidemia there is instead an activation of the sympathetic nervous system, which increases islet blood flow mainly through β_3-adrenoceptors (unpublished observation). However, within a few minutes glucose-stimulation of the β-cell metabolism generates islet blood flow increase, with adenosine as the main vasodilator [17]. We have recently been able to perfuse single islets with intact afferent arterioles (Fig. 12.2) [51, 52, 67]. The results confirm that adenosine has a direct dilatory effect on arterioles [51]. So far preliminary studies on intraluminally administered ATP have demonstrated a vasoconstrictive effect when given at 100 μM (Fig. 12.3). This contrasts to the ATP-effects in skeletal muscle circulation, where it mediates exercise hyperaemia [21, 61].

It is known that endothelial cells in capillary networks communicate via gap junctions [42]. As a consequence pulses of depolarization are conducted both up and downstream in the capillaries [62] and propagate to the arterioles and/or venules [3]. It can be envisaged from the islet microvasculature that electrical coupling between endothelial cells enables the capillaries to regulate the islet blood flow by conducted responses to the arterioles. If this holds true, external ATP may affect islet blood flow differently depending on whether it exerts the predominant action on endothelial cells or VSM. As mentioned above, we have observed adenosine-induced vasodilation in isolated islet arteriolar VSM [51].

Purinoceptor regulation of islet microcirculation may involve release of ATP from erythrocytes by mechanical deformation [77] and deoxygenation of haemoglobin [43]. Together with ATP release from granulocytes [23], this is likely to be a major determinant for the deranged microcirculation induced by hypoxia and non-specific inflammation early after islet transplantations. Similar responses are seen in damaged skeletal muscle [27, 76]. Non-specific inflammatory responses, associated with local hyperemia, are amplified by immunological processes associated with rejection. Accordingly, ATP may influence islet microcirculation in the immediate post-transplantation period. Infiltration of leukocytes is frequently seen with insulitis in the initial stages of type 1 diabetes. This insulitis is coupled to a marked increase in islet blood flow [14, 18], and may well reflect the local release of ATP.

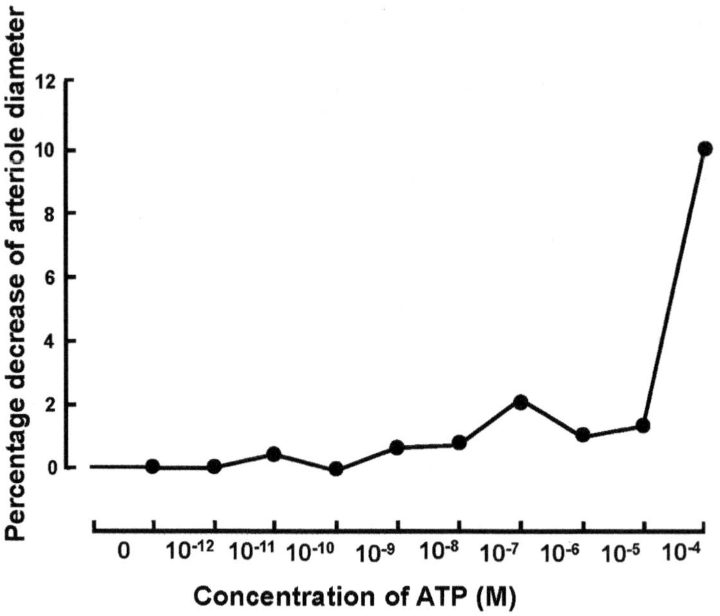

Fig. 12.3 Percentage alterations of the diameter in an arteriole, associated with an isolated islet, during step-by-step increase of ATP in 2-min intervals. The experiment was performed as previously described [51]

12.4 Interactions Between Endocrine Cells and Islet Endothelium

12.4.1 External ATP is a Regulator of Ca^{2+} Rises Triggering Pulsatile Insulin Release

Each β-cell is a biological oscillator responding to a glucose stimulus with periodic depolarization and accompanying entry of Ca^{2+}. It is generally accepted that the resulting $[Ca^{2+}]_i$ oscillations are the main trigger for pulses of insulin release [36]. A key factor for the depolarization is closure of K^+ channels mediated by the increase of cytoplasmic ATP. Within the pancreas, the β-cells will be entrained into a common rhythm both by gap junctions [5] and diffusible messengers, such as ATP [30, 35]. The coordinating action of external ATP is supposed to be mediated by generation of IP_3-induced increases (transients) of $[Ca^{2+}]_i$, which temporarily interrupt the entry of Ca^{2+} by activating a repolarizing K^+ current [22]. Pancreatic β-cells are both recipients of coordinating ATP signals, and release this nucleotide intermittently through exocytosis [32, 75]. The prerequisite for cyclic variations of circulating insulin is that $[Ca^{2+}]_i$ oscillations in the β-cells appear in the same phase. It is likely that neural activity with discharge of ATP accounts for the entrainment of the islets into common rhythm [28].

12.4.2 Purinoceptor-Mediated Ca^{2+} Rise in Isolated Islet Endothelial Cells and β-Cells

It was possible to compare the Ca^{2+} responses of islet endothelial cells and β-cells after dissociation of β-cell rich ob/ob mouse islets [34] into single cells and small aggregates [37]. Suspensions of islet cells were allowed to attach to coverslips during culture in medium supplemented with endothelial cell growth factor (ECGF), to maintain the phenotype of the endothelial cells [59]. $[Ca^{2+}]_i$ was measured during perifusion with ratiometric fura-2 technique [37, 59]. Endothelial cells, preliminary identified by addition of Dynabeads coated with the *Bandeira simplicifolia 1* lectin [60], responded in a similar way as those stained with CD31 antibodies after measurements of $[Ca^{2+}]_i$.

Endothelial cells differed from β-cells by lacking spontaneous fluctuations of $[Ca^{2+}]_i$ The ability of different test substances to generate $[Ca^{2+}]_i$ rises in the presence of 5 mM glucose is shown in Table 12.2 and Fig. 12.4. Both endothelial cells and β-cells responded to 10 μM ATP with rise of $[Ca^{2+}]_i$. In the endothelial cells 10 μM UTP was equally effective as ATP in raising $[Ca^{2+}]_i$ but equimolar ADP, UTP, adenosine or acetylcholine lacked effect. Studying the responses of the β-cells it was found that 10 μM ADP and acetylcholine initiated rise of $[Ca^{2+}]_i$, but UTP, UDP, adenosine and acetylcholine were without effect.

Subsequent fura-2 measurements confirmed that UTP was equally effective as ATP in raising $[Ca^{2+}]_i$ in islet endothelial cells (Figs. 12.5 and 12.6). The observation that ATP lacks effects in the presence of UTP (Fig. 12.5a) and that both nucleotides elicit $[Ca^{2+}]_i$ rises with similar kinetics, suppressed with 100 μM suramin (Fig. 12.5b), but not with methoxyverapamil (Fig. 12.6), suggest that most of their actions are mediated by the same type of purinoceptor, probably $P2Y_2$. Additional binding to $P2Y_4$ and $P2Y_6$ receptors seems unlikely, since these receptors are relatively resistant to suramin inhibition [19]. Moreover, there was no increase of $[Ca^{2+}]_i$ when the endothelial cells were exposed to UDP, the principal activator of $P2Y_6$ receptors. In many types of endothelial cells the ATP- induced increase of $[Ca^{2+}]_i$ is mediated by activation of $P2Y_1$ receptors [24, 63]. However, there was no $[Ca^{2+}]_i$ response to ADP, indicating that mouse islet endothelium lacks

Table 12.2 Rise of cytoplasmic Ca^{2+} in mouse islet endothelial cells and β-cells in the presence of 5 mM D-glucose

Cell	Endothelial cell	Beta-cell
ATP; 10 μM	+	+
ADP; 10 μM	0	+
UTP; 10 μM	+	0
UDP; 10 μM	0	0
Adenosine; 10 μM	0	0
Acetylcholine; 10 μM	0	+

Results from reference [37]

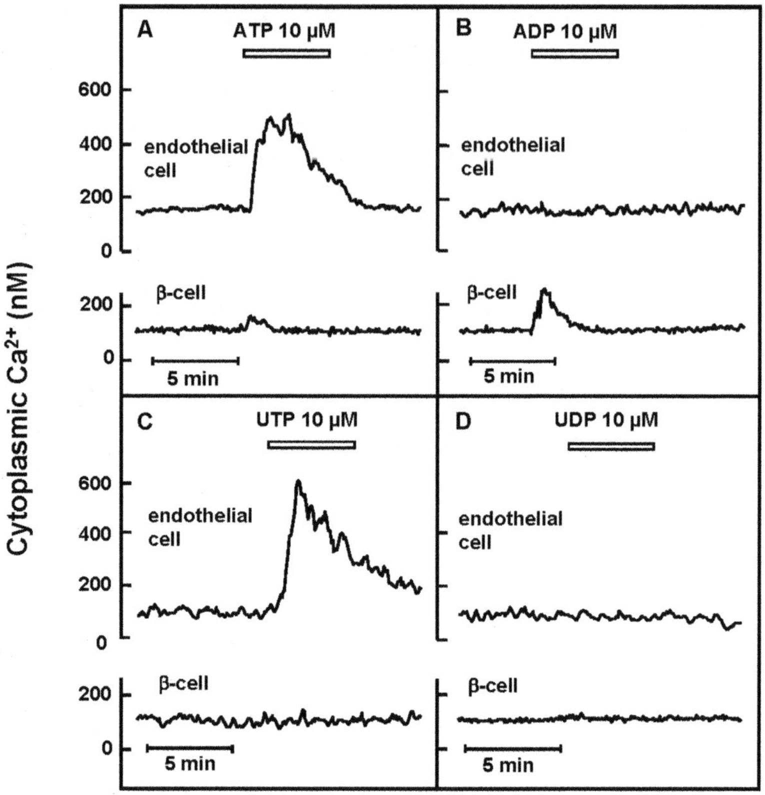

Fig. 12.4 Effects of 10 μM of ATP, ADP, UTP and UDP on $[Ca^{2+}]_i$ in islet endothelial cells and β-cells superfused with a medium containing 5 mM glucose. The test substances were added during 5-min periods. Representative for 5 experiments. From [37]

functionally active $P2Y_1$ receptors. Endothelial cells sometimes express P2X receptors, which mediate vasodilation [64, 79, 85]. Since UTP was equally effective as ATP in raising $[Ca^{2+}]_i$, it is unlikely that islet endothelial cells have functionally active P2X receptors.

Evidence has been provided that glucose stimulation of the islet blood flow involves adenosine P1 receptors [17]. However there was no increase of $[Ca^{2+}]_i$ when islet endothelial cells were exposed to 10 μM adenosine. The latter finding supports the idea that the vasodilator action of adenosine involves arterioles outside the islets rather than the intra-islet capillaries. Acetylcholine is widely employed for initiating release of endothelial factors regulating the vascular tone [88]. Nevertheless 10 μM acetylcholine elicited $[Ca^{2+}]_i$ increase only in the β-cells. The observation that endothelial cells respond with rise of $[Ca^{2+}]_i$ to purinergic, but not to cholinergic stimuli, is not unique for islet capillaries but is seen also in brain [44].

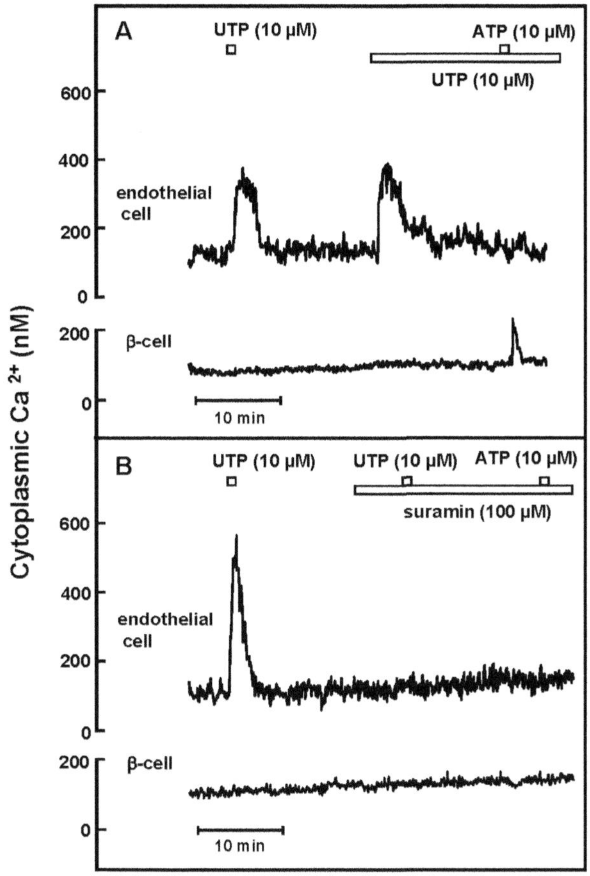

Fig. 12.5 Effects of 60 s pulses of 10 μM UTP and ATP on $[Ca^{2+}]_i$ in islet endothelial cells and β-cells superfused with a medium containing 5 mM glucose. The experiments were performed in the absence and presence of 10 μM UTP (**a**) or 100 μM suramin (**b**). Representative for 5 experiments. From [37]

12.4.3 ATP-Induced Ca^{2+} Signalling Between Islet Endothelium and β-Cells

Increase of $[Ca^{2+}]_i$ triggers a number of physiological processes, including release of secretory products both from β-cells and the islet endothelial cells. As already referred to (see above), there are reasons to believe that regenerative ATP release propagates Ca^{2+} signals both from one β-cell to another as well as between the islet endothelial cells. Accumulating data supports the idea that ATP released from the β-cells has regulatory effects on endothelial cells and vice versa. In support for the idea that ATP release, related to exocytosis of insulin secretory granules, affects the secretory activity of the endothelial cells, the latter cells responded to ATP with pronounced and extended rises of $[Ca^{2+}]_i$. The presence of this ATP effect is consistent with reports that glucose stimulation of β-cells promotes the blood flow within the islets and the endothelial cell production of pro-angiogenic and angiostatic factors [9]. ATP is released intermittently not only from nerves but also from

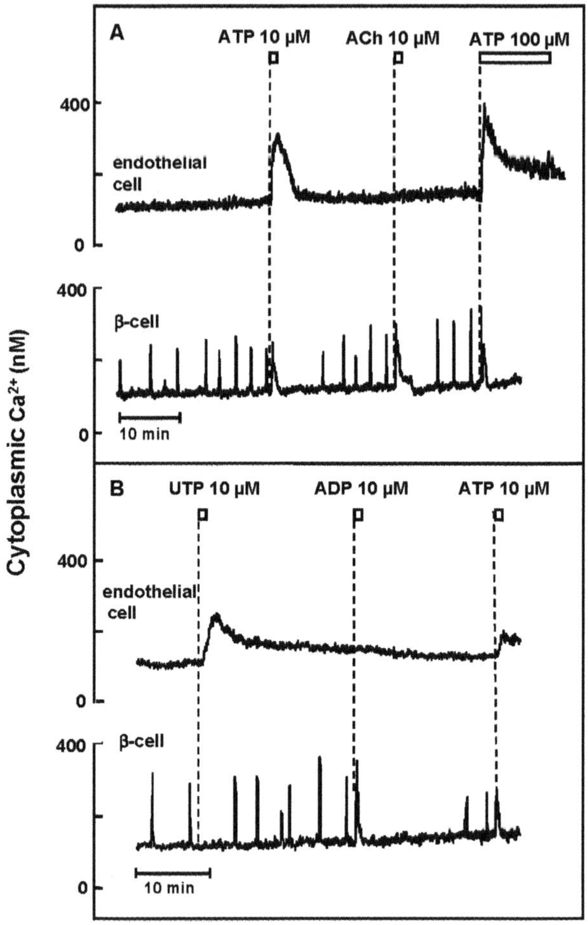

Fig. 12.6 Effects of 60 s pulses of 10 μM ATP, ADP, UTP and acetylcholine (ACh) on $[Ca^{2+}]_i$ in islet endothelial cells and β-cells superfused with a medium containing 20 mM glucose, 20 nM glucagon and 50 μM methoxyverapamil. The endothelial cells respond with extended rises of $[Ca^{2+}]_i$ to ATP and UTP, but not to ADP or acetylcholine. Spontaneous transients are seen in β-cells, but not in endothelial cells. Transients were promptly elicited in β-cells by ATP and ADP. Representative for 5 (**a**) and 4 (**b**) experiments. From [37]

β-cells [29, 32]. Prolonged exposure of β-cells to external ATP or ADP results in disappearance of the $[Ca^{2+}]_i$ transients, which coordinate the hormone pulses [28]. Administered as 60-s pulses it was found that ADP temporarily suppressed glucose-induced transients of $[Ca^{2+}]_i$ (Fig. 12.6). Since the islet endothelial cells respond to P2Y receptor activation with extended increase of $[Ca^{2+}]_i$, it is likely that the accompanying ATP release is prolonged. Accordingly, islet endothelial cells may have a tonic inhibitory action on the β-cell transients of $[Ca^{2+}]_i$, resulting in impaired synchronization of the insulin release pulses.

12.5 Conclusion

The pancreatic islets constitute a unique micro-organ, where both microvascular endothelium and the endocrine cells can release and react upon ATP. A crucial question is whether ATP released from the β-cells has regulatory effects on islet endothelial cells and vice versa. In support for the idea that intermittent release of ATP, related to exocytosis of insulin secretory granules, affects the secretory activity of the capillary endothelium, isolated endothelial cells responded to activation of $P2Y_2$ receptors with pronounced and extended rise of $[Ca^{2+}]_i$. This $[Ca^{2+}]_i$ increase subsequently triggers release of ATP from the endothelial cells. However, contrary to what is seen in β-cells, the extrusion of ATP from islet endothelial cells is not intermittent and results in accumulation of the nucleotide within the islets. Accordingly, the endothelial cells may have a tonic inhibitory action on β-cell coordination, mediated by down-regulation of the $P2Y_1$ receptors responsible for the generation of $[Ca^{2+}]_i$ transients.

Acknowledgments Financial support for our own studies referred to in the article was received from the Swedish Research Council, the Swedish Diabetes Association, the Juvenile Diabetes Research Foundation and the Family Ernfors Fund.

References

1. Ahren B. (2000) Autonomic regulation of islet hormone secretion – implications for health and disease. Diabetologia 43:393–410.
2. Anneren C, Welsh M, Jansson L. (2007) Glucose intolerance and reduced islet blood flow in transgenic mice expressing the FRK tyrosine kinase under the control of the rat insulin promoter. Am J Physiol Endocrinol Metab 292:E1183–90.
3. Beach JM, McGahren ED, Duling BR. (1998) Capillaries and arterioles are electrically coupled in hamster cheek pouch. Am J Physiol 275: H1489–96.
4. Bertrand G, Chapal J, Puech R, Loubatieres-Mariani MM. (1991) Adenosine-5′-O-(2-thiodiphosphate) is a potent agonist at P2 purinoceptors mediating insulin secretion from perfused rat pancreas. Br J Pharmacol 102:627–30.
5. Bertuzzi F, Davalli AM, Nano R, et al.. (1999) Mechanisms of coordination of Ca^{2+} signals in pancreatic islet cells. Diabetes 48:1971–8.
6. Bishara NB, Triggle CR, Hill MA. (2005) Cytochrome P450 products and arachidonic acid-induced, non-store-operated, Ca^{2+} entry in cultured bovine endothelial cells. Endothelium 12:153–61.
7. Bonner-Weir S. (1988) Morphological evidence for pancreatic polarity of beta-cell within islets of Langerhans. Diabetes 37:616–21.
8. Bonner-Weir S, Orci L. (1982) New perspectives on the microvasculature of the islets of Langerhans in the rat. Diabetes 31:883–9.
9. Brunicardi FC, Stagner J, Bonner-Weir S, et al. (1996) Microcirculation of the islets of Langerhans. Long Beach veterans administration regional medical education center symposium. Diabetes 45:385–92.
10. Burnstock G. (2002) Purinergic signaling and vascular cell proliferation and death. Arterioscler Thromb Vasc Biol 22:364–73.
11. Burnstock G. (2007) Physiology and pathophysiology of purinergic neurotransmission. Physiol Rev 87:659–797.
12. Burnstock G, Sneddon P. (1985) Evidence for ATP and noradrenaline as cotransmitters in sympathetic nerves. Clin Sci (Lond) 68(Suppl 10): 89s–92s.
13. Buxton IL, Kaiser RA, Oxhorn BC, Cheek DJ. (2001) Evidence supporting the Nucleotide Axis Hypothesis: ATP release and metabolism by coronary endothelium. Am J Physiol Heart Circ Physiol 281:H1657–66.
14. Carlsson PO, Flodstrom M, Sandler S. (2000) Islet blood flow in multiple low dose streptozotocin-treated wild-type and inducible nitric oxide synthase-deficient mice. Endocrinology 141:2752–57.

15. Carlsson PO, Iwase M, Jansson L. (1999) Stimulation of intestinal glucoreceptors in rats increases pancreatic islet blood flow through vagal mechanisms. Am J Physiol 276: R233–6.
16. Carlsson PO, Iwase M, Jansson L. (2000) Intraportal glucose infusion and pancreatic islet blood flow in anesthetized rats. Am J Physiol Regul Integr Comp Physiol 279:R1224–9.
17. Carlsson PO, Olsson R, Källskog Ö, Bodin B, Andersson A, Jansson L. (2002) Glucose-induced islet blood flow increase in rats: interaction between nervous and metabolic mediators. Am J Physiol Endocrinol Metab 283: E457–64.
18. Carlsson PO, Sandler S, Jansson L. (1998) Pancreatic islet blood perfusion in the nonobese diabetic mouse: diabetes-prone female mice exhibit a higher blood flow compared with male mice in the prediabetic phase. Endocrinology 139: 3534–41.
19. Castro E, Mateo J, Tome AR, Barbosa RM, Miras-Portugal MT, Rosario LM. (1995) Cell-specific purinergic receptors coupled to Ca^{2+} entry and Ca^{2+} release from internal stores in adrenal chromaffin cells. Differential sensitivity to UTP and suramin. J Biol Chem 270: 5098–106.
20. Cioffi DL, Stevens T. (2006) Regulation of endothelial cell barrier function by store-operated calcium entry. Microcirculation 13:709–23.
21. Collins DM, McCullough WT, Ellsworth ML. (1998) Conducted vascular responses: communication across the capillary bed. Microvasc Res 56:43–53.
22. Dryselius S, Grapengiesser E, Hellman B, Gylfe E. (1999) Voltage-dependent entry and generation of slow Ca^{2+} oscillations in glucose-stimulated pancreatic β-cells. Am J Physiol 276:E512–8.
23. Eltzschig HK, Macmanus CF, Colgan SP. (2008) Neutrophils as sources of extracellular nucleotides: functional consequences at the vascular interface. Trends Cardiovasc Med 18:103–7.
24. Erlinge D, Burnstock G. (2008) P2 receptors in cardiovascular regulation and disease. Purinergic Signal 4:1–20.
25. Ferner H. (1957) Dissemination of testicular interstitial cells and of the islets of Langerhans as a functional principle for the seminiferous tubules and the exocrine pancreas. Z Mikrosk Anat Forsch 63:35–52.
26. Gerasimovskaya EV, Woodward HN, Tucker DA, Stenmark KR. (2008) Extracellular ATP is a pro-angiogenic factor for pulmonary artery vasa vasorum endothelial cells. Angiogenesis 11:169–82.
27. Gonzalez-Alonso J, Olsen DB, Saltin B. (2002) Erythrocyte and the regulation of human skeletal muscle blood flow and oxygen delivery: role of circulating ATP. Circ Res 91: 1046–55.
28. Grapengiesser E, Dansk H, Hellman B. (2004) Pulses of external ATP aid to the synchronization of pancreatic β-cells by generating premature Ca^{2+} oscillations. Biochem Pharmacol 68:667–74.
29. Grapengiesser E, Gylfe E, Dansk H, Hellman B. (2003) Stretch activation of Ca^{2+} transients in pancreatic β cells by mobilization of intracellular stores. Pancreas 26: 82–6.
30. Grapengiesser E, Gylfe E, Hellman B. (1999) Synchronization of glucose-induced Ca^{2+} transients in pancreatic β-cells by a diffusible factor. Biochem Biophys Res Commun 254:436–9.
31. Hansen B, Longati P, Elvevold K, et al.. (2005) Stabilin-1 and stabilin-2 are both directed into the early endocytic pathway in hepatic sinusoidal endothelium via interactions with clathrin/AP-2, independent of ligand binding. Exp Cell Res 303:160–73.
32. Hazama A, Hayashi S, Okada Y. (1998) Cell surface measurements of ATP release from single pancreatic beta cells using a novel biosensor technique. Pflugers Arch 437: 31–5.
33. Hellman B. (1959) Methodological aspects on the differential cell count of the islet tissue in the rat. Acta Pathol Microbiol Scand 45:336–46.

34. Hellman B. (1965) Studies in obese-hyperglycemic mice. Ann NY Acad Sci 131:541–58.
35. Hellman B, Dansk H, Grapengiesser E. (2004) Pancreatic β-cells communicate via intermittent release of ATP. Am J Physiol Endocrinol Metab 286:E759–65.
36. Hellman B, Gylfe E, Grapengiesser E, Lund PE, Berts A. (1992) Cytoplasmic Ca^{2+} oscillations in pancreatic β-cells. Biochim Biophys Acta 1113: 295–305.
37. Hellman B, Jansson L, Dansk H, Grapengiesser E. (2007) Effects of external ATP on Ca^{2+} signalling in endothelial cells isolated from mouse islets. Endocrine 32:33–40.
38. Hellman B, Wallgren A, Petersson B. (1962) Cytological characteristics of the exocrine pancreatic cells with regard to their position in relation to the islets of Langerhans. A study in normal and obese-hyperglycaemic mice. Acta Endocrinol. (Copenh) 39:465–73.
39. Hillaire-Buys D, Dietz S, Chapal J, Petit P, Loubatieres-Mariani MM. (1999) Involvement of P2X and P2U receptors in the constrictor effect of ATP on the pancreatic vascular bed. J Soc Biol 193:57–61.
40. Himmel HM, Whorton AR, Strauss HC. (1993) Intracellular calcium, currents, and stimulus-response coupling in endothelial cells. Hypertension 21:112–27.
41. Holmqvist K, Cross MJ, Rolny C, et al. (2004) The adaptor protein shb binds to tyrosine 1175 in vascular endothelial growth factor. (VEGF) receptor-2 and regulates VEGF-dependent cellular migration. J Biol Chem 279:22267–75.
42. Isakson BE, Damon DN, Day KH, Liao Y, Duling BR. (2006) Connexin 40 and connexin 43 in mouse aortic endothelium: evidence for coordinated regulation. Am J Physiol Heart Circ Physiol 290:H1199–205.
43. Jagger JE, Bateman RM, Ellsworth ML, Ellis CG. (2001) Role of erythrocyte in regulating local O_2 delivery mediated by hemoglobin oxygenation. Am J Physiol Heart Circ Physiol 280: H2833–9.
44. Jakovcevic D, Harder DR. (2007) Role of astrocytes in matching blood flow to neuronal activity. Curr Top Dev Biol 79:75–97.
45. Jansson L. (1994) The regulation of pancreatic islet blood flow. Diabetes Metab Rev 10: 407–16.
46. Jansson L, Hellerstrom C. (1986) Glucose-induced changes in pancreatic islet blood flow mediated by central nervous system. Am J Physiol 251: E644–7.
47. Johansson M, Andersson A, Carlsson PO, Jansson L. (2006) Perinatal development of the pancreatic islet microvasculature in rats. J Anat 208: 191–6.
48. Johansson M, Mattsson G, Andersson A, Jansson L, Carlsson PO. (2006) Islet endothelial cells and pancreatic beta-cell proliferation: studies in vitro and during pregnancy in adult rats. Endocrinology 147:2315–24.
49. Kaido T, Yebra M, Cirulli V, Montgomery AM. (2004) Regulation of human beta-cell adhesion, motility, and insulin secretion by collagen IV and its receptor α1β1. J Biol Chem 279:53762–9.
50. Konstantinova I, Lammert E. (2004) Microvascular development: learning from pancreatic islets. Bioessays 26:1069–75.
51. Lai EY, Jansson L, Patzak A, Persson AE. (2007) Vascular reactivity in arterioles from normal and alloxan-diabetic mice: studies on single perfused islets. Diabetes 56:107–12.
52. Lai EY, Persson AE, Bodin B, et al. (2007) Endothelin-1 and pancreatic islet vasculature: studies in vivo and on isolated, vascularly perfused pancreatic islets. Am J Physiol Endocrinol Metab 292: E1616–23.
53. Lammert E. (2008) The vascular trigger of type II diabetes mellitus. Exp Clin Endocrinol Diabetes 116 (Suppl 1):S21–5.
54. Lammert E, Cleaver O, Melton D. (2001) Induction of pancreatic differentiation by signals from blood vessels. Science 294:564–7.
55. Lammert E, Gu G, McLaughlin M, et al. (2003) Role of VEGF-A in vascularization of pancreatic islets. Curr Biol 13: 1070–4.
56. Langenkamp E, Molema G. (2009) Microvascular endothelial cell heterogeneity: general concepts and pharmacological consequences for anti-angiogenic therapy of cancer. Cell Tissue Res 335:205–22.

57. Lopez-Jaramillo P, Gonzalez MC, Palmer RM, Moncada S. (1990) The crucial role of physiological Ca^{2+} concentrations in the production of endothelial nitric oxide and the control of vascular tone. Br J Pharmacol 101:489–93.
58. Luo X, Zheng W, Yan M, Lee MG, Muallem S. (1999) Multiple functional P2X and P2Y receptors in the luminal and basolateral membranes of pancreatic duct cells. Am J Physiol 277:C205 15.
59. Mattsson G. (2005) The endothelial cells in islets of Langerhans. Ups J Med Sci 110:1–15.
60. Mattsson G, Carlsson PO, Olausson K, Jansson L. (2002) Histological markers for endothelial cells in endogenous and transplanted rodent pancreatic islets. Pancreatology 2: 155–62.
61. McCullough WT, Collins DM, Ellsworth ML. (1997) Arteriolar responses to extracellular ATP in striated muscle. Am J Physiol 272:H1886–91.
62. McGahren ED, Beach JM, Duling BR. (1998) Capillaries demonstrate changes in membrane potential in response to pharmacological stimuli. Am J Physiol 274:H60–5.
63. Moccia F, Baruffi S, Spaggiari S, et al. (2001) P2y1 and P2y2 receptor-operated Ca2+ signals in primary cultures of cardiac microvascular endothelial cells. Microvasc Res 61:240–52.
64. Morato M, Sousa T, Albino-Teixeira A. (2008) Purinergic receptors in the splanchnic circulation. Purinergic Signal 4:267–85.
65. Nikolova G, Jabs N, Konstantinova I, et al. (2006) The vascular basement membrane: a niche for insulin gene expression and β cell proliferation. Dev Cell 10:397–405.
66. Noble MD, Liddle RA. (2005) Neurohormonal control of exocrine pancreatic secretion. Curr Opin Gastroenterol 21:531–7.
67. Nordquist L, Lai EY, Sjoquist M, Jansson L, Persson AE. (2008) C-peptide constricts pancreatic islet arterioles in diabetic, but not normoglycaemic mice. Diabetes Metab Res Rev 24:165–8.
68. Novak I. (2008) Purinergic receptors in the endocrine and exocrine pancreas. Purinergic Signal 4:237–53.
69. Nyman LR, Wells KS, Head WS, et al. (2008) Real-time, multidimensional in vivo imaging used to investigate blood flow in mouse pancreatic islets. J Clin Invest 118:3790–97.
70. Olerud J, Johansson M, Lawler J, Welsh N, Carlsson PO. (2008) Improved vascular engraftment and graft function after inhibition of the angiostatic factor thrombospondin-1 in mouse pancreatic islets. Diabetes 57:1870–7.
71. Olsson R, Carlsson PO. (2005) Better vascular engraftment and function in pancreatic islets transplanted without prior culture. Diabetologia 48:469–76.
72. Olsson R, Carlsson PO. (2006) The pancreatic islet endothelial cell: emerging roles in islet function and disease. Int J Biochem Cell Biol 38:710–4.
73. Parr EL, Bowen KM, Lafferty KJ. (1980) Cellular changes in cultured mouse thyroid glands and islets of Langerhans. Transplantation 30: 135–41.
74. Persson-Sjogren S, Forsgren S, Taljedal IB. (2000) Peptides and other neuronal markers in transplanted pancreatic islets. Peptides 21:741–52.
75. Petit P, Lajoix AD, Gross R. (2009) P2 purinergic signalling in the pancreatic β-cell: control of insulin secretion and pharmacology. Eur J Pharm Sci 37:67–75.
76. Pittman RN. (2000) Oxygen supply to contracting skeletal muscle at the microcirculatory level: diffusion vs. convection. Acta Physiol Scand 168:593–602.
77. Sprague RS, Ellsworth ML, Stephenson AH, Lonigro AJ. (2001) Participation of cAMP in a signal-transduction pathway relating erythrocyte deformation to ATP release. Am J Physiol Cell Physiol 281:C1158–64.
78. Stagner JI, Samols E, Bonner-Weir S. (1988) beta – alpha – delta pancreatic islet cellular perfusion in dogs. Diabetes 37:1715–21.
79. Surprenant A, North RA. (2009) Signaling at Purinergic P2X Receptors. Annu Rev Physiol 71:333–59.
80. Svensson AM, Ostenson CG, Sandler S, Efendic S, Jansson L. (1994) Inhibition of nitric oxide synthase by N^G-nitro-L-arginine causes a preferential decrease in pancreatic islet blood flow in normal rats and spontaneously diabetic GK rats. Endocrinology 135:849–53.

81. Tang EH, Ku DD, Tipoe GL, Feletou M, Man RY, Vanhoutte PM. (2005) Endothelium-dependent contractions occur in the aorta of wild-type and $COX2^{-/-}$ knockout but not $COX1^{-/-}$ knockout mice. J Cardiovasc Pharmacol 46:761–5.
82. Teitelman G, Guz Y, Ivkovic S, Ehrlich M. (1998) Islet injury induces neurotrophin expression in pancreatic cells and reactive gliosis of peri-islet Schwann cells. J Neurobiol 34:304–18.
83. Tran QK, Ohashi K, Watanabe H. (2000) Calcium signalling in endothelial cells. Cardiovasc Res 48:13–22.
84. Virtanen I, Banerjee M, Palgi J, et al. (2008) Blood vessels of human islets of Langerhans are surrounded by a double basement membrane. Diabetologia 51:1181–91.
85. Yamamoto K, Korenaga R, Kamiya A, Qi Z, Sokabe M, Ando J. (2000) $P2X_4$ receptors mediate ATP-induced calcium influx in human vascular endothelial cells. Am J Physiol Heart Circ Physiol 279:H285–92.
86. Yamamoto K, Sokabe T, Matsumoto T, et al. (2006) Impaired flow-dependent control of vascular tone and remodeling in P2X4-deficient mice. Nat Med 12:133–7.
87. Yegutkin GG. (2008) Nucleotide- and nucleoside-converting ectoenzymes: Important modulators of purinergic signalling cascade. Biochim Biophys Acta 1783:673–94.
88. Zhou Y, Varadharaj S, Zhao X, Parinandi N, Flavahan NA, Zweier JL. (2005) Acetylcholine causes endothelium-dependent contraction of mouse arteries. Am J Physiol Heart Circ Physiol 289:H1027–32.

Index

E. Gerasimovskaya, E. Kaczmarek (eds.), *Extracellular ATP and Adenosine as Regulators of Endothelial Cell Function*, DOI 10.1007/978-90-481-3435-9,